Test Bank

Test Bank

for

RAVEN • EVERT • EICHHORN

Biology of Plants

SIXTH EDITION

Robert C. Evans
Rutgers University, Camden

W. H. FREEMAN AND COMPANY
WORTH PUBLISHERS

Test Bank

for

Biology of Plants, Sixth Edition

Printed in the United States of America

ISBN: 1-57259-607-4

Printing: 1 2 3 4 5 – 03 02 01 00 99

W. H. FREEMAN AND COMPANY
41 Madison Avenue
New York, New York 10010
http://www.whfreeman.com

Contents

Preface

This Test Bank is designed to help instructors evaluate their students' understanding of the material presented in *Biology of Plants*, Sixth Edition, by Peter H. Raven, Ray F. Evert, and Susan E. Eichhorn. The Test Bank contains more than 2800 questions—multiple-choice, true-false, and essay—for the textbook's 34 chapters.

The following question identification information is included with each question to help make selections easier:

Textbook section heading This enables you to select questions from sections you have assigned. Because boxed essays are often considered optional reading, questions on this material have not been included in the Test Bank.

Difficulty level Each item is designated easy, moderate, or difficult. Generally, "easy" means easily abstracted from the text—the wording of the question and correct answer is almost identical to the words in the chapter. A "moderate" question is not worded identically and requires some extrapolation by the student. A "difficult" question requires a broader reading of the text and/or a greater degree of extrapolation.

Answer The correct answer is provided for multiple-choice and true-false questions.

Most of the questions are multiple-choice, but true-false and essay questions have also been included for those instructors who want a variety of testing formats. All multiple-choice questions have five alternative answers, and we have taken care to avoid those that students might find tricky or ambiguous. Also, we have avoided using alternatives such as "all/none of the above" and "both answers c and d are correct." Finally, all questions are written so they can "stand alone." Thus, an instructor can prepare an exam that consists of randomly arranged questions from a variety of chapters.

These Test Bank questions form the database for programs for use on Windows or Macintosh computers. For more information on the computerized test-generation system, contact W. H. Freeman or your SASMP sales representative.

Acknowledgments

I am very grateful to all the individuals who helped put this Test Bank together. In particular I would like to thank Linda Strange, who made numerous suggestions for improving my questions and who wrote many questions herself, Michael Gittings and Yuna Lee for helping with the final touches, and Julie Kerr of Worth Publishers who coordinated the entire project. Julie's cheerful supervision, efficient assistance, and overall good will helped make the project so enjoyable to work on. Finally, I thank the authors of the textbook for producing such an impressive revision. It was a delight to read the book as thoroughly and carefully as was necessary to produce these questions.

Robert C. Evans
May 1998

Test Bank

Chapter 1

An Introduction

Multiple-Choice Questions

1. **Botany: An Introduction; p. 2; easy; ans: c**
   The process of photosynthesis results in the formation of two substances
   essential to our existence:
   a. chlorophyll and water.
   b. electrons and protons.
   c. sugar and oxygen.
   d. sugar and water.
   e. chlorophyll and oxygen.

2. **Botany: An Introduction; p. 2; easy; ans: a**
   In the process of photosynthesis, a particle of light raises an electron of
   a(n)_____ molecule to a higher energy level.
   a. chlorophyll
   b. oxygen
   c. sugar
   d. carbon dioxide
   e. water

3. **Evolution of Plants; p. 2; moderate; ans: d**
   Life existed on Earth as early as _____ years ago.
   a. 300 to 400 thousand
   b. 3 to 4 million
   c. 300 to 400 million
   d. 3 to 4 billion
   e. 300 to 400 billion

4. **Evolution of Plants; p. 3; difficult; ans: c**
   Oparin proposed that:
   a. living organisms were brought to Earth on asteroids.
   b. the collision of comets with the Earth resulted in the formation of the
      basic chemical building blocks of life.
   c. organic compounds were formed from volcanic gases in the atmosphere.
   d. a variety of complex molecules could be formed in the laboratory if a
      spark were applied to a mixture of gases.
   e. cells arose from proteinoid microspheres.

5. **Evolution of Plants; p. 3; moderate; ans: b**
    In addition to water vapor, the major gases present in the Earth's first
    atmosphere were probably:

    a. oxygen and carbon dioxide.
    b. carbon dioxide and nitrogen.
    c. methane and ammonia.
    d. methane and nitrogen.
    e. nitrogen and oxygen.

6. **Evolution of Plants; p. 4; difficult; ans: a**
    The Hale-Bopp comet is important from a biological perspective because it:

    a. provides evidence that comets can provide a rich array of important
        organic molecules.
    b. creates electrical discharges that cause biologically important molecules to
        form in the Earth's atmosphere.
    c. warms the primitive oceans, thus stimulating the formation of organic
        molecules.
    d. stimulates volcanic eruptions, thus increasing the discharge of important
        gases.
    e. contains primitive cells.

7. **Evolution of Plants; pp. 4-5; difficult; ans: b**
    Which of the following statements concerning primitive cells is FALSE?

    a. They used organic molecules to satisfy their energy requirements.
    b. They constructed new cells from organic molecules made via
        photosynthesis.
    c. They acquired the ability to grow.
    d. They acquired the ability to reproduce.
    e. They acquired the ability to pass on their characteristics to subsequent
        generations.

8. **Evolution of Plants; p. 5; moderate; ans: a**
    Which of the following statements about photosynthetic autotrophs is FALSE?

    a. They obtain their required organic compounds from external sources.
    b. They channel radiant energy into the biosphere.
    c. The word "autotroph" means "self-feeder."
    d. They have a complex pigment system.
    e. An example of an autotroph is a plant.

9. **Evolution of Plants; p. 5; difficult; ans: e**
    Which of the following statements concerning the earliest photosynthetic
    organisms is FALSE?

    a. They were simple in comparison to plants.
    b. They were more complex than primitive heterotrophs.
    c. They had a complex pigment system.
    d. They had a way of storing energy in an organic molecule.
    e. They have been found in rocks 3.8 billion years old.

10. Evolution of Plants; p. 5; easy; ans: b
The oxygen gas released in photosynthesis originates from:

    a.   carbon dioxide.
    b.   water.
    c.   ozone.
    d.   sugar.
    e.   nitrates.

11. Evolution of Plants; p. 6; easy; ans: d
Ozone in the outer layer of the atmosphere has important consequences for living things in that it:

    a.   is a pollutant.
    b.   is involved directly in respiration.
    c.   aids in the aggregation of molecules.
    d.   absorbs ultraviolet rays from sunlight.
    e.   is used by autotrophs to make sugars.

12. Evolution of Plants; p. 6; difficult; ans: e
The first cells on Earth were most likely:

    a.   bacteria.
    b.   autotrophs.
    c.   eukaryotes.
    d.   proteinoid microspheres.
    e.   archaeans.

13. Evolution of Plants; pp. 6-7; difficult; ans: e
Which of the following is NOT an adaptation of photosynthetic organisms to rocky coasts?

    a.   a multicellular body
    b.   strong cell walls
    c.   structures to anchor their bodies
    d.   food-conducting tissues
    e.   pigment systems

14. Evolution of Plants; p. 7; easy; ans: d
One function of the cuticle is to:

    a.   transport water.
    b.   transport food.
    c.   add cells to the plant body.
    d.   retard water loss.
    e.   serve as a protective layer of cells.

15. Evolution of Plants; p. 7; moderate; ans: e
In perennials, the _____ is most similar in function to the cuticle-covered epidermis of annuals.

    a.   xylem
    b.   phloem
    c.   stoma
    d.   vascular cambium
    e.   cork

16. **Evolution of Plants; p. 8; easy; ans: c**
    Water is transported upward through the plant body in the:

    a. epidermis.
    b. cork.
    c. xylem.
    d. phloem.
    e. apical meristems.

17. **Evolution of Plants; p. 8; easy; ans: d**
    The food manufactured by photosynthesis is transported throughout the plant
    body in the:

    a. epidermis.
    b. cork.
    c. xylem.
    d. phloem.
    e. apical meristems.

18. **Evolution of Plants; p. 8; easy; ans: a**
    The _____ are regions at the shoot and root tips that are capable of adding
    cells indefinitely to the plant body.

    a. apical meristems
    b. lateral meristems
    c. stomata
    d. vascular systems
    e. seed coats

19. **Evolution of Plants; p. 8; easy; ans: d**
    Secondary growth refers to growth:

    a. that is of secondary importance to the plant.
    b. that results in the extension of roots and stems.
    c. originating from apical meristems.
    d. originating from lateral meristems.
    e. originating from the epidermis.

20. **Evolution of Plants; p. 8; easy; ans: d**
    The activity of the _____ results in a thickening of stems, branches, and
    roots.

    a. xylem and phloem regions
    b. epidermal regions
    c. vascular systems
    d. lateral meristems
    e. apical meristems

21. **Evolution of Plants; p. 9; moderate; ans: c**
    A seed is composed of three parts:

    a. root, stem, and leaves.
    b. xylem, phloem, and seed coat.
    c. seed coat, embryo, and food supply.
    d. apical meristems, lateral meristems, and seed coat.
    e. spore coat, embryo, and vascular system.

22. **Evolution of Communities; p. 9; moderate; ans: c**
Natural communities of organisms of wide extent, characterized by distinctive, climatically controlled groups of plants and animals, are called:

    a. biospheres.
    b. ecosystems.
    c. biomes.
    d. species.
    e. aggregations.

23. **Evolution of Communities; p. 9; easy; ans: c**
In all ecosystems, heterotrophs are completely dependent on the productivity of all the following groups of organisms EXCEPT:

    a. photosynthetic bacteria.
    b. autotrophs.
    c. animals.
    d. plants.
    e. algae.

24. **Appearance of Human Beings; p. 11; easy; ans: c**
Humans first appeared about _____ years ago.

    a. 2000
    b. 200,000
    c. 2 million
    d. 20 million
    e. 200 million

25. **Appearance of Human Beings; p. 11; easy; ans: b**
The development of agriculture started at least _____ years ago.

    a. 1000
    b. 11,000
    c. 111,000
    d. 1 million
    e. 11 million

26. **Appearance of Human Beings; p. 11; moderate; ans: e**
Cytology is the study of:

    a. energy transformations.
    b. plant form.
    c. heredity.
    d. fossil plants.
    e. cell structure, function, and life histories.

27. **Appearance of Human Beings; p. 13; moderate; ans: d**
The greenhouse effect refers to the:

    a. depletion of the ozone layer.
    b. increased incidence of skin cancer.
    c. problem of feeding the world's population.
    d. trapping of heat radiated from Earth.
    e. disappearance of species.

True-False Questions

1. **Evolution of Plants; p. 2; moderate; ans: T**
    The Earth's earliest geologic period ended at the close of a massive meteor bombardment, 3.8 billion years ago.

2. **Evolution of Plants; p. 3; moderate; ans: T**
    Organic molecules similar to the building blocks of life can be produced in the laboratory by exposing a mixture of gases over heated water to electric sparks.

3. **Evolution of Plants; p. 4; easy; ans: T**
    Dust from comets and asteroids contains organic molecules.

4. **Evolution of Plants; pp. 4-5; moderate; ans: F**
    The first cell-like structures were able to use inorganic compounds as a source of energy.

5. **Evolution of Plants; p. 5; easy; ans: F**
    Most likely, autotrophs evolved before heterotrophs.

6. **Evolution of Plants; p. 6; easy; ans: F**
    Eukaryotic cells evolved before prokaryotic cells.

7. **Evolution of Plants; p. 6; easy; ans: F**
    According to the fossil record, the appearance of prokaryotic cells was associated with an increase in free oxygen in the atmosphere.

8. **Evolution of Plants; p. 6; easy; ans: T**
    Archaeans are a group of prokaryotes.

9. **Evolution of Plants; pp. 6-7; moderate; ans: T**
    As multicellular organisms evolved in coastal environments, food-conducting tissues developed to connect photosynthesizing and nonphotosynthesizing body parts.

10. **Evolution of Plants; p. 7; moderate; ans: T**
    In plants, water moves in a continuous stream from roots to stems to leaves.

11. **Evolution of Plants; p. 7; moderate; ans: F**
    Stomata are specialized guard cells in the epidermis of leaves.

12. **Evolution of Plants; p. 8; easy; ans: T**
    Plants that contain xylem and phloem are called vascular plants.

13. **Evolution of Plants; p. 8; difficult; ans: T**
    A plant must first exhibit primary growth before it can exhibit secondary growth.

14. **Evolution of Communities; p. 9; difficult; ans: F**
    An example of an ecosystem is a tree and all the animals that live on and in it.

15. **Evolution of Communities; p. 9; moderate; ans: F**
    In an ecosystem, both elements and energy are recycled.

16. **Evolution of Communities; pp. 9, 11; moderate; ans: F**
The oxygen released by photosynthesis is required by heterotrophs, but not by autotrophs, for energy-producing metabolic activities.

17. **Appearance of Human Beings; p. 12; easy; ans: F**
Plants, algae, and bacteria can no longer supply industrial civilization with renewable energy.

18. **Appearance of Human Beings; p. 13; easy; ans: T**
One effect of chlorofluorocarbons has been to deplete the ozone layer.

Essay Questions

1. **Evolution of Plants; p. 2; moderate**
What is the evidence that life may have originated elsewhere in the universe and reached Earth by traveling through space?

2. **Evolution of Plants; pp. 4-5; difficult**
List the four main properties that characterize living things, and explain the significance of each.

3. **Evolution of Plants; p. 5; easy**
What is the difference between an autotroph and a heterotroph? In what way was the evolution of autotrophs crucial to the survival of life on Earth?

4. **Evolution of Plants; pp. 5-6; moderate**
In what two ways did photosynthesis alter the Earth's early atmosphere, and what was the significance of each for life on Earth?

5. **Evolution of Plants; pp. 6-7; moderate**
What environmental factors at the seashore favored the evolution of photosynthetic organisms? What plant structures evolved in response to this environment?

6. **Evolution of Plants; p. 8; easy**
What is the difference between primary growth and secondary growth? Which types of meristems are involved in each?

7. **Evolution of Plants; pp. 7-9; moderate**
Discuss the principal characteristics that helped plants adapt to life on land.

8. **Evolution of Communities; p. 9; difficult**
Explain how an ecosystem can be stable but not static.

9. **Appearance of Human Beings; p. 11; moderate**
Which groups of organisms are studied under the umbrella of botany, and why?

10. **Appearance of Human Beings; p. 12; moderate**
Discuss the ways in which plants are involved in many of the environmental issues facing today's world.

11. **Appearance of Human Beings; pp. 12-13; easy**
    List some detrimental effects that human activities have had on the
    environment.

Chapter 2

The Molecular Composition of Plant Cells

Multiple-Choice Questions

1. **The Molecular Composition of Plant Cells; p. 18; easy; ans: e**
    Which substance makes up more than half of all living matter and more than 90 percent of the weight of most plant tissues?

    a. protein
    b. cellulose
    c. starch
    d. triglyceride
    e. water

2. **The Molecular Composition of Plant Cells; p. 18; easy; ans: c**
    By definition, an organic molecule contains:

    a. hydrogen.
    b. oxygen.
    c. carbon.
    d. calcium.
    e. double bonds.

3. **Organic Molecules; p. 18; easy; ans: a**
    Carbohydrates, lipids, proteins, and nucleic acids consist mainly of:

    a. carbon and hydrogen.
    b. nitrogen and phosphorus.
    c. nitrogen and sulfur.
    d. oxygen and nitrogen.
    e. carbon and phosphorus.

4. **Organic Molecules; p. 18; moderate; ans: d**
    Nucleic acids are different from proteins in that nucleic acids contain:

    a. carbon.
    b. hydrogen.
    c. nitrogen.
    d. phosphorus.
    e. sulfur.

5. **Carbohydrates; p. 18; easy; ans: b**
The most abundant organic molecules in nature are:

   a. proteins.
   b. carbohydrates.
   c. lipids.
   d. nucleic acids.
   e. water molecules.

6. **Carbohydrates; p. 18; easy; ans: e**
Which of the following is a polymer?

   a. fructose
   b. sucrose
   c. maltose
   d. ribose
   e. starch

7. **Carbohydrates; p. 19; moderate; ans: d**
The primary source of chemical energy in plants and animals is:

   a. sucrose.
   b. starch.
   c. fructose.
   d. glucose.
   e. ribose.

8. **Carbohydrates; p. 20; moderate; ans: b**
The common transport form of sugar in animals is _____ and in plants is

   _____.

   a. sucrose; glucose
   b. glucose; sucrose
   c. starch; sucrose
   d. maltose; lactose
   e. fructose; maltose

9. **Carbohydrates; p. 20; moderate; ans: e**
The formation of _____ from _____ occurs by dehydration synthesis.

   a. glucose and fructose; sucrose
   b. glucose; starch
   c. monomers; polymers
   d. glucose; cellulose
   e. sucrose; glucose and fructose

10. **Carbohydrates; p. 20; moderate; ans: b**
Which of the following statements about hydrolysis reactions is FALSE?

   a. They are energy-yielding reactions.
   b. They are a type of condensation reaction.
   c. They involve the addition of a molecule of water.
   d. An example is the conversion of a polymer to its monomers.
   e. An example is the conversion of disaccharides to monosaccharides.

11. Carbohydrates; p. 20; moderate; ans: a
Starch is a _____ composed of _____ subunits.

    a.   polysaccharide; glucose
    b.   disaccharide; glucose and fructose
    c.   monosaccharide; fructan
    d.   polysaccharide; cellulose
    e.   disaccharide; sucrose

12. Carbohydrates; p. 20; easy; ans: b
Which molecules are found in the starch grains of plant cells?

    a.   fructans
    b.   amylose and amylopectin
    c.   glycogen and sucrose
    d.   oligosaccharins and chitin
    e.   cellulose and maltose

13. Carbohydrates; p. 20; moderate; ans: e
In wheat, rye, and barley, the principal storage polysaccharide of leaves and stems is:

    a.   amylose.
    b.   amylopectin.
    c.   starch.
    d.   glycogen.
    e.   fructan.

14. Carbohydrates; p. 20; moderate; ans: d
When amylose is hydrolyzed, one product is:

    a.   amylopectin.
    b.   sucrose.
    c.   fructose.
    d.   glucose.
    e.   cellulose.

15. Carbohydrates; p. 21; moderate; ans: c
The principal polysaccharide in the plant cell wall is:

    a.   starch.
    b.   fructan.
    c.   cellulose.
    d.   glycogen.
    e.   sucrose.

16. Carbohydrates; p. 21; easy; ans: a
What is the most abundant organic compound known?

    a.   cellulose
    b.   sucrose
    c.   DNA
    d.   phospholipid
    e.   starch

17. Carbohydrates; p. 21; moderate; ans: d
Cotton fibers and wood consist mainly of:

    a.   protein.
    b.   lipid.
    c.   starch.
    d.   cellulose.
    e.   chitin.

18. Carbohydrates; p. 22; moderate; ans: d
Which of the following consists of beta-glucose subunits?

    a.   amylose
    b.   amylopectin
    c.   chitin
    d.   cellulose
    e.   glycogen

19. Carbohydrates; p. 22; easy; ans: b
The principal component of the cell walls of fungi is:

    a.   cellulose.
    b.   chitin.
    c.   starch.
    d.   protein.
    e.   phospholipid.

20. Carbohydrates; p. 22; moderate; ans: b
The matrix of the plant cell wall contains:

    a.   microfibrils and pectins.
    b.   pectins and hemicelluloses.
    c.   hemicelluloses and oligosaccharins.
    d.   cellulose and chitin.
    e.   chitin and glycogen.

21. Carbohydrates; p. 22; easy; ans: e
"Oligosaccharins" are cell wall components that may function as:

    a.   microfibrils.
    b.   enzymes.
    c.   storage polysaccharides.
    d.   transport forms of carbohydrate.
    e.   hormones.

22. Lipids; p. 22; easy; ans: c
Which of the following groups of substances is generally hydrophobic?

    a.   carbohydrates
    b.   proteins
    c.   lipids
    d.   nucleic acids
    e.   sugars

23. Lipids; p. 22; moderate; ans: a
Which of the following yields the greatest amount of energy per gram?

 a. fats
 b. starch
 c. glycogen
 d. protein
 e. cellulose

24. Lipids; p. 22; easy; ans: a
A fat consists of one _____ molecule bonded to three _____ molecules.

 a. glycerol; fatty acid
 b. glucose; fatty acid
 c. amino acid; glycerol
 d. fatty acid; glycerol
 e. sugar; amino acid

25. Lipids; p. 24; moderate; ans: d
A phospholipid differs from a triglyceride in that a phospholipid contains:

 a. fatty acids.
 b. glycerol.
 c. double bonds.
 d. a hydrophilic group.
 e. a hydrophobic group.

26. Lipids; p. 24; moderate; ans: b
The main function of cutin and suberin is to:

 a. serve as structural components of cellular membranes.
 b. prevent water loss.
 c. catalyze chemical reactions.
 d. provide strength to the cell wall.
 e. serve as hormones.

27. Lipids; p. 25; easy; ans: c
A major lipid component of the walls of cork cells is:

 a. phospholipid.
 b. cutin.
 c. suberin.
 d. steroid.
 e. epicuticular wax.

28. Lipids; p. 25; moderate; ans: d
Sitosterol and ergosterol are examples of _____ present in cell membranes.

 a. polysaccharides
 b. proteins
 c. nucleic acids
 d. lipids
 e. triglycerides

29. Lipids; p. 25; moderate; ans: e
In all organisms except prokaryotes, an important role of sterols is to:

a. serve as storage forms of energy.
b. prevent water loss.
c. catalyze chemical reactions.
d. provide strength to the cell wall.
e. stabilize the phospholipid tails in cell membranes.

30. Proteins; p. 26; easy; ans: b
The monomers of proteins are:

a. monosaccharides.
b. amino acids.
c. glycerol and fatty acids.
d. nucleotides.
e. fused hydrocarbon rings.

31. Proteins; p. 26; moderate; ans: d
Which of the following plant structures contains the highest concentration of proteins?

a. leaves
b. stems
c. roots
d. seeds
e. flowers

32. Proteins; p. 28; easy; ans: c
How many different kinds of amino acids are used to build proteins?

a. 5
b. 10
c. 20
d. 50
e. 100

33. Proteins; p. 28; easy; ans: a
A peptide bond occurs between the _____ groups of adjacent amino acids in polypeptides.

a. amino and carboxyl
b. sulfhydryl
c. amino and phosphate
d. carboxyl and hydroxyl
e. amino

34. Proteins; p. 28; easy; ans: a
The linear sequence of amino acids is called the _____ structure of a protein.

a. primary
b. secondary
c. tertiary
d. quaternary
e. helix

35. Proteins; p. 28; moderate; ans: a
A common secondary structure in proteins is:

a. the alpha helix.
b. the peptide bond.
c. the disulfide bridge.
d. interaction fostered by molecular chaperones.
e. interaction between two or more polypeptide chains.

36. Proteins; p. 28 and Fig. 2-17; difficult; ans: b
A common _____ structure of proteins involves a hydrogen bond between the amino group of one amino acid and the carboxyl group of an amino acid farther along the peptide chain.

a. primary
b. secondary
c. tertiary
d. quaternary
e. pentenary

37. Proteins; p. 29; difficult; ans: e
The tertiary structure of a protein is a result of all of the following EXCEPT:

a. the folding of the secondary structure.
b. the linear sequence of amino acids.
c. the formation of disulfide bridges.
d. interactions among the R groups of a single polypeptide.
e. interactions between two or more polypeptide chains.

38. Proteins; p. 29; difficult; ans: a
The highest level of organization in a fibrous protein is _____ structure and in a globular protein with one polypeptide is _____ structure.

a. secondary; tertiary
b. tertiary; secondary
c. secondary; quaternary
d. tertiary; quaternary
e. quaternary; tertiary

39. Proteins; p. 29; moderate; ans: c
Molecular chaperones function by:

a. catalyzing the hydrolysis of peptide bonds.
b. catalyzing the dehydration synthesis of a protein.
c. facilitating protein folding.
d. denaturing proteins.
e. escorting proteins across membranes.

40. Proteins; p. 29; moderate; ans: c
Disulfide bridges help stabilize the _____ structure of a polypeptide.

a. primary
b. secondary
c. tertiary
d. quaternary
e. pentenary

41. Proteins; p. 30; moderate; ans: d
Which of the following statements concerning the denaturation of proteins is FALSE?

a. It can be caused by heat.
b. It can be caused by increased acidity.
c. It involves a disruption of tertiary structure.
d. It involves a disruption of primary structure.
e. It results in a loss of the protein's biological activity.

42. Proteins; p. 30; easy; ans: d
The _____ structure of a protein involves interactions between two or more polypeptide chains.

a. primary
b. secondary
c. tertiary
d. quaternary
e. pentenary

43. Proteins; p. 30; easy; ans: e
Which of the following statements about enzymes is FALSE?

a. They are globular proteins.
b. They enable cells to carry out chemical reactions at room temperature.
c. They lower the energy of activation.
d. They can be used over and over again.
e. They are typically effective only at high concentrations.

44. Nucleic Acids; p. 30; moderate; ans: b
The monomers of nucleic acids are:

a. DNA and RNA.
b. nucleotides.
c. ribose and deoxyribose.
d. ATP and ADP.
e. genes.

45. Nucleic Acids; p. 30; easy; ans: c
The subunits of a nucleotide are a(n):

a. five-carbon sugar and an amino acid.
b. amino acid, glycerol, and a fatty acid.
c. five-carbon sugar, a nitrogenous base, and a phosphate group.
d. nitrogenous base, an amino acid, and a monosaccharide.
e. phosphate group, a disaccharide, and a fatty acid.

46. Nucleic Acids; p. 31; moderate; ans: d
_____ contains genetic information organized into genes.

a. A ribozyme
b. RNA
c. An enzyme
d. DNA
e. A protein

47. Nucleic Acids; p. 31; moderate; ans: e
Adenosine triphosphate is a type of:

  a. fatty acid.
  b. amino acid.
  c. enzyme.
  d. steroid.
  e. nucleotide.

48. Nucleic Acids; p. 32; easy; ans: b
The principal role of ATP in the cell is:

  a. catalyzing chemical reactions.
  b. providing energy.
  c. serving as structural support.
  d. functioning as a hormone.
  e. keeping the phospholipid tails of membranes aligned.

49. Secondary Metabolites; p. 32; moderate; ans: a
Which of the following statements about secondary metabolites is FALSE?

  a. They are found in all cells of a plant.
  b. Some function as chemical signals or in the defense of the plant against
      herbivores.
  c. They frequently are synthesized in one part of the plant and stored in
      another.
  d. Some are produced only after the plant has been damaged.
  e. Their concentration in a plant can vary greatly over a 24-hour period.

50. Secondary Metabolites; p. 32; easy; ans: e
The major classes of secondary plant metabolites are:

  a. sugars and proteins.
  b. alkaloids and sugars.
  c. nucleic acids, alkaloids, and phenolics.
  d. terpenoids, phenolics, and proteins.
  e. alkaloids, phenolics, and terpenoids.

51. Secondary Metabolites; p. 32; moderate; ans: c
An alkaloid used as an analgesic and cough suppressant is:

  a. caffeine.
  b. nicotine.
  c. morphine.
  d. salicylic acid.
  e. cocaine.

52. Secondary Metabolites; p. 33; easy; ans: b
_____ is an alkaloid used to dilate pupils in eye examinations.

  a. Cocaine
  b. Atropine
  c. Caffeine
  d. Morphine
  e. Nicotine

53. Secondary Metabolites; p. 33; easy; ans: a
The terpenoid that is responsible for the bluish haze in hills and mountains is:

a.   isoprene.
b.   taxol.
c.   anthocyanin.
d.   salicylic acid.
e.   tannin.

54. Secondary Metabolites; p. 34; easy; ans: d
Essential oils are types of:

a.   tannins.
b.   flavonoids.
c.   alkaloids.
d.   terpenoids.
e.   anthocyanins.

55. Secondary Metabolites; p. 34; moderate; ans: d
Which of the following secondary metabolites consists of isoprene units?

a.   morphine
b.   atropine
c.   anthocyanin
d.   taxol
e.   salicylic acid

56. Secondary Metabolites; p. 34; easy; ans: e
_____ is a terpenoid consisting of as many as 100,000 isoprene units.

a.   Taxol
b.   Menthol
c.   Cardiac glycoside
d.   Anthocyanin
e.   Rubber

57. Secondary Metabolites; p. 34; moderate; ans: c
One of the hypothesized roles of cardiac glycosides is to:

a.   help the plant cope with heat.
b.   attract insect pollinators.
c.   protect the plant against predators.
d.   function as hormones.
e.   serve as a photosynthetic pigment.

58. Secondary Metabolites; pp. 34–35; easy; ans: c
The largest group of plant phenolics are the:

a.   tannins.
b.   cardiac glycosides.
c.   flavonoids.
d.   lignins.
e.   salicylic acids.

59. Secondary Metabolites; p. 35; difficult; ans: e
Co-pigmentation in flowers involves the formation of complexes of colored and colorless _____ with metal ions.

a. tannins
b. essential oils
c. cardiac glycosides
d. terpenoids
e. flavonoids

60. Secondary Metabolites; p. 35; difficult; ans: d
Phenolics used by humans to denature the protein of leather are:

a. lignins.
b. salicylic acids.
c. flavonoids.
d. tannins.
e. anthocyanins.

61. Secondary Metabolites; p. 36; easy; ans: b
_____ is a secondary metabolite responsible for adding compressive strength, stiffness, and waterproofing to the plant cell wall.

a. Tannin
b. Lignin
c. Cellulose
d. Flavone
e. Flavonol

62. Secondary Metabolites; p. 37; moderate; ans: a
The active ingredient in aspirin is a(n) _____ that has an essential role in plant resistance to pathogens.

a. phenolic
b. tannin
c. terpenoid
d. alkaloid
e. flavonol

True-False Questions

1. The Molecular Composition of Plant Cells; p. 18; easy; ans: T
Only six elements make up 99 percent of the weight of all living matter.

2. Organic Molecules; p. 18; moderate; ans: F
All organic molecules contain carbon, hydrogen, and oxygen.

3. Carbohydrates; p. 18; easy; ans: F
In the ring form but not the chain form, pentoses and hexoses have a carbonyl group.

4. Carbohydrates; p. 20; easy; ans: F
Plants break down their carbohydrate reserves and transport starch to the site where it is needed.

5. **Carbohydrates; p. 22; easy; ans: T**
Starch and glycogen consist of alpha-glucose subunits, whereas cellulose is made up entirely of beta-glucose.

6. **Carbohydrates; p. 22; moderate; ans: T**
Pectins and hemicelluloses comprise the matrix of the plant cell wall in which cellulose microfibrils are embedded.

7. **Lipids; p. 22; easy; ans: F**
On average, proteins contain more chemical energy than fats.

8. **Lipids; p. 24; moderate; ans: T**
In a phospholipid, the phosphate group is attached directly to the glycerol backbone.

9. **Lipids; p. 26; moderate; ans: T**
Steroids have hormonal functions in plants and animals.

10. **Proteins; p. 28; easy; ans: T**
A peptide bond is a linkage between an amino group and a carboxyl group.

11. **Proteins; p. 30; moderate; ans: F**
All proteins have a primary, secondary, tertiary, and quaternary structure.

12. **Proteins; p. 30; easy; ans: T**
The denaturation of a protein involves a disruption in its tertiary structure.

13. **Nucleic Acids; p. 31; easy; ans: T**
DNA molecules are the largest macromolecules found in cells.

14. **Nucleic Acids; p. 31; difficult; ans: F**
In the reaction in which a phosphate group is linked to ADP forming ATP, energy is released.

15. **Secondary Metabolites; p. 33; easy; ans: T**
Allelopathy is the process in which a chemical produced by one plant inhibits the growth of competitors.

16. **Secondary Metabolites; p. 34; difficult; ans: F**
Essential oils, components of plant fragrances, are phenolic compounds.

17. **Secondary Metabolites; p. 37; moderate; ans: F**
It is thought that lignin first evolved as a support molecule and only later functioned as an antibacterial and antifungal agent.

Essay Questions

1. **Carbohydrates; p. 22; moderate**
What features do the structural polysaccharides and energy-storage polysaccharides of plants have in common? How are they different?

2. **Carbohydrates; p. 22; difficult**
   Describe the molecular structure of the plant cell wall in relation to its function.

3. **Lipids; pp. 22, 24; moderate**
   Discuss the relationship between (a) the structure and the solubility of saturated and unsaturated fats; (b) triglycerides and phospholipids.

4. **Lipids; pp. 24–25; moderate**
   Describe the molecules and structures involved in the prevention of water loss in plants.

5. **Proteins; pp. 28–30; difficult**
   Explain how each level of protein organization is influenced by the previous level.

6. **Proteins; p. 30; moderate**
   Discuss the importance of enzymes in the plant cell.

7. **Nucleic Acids; p. 31; easy**
   How do DNA and RNA differ in structure and function?

8. **Nucleic Acids; p. 31; moderate**
   In what way is ATP the cell's energy currency?

9. **Secondary Metabolites; p. 32; easy**
   What is the difference between a primary metabolite and a secondary metabolite? Name the three main classes of secondary metabolites and give an example of each.

10. **Secondary Metabolites; pp. 32–37; difficult**
    It has been said that secondary metabolites are of secondary importance to the plant. Do you agree with this statement? Use examples to support your answer.

Chapter 3

Introduction to the Plant Cell

Multiple-Choice Questions

1. **Development of the Cell Theory; p. 41; easy; ans: c**
   The word "cell" was first used in a biological sense by:

   a. Charles Darwin.
   b. Matthias Schleiden.
   c. Robert Hooke.
   d. Rudolf Virchow.
   e. Theodor Schwann.

2. **Development of the Cell Theory; p. 41; easy; ans: e**
   _____ proposed that cells can arise only from preexisting cells.

   a. Matthias Schleiden
   b. Theodor Schwann
   c. Robert Hooke
   d. Charles Darwin
   e. Rudolf Virchow

3. **Development of the Cell Theory; p. 41; difficult; ans: b**
   Which of the following statements is NOT part of the modern cell theory?

   a. Cells arise from other cells.
   b. Cells are either prokaryotic or eukaryotic.
   c. Cells contain hereditary information that can be passed from parent cell to daughter cell.
   d. The chemical reactions of a living organism take place within cells.
   e. All living organisms are composed of one or more cells.

4. **Prokaryotic Cells and Eukaryotic Cells; p. 41; easy; ans: a**
   Eukaryotic cells differ from prokaryotic cells in that eukaryotic cells have:

   a. a nuclear envelope.
   b. a cytoplasm.
   c. a plasma membrane.
   d. genetic material.
   e. ribosomes.

5. Prokaryotic Cells and Eukaryotic Cells; p. 41; difficult; ans: e
Which of the following is/are NOT found in BOTH prokaryotic cells and eukaryotic cells?

   a.   cytoplasm
   b.   genetic material
   c.   plasmalemma
   d.   ribosomes
   e.   organelles

6. Prokaryotic Cells and Eukaryotic Cells; pp. 41, 43; moderate; ans: c
Which of the following statements about eukaryotes and prokaryotes is FALSE?

   a.   Eukaryotic cells are usually larger than prokaryotic cells.
   b.   Eukaryotic cells possess a cytoskeleton, but prokaryotic cells do not.
   c.   Eukaryotic cells have a nucleoid, but prokaryotic cells do not.
   d.   The DNA of eukaryotic cells is bound to histones, but prokaryotic DNA is not.
   e.   The cell walls of eukaryotic cells and prokaryotic cells differ in composition.

7. Prokaryotic Cells and Eukaryotic Cells; p. 43; moderate; ans: b
Which of the following statements about eukaryotes and/or prokaryotes is TRUE?

   a.   The earliest cells resembled present-day eukaryotes.
   b.   Prokaryotes are represented by archaeans and bacteria.
   c.   Multicellular organisms appeared approximately 7.5 million years ago.
   d.   The transition from prokaryotic cell to eukaryotic cell is considered a relatively insignificant evolutionary event.
   e.   The alga *Chlamydomonas* is an example of a modern prokaryote.

8. The Plant Cell: An Overview; p. 45; moderate; ans: e
Which of the following statements concerning multicellular organisms is FALSE?

   a.   The first multicellular organisms appeared in the fossil record approximately 750 million years ago.
   b.   Their cells closely resemble those of single-celled eukaryotes.
   c.   Each of their cells is specialized to carry out a relatively limited function.
   d.   Each of their cells is self-sustaining.
   e.   An example is *Chlamydomonas*.

9. The Plant Cell: An Overview; pp. 45–46; moderate; ans: e
Which of the following is/are NOT part of the protoplast?

   a.   cytoplasm
   b.   nucleus
   c.   organelles
   d.   ribosomes
   e.   cell wall

10. **The Plant Cell: An Overview; p. 46; moderate; ans: a**
Cyclosis refers to:

    a.   the constant streaming of the cytoplasm.
    b.   the process of cell division.
    c.   that portion of the cytoplasm outside the plasma membrane.
    d.   that portion of the cytoplasm inside the plasma membrane.
    e.   the liquid inside the vacuole.

11. **Plasma Membrane; p. 46; moderate; ans: e**
Which of the following statements about the plasma membrane is FALSE?

    a.   It receives and transmits hormonal and environmental signals.
    b.   It mediates the transport of substances into and out of the cell.
    c.   It coordinates the synthesis and assembly of cellulose microfibrils.
    d.   It has a three-layered appearance when viewed with the electron microscope.
    e.   It has a unique structure, unlike that of the internal cellular membranes.

12. **Nucleus; p. 47; moderate; ans: c**
Nuclear pores are lined with:

    a.   nucleoplasm.
    b.   chromatin.
    c.   portions of the nuclear envelope.
    d.   nucleoli.
    e.   portions of the plasma membrane.

13. **Nucleus; p. 47; moderate; ans: d**
The nuclear envelope may be considered a portion of the:

    a.   nucleoplasm.
    b.   chromatin.
    c.   nucleoli.
    d.   endoplasmic reticulum.
    e.   ribosomes.

14. **Nucleus; p. 47; easy; ans: a**
The dispersed chromatin of nondividing cells seems to be attached to the:

    a.   nuclear envelope.
    b.   nucleolus.
    c.   nuclear pores.
    d.   endoplasmic reticulum.
    e.   nucleoplasm.

15. **Nucleus; p. 47; moderate; ans: b**
If a plant has a diploid chromosome number of 60, how many chromosomes are present in its gametes?

    a.   20
    b.   30
    c.   40
    d.   60
    e.   80

16. **Nucleus; p. 48; easy; ans: c**
    The nucleolus is the structure in which _____ are constructed.

    a. nuclear pores
    b. chromosomes
    c. ribosomes
    d. units of endoplasmic reticulum
    e. portions of the nuclear envelope

17. **Chloroplasts and Other Plastids; p. 48; easy; ans: e**
    The principal plastids involved in photosynthesis are the:

    a. etioplasts.
    b. chromoplasts.
    c. leucoplasts.
    d. proplastids.
    e. chloroplasts.

18. **Chloroplasts and Other Plastids; p. 49; easy; ans: b**
    Grana are stacks of _____ within chloroplasts.

    a. prolamellar bodies
    b. thylakoids
    c. stroma
    d. carotenoids
    e. etioplasts

19. **Chloroplasts and Other Plastids; p. 49; easy; ans: c**
    Chlorophylls and carotenoid pigments are embedded in the:

    a. stroma.
    b. outer chloroplast membrane.
    c. thylakoid membranes.
    d. plasmalemma.
    e. nucleoids.

20. **Chloroplasts and Other Plastids; p. 49; easy; ans: e**
    Plastoglobuli are _____ found in the chloroplasts of plants and green algae.

    a. homogeneous portions of stroma
    b. semicrystalline bodies
    c. starch grains
    d. photosynthetic pigments
    e. small oil bodies

21. **Chloroplasts and Other Plastids; p. 49; moderate; ans: d**
    Which of the following statements about chloroplasts is FALSE?

    a. They can reorient themselves under the influence of light.
    b. Stroma thylakoids interconnect the grana.
    c. Both nuclear and plastid DNA are involved in the formation of
        chloroplasts.
    d. Chloroplasts of plants kept in the dark usually contain starch grains.
    e. Chloroplasts are involved in amino acid and fatty acid synthesis.

22. Chloroplasts and Other Plastids; p. 51; easy; ans: c
Leucoplasts that synthesize starch are known as:

a. etioplasts.
b. chromoplasts.
c. amyloplasts.
d. proplastids.
e. chloroplasts.

23. Chloroplasts and Other Plastids; p. 51; easy; ans: d
_____ are colorless plastids that are the precursors of other, more highly differentiated plastids.

a. Leucoplasts
b. Chromoplasts
c. Amyloplasts
d. Proplastids
e. Chloroplasts

24. Chloroplasts and Other Plastids; p. 51; moderate; ans: a
Etioplasts, by definition, are plastids containing:

a. prolamellar bodies.
b. thylakoids.
c. pigments.
d. plastoglobuli.
e. starch.

25. Chloroplasts and Other Plastids; pp. 51–52; moderate; ans: c
Which of the following statements about plastids is FALSE?

a. They contain one or more nucleoids.
b. They reproduce by fission.
c. They have circular DNA with associated histones.
d. Their ribosomes are sites of protein synthesis.
e. Protein synthesis on their ribosomes is inhibited by antibiotics that have no effect on other eukaryotic ribosomes.

26. Mitochondria; p. 53; moderate; ans: d
The release of energy from fuel molecules occurs in:

a. chloroplasts.
b. the nucleus.
c. the endoplasmic reticulum.
d. mitochondria.
e. peroxisomes.

27. Mitochondria; p. 53; moderate; ans: b
Which of the following statements about mitochondria is FALSE?

a. They are the sites of respiration.
b. They are generally larger than plastids.
c. They can divide by fission.
d. Their DNA is circular.
e. They congregate in parts of the cell where energy is required.

28. Mitochondria; p. 53; moderate; ans: c
In plant cells, genetic information is found in the:

    a.   nucleus only.
    b.   nucleus and mitochondria only.
    c.   nucleus and plastids only.
    d.   nucleus, plastids, and mitochondria only.
    e.   nucleus, plastids, mitochondria, and peroxisomes only.

29. Peroxisomes; p. 54; easy; ans: b
Which organelle plays an important role in photorespiration?

    a.   etioplast
    b.   peroxisome
    c.   chromoplast
    d.   vacuole
    e.   oil body

30. Peroxisomes; p. 54; moderate; ans: c
Peroxisomes are organelles that:

    a.   evolved from the endoplasmic reticulum.
    b.   possess their own DNA.
    c.   self-replicate.
    d.   are bounded by a double membrane.
    e.   possess their own ribosomes.

31. Vacuoles; p. 54; difficult; ans: e
Crystals of _____ are especially common in vacuoles.

    a.   water
    b.   inorganic ions
    c.   sugar
    d.   amino acids
    e.   calcium oxalate

32. Vacuoles; p. 54; easy; ans: a
The _____ in the cell sap are responsible for the red and blue colors of many fruits and vegetables.

    a.   anthocyanins
    b.   carotenoids
    c.   chlorophylls
    d.   calcium oxalate crystals
    e.   oil bodies

33. Vacuoles; p. 54; moderate; ans: b
The vacuoles of plant cells are comparable in function to the _____ of animal cells.

    a.   mitochondria
    b.   lysosomes
    c.   microbodies
    d.   vacuoles
    e.   glyoxysomes

34. **Oil Bodies; p. 55; moderate; ans: d**
    Which of the following statements about oil bodies is FALSE?

    a.   They are lipid droplets.
    b.   They give the cytoplasm a granular appearance.
    c.   They probably arise in the endoplasmic reticulum.
    d.   They are membrane-bounded organelles.
    e.   They are most widely distributed in fruits and seeds.

35. **Ribosomes; p. 55; easy; ans: c**
    Ribosomes are the sites of _____ synthesis.

    a.   lipid
    b.   DNA
    c.   protein
    d.   carbohydrate
    e.   calcium oxalate

36. **Ribosomes; p. 55; easy; ans: e**
    Ribosomes are found both free in the cytoplasm and attached to the:

    a.   plasma membrane.
    b.   oil bodies.
    c.   tonoplast.
    d.   glyoxysomes.
    e.   endoplasmic reticulum.

37. **Ribosomes; pp. 55–56; moderate; ans: c**
    By definition, polysomes are:

    a.   free in the cytoplasm.
    b.   attached to a membrane.
    c.   actively involved in protein synthesis.
    d.   involved with calcium-requiring processes.
    e.   facilitating cellular communication.

38. **Endoplasmic Reticulum; p. 56; easy; ans: c**
    Smooth endoplasmic reticulum is involved in _____ synthesis.

    a.   protein
    b.   glycoprotein
    c.   lipid
    d.   ribosome
    e.   carbohydrate

39. **Endoplasmic Reticulum; p. 56; moderate; ans: b**
    Cisternal endoplasmic reticulum is typically:

    a.   tubular ER.
    b.   rough ER.
    c.   smooth ER.
    d.   lacking a lumen.
    e.   lacking parallel membranes.

40. Endoplasmic Reticulum; p. 56; moderate; ans: a
The most likely function of cortical endoplasmic reticulum is:

   a.   regulating calcium ion levels in the cytoplasm.
   b.   synthesizing proteins on polysomes.
   c.   stabilizing the cytoskeleton.
   d.   forming new nuclear envelopes following nuclear division.
   e.   synthesizing lipids.

41. Endoplasmic Reticulum; p. 56; easy; ans: b
Plasmodesmata are:

   a.   carbohydrates that protrude from the plasma membrane.
   b.   cytoplasmic threads connecting adjacent cells.
   c.   vesicles that move from one portion of the cytoplasm to another.
   d.   portions of the *trans*-Golgi network.
   e.   crystalline materials stored in the vacuole.

42. Golgi Complex; p. 57; moderate; ans: d
Golgi bodies have the appearance of:

   a.   a system of parallel membranes with a narrow space between them.
   b.   a smooth outer membrane and an inner membrane with numerous
        invaginations.
   c.   spherical organelles bounded by a single membrane.
   d.   stacks of flattened, disk-shaped sacs.
   e.   small particles, 17 to 23 nm in diameter.

43. Golgi Complex; p. 58; difficult; ans: c
Which of the following lists the correct sequence in which glycoproteins travel
through the Golgi complex?

   a.   *trans*-Golgi network, maturing face, shuttle vesicles, forming face
   b.   shuttle vesicles, *trans*-Golgi network, forming face, maturing face
   c.   forming face, shuttle vesicles, maturing face, *trans*-Golgi network
   d.   maturing face, shuttle vesicles, *trans*-Golgi network, forming face
   e.   forming face, maturing face, shuttle vesicles, *trans*-Golgi network

44. Golgi Complex; p. 58; moderate; ans: b
Coated vesicles transport proteins from the Golgi bodies to the:

   a.   plasma membrane.
   b.   vacuole.
   c.   mitochondria.
   d.   chloroplasts.
   e.   leucoplasts.

45. Golgi Complex; p. 58; moderate; ans: b
Which of the following is NOT part of the endomembrane system?

   a.   endoplasmic reticulum
   b.   peroxisome membrane
   c.   plasma membrane
   d.   Golgi complex
   e.   nuclear envelope

46. Cytoskeleton; p. 58; easy; ans: e
Tubulin is the protein that:

a. covers coated vesicles.
b. coats microfibrils.
c. lines the tubular endoplasmic reticulum.
d. is the subunit of actin filaments.
e. is the subunit of microtubules.

47. Cytoskeleton; p. 58; moderate; ans: c
Which of the following statements about microtubules is FALSE?

a. They are components of the cytoskeleton.
b. Their subunits are assembled at microtubule organizing centers.
c. Their subunits are arranged in a solid cylinder.
d. They are polar structures, with plus and minus ends.
e. They exhibit dynamic instability.

48. Cytoskeleton; p. 59; easy; ans: a
The alignment of cellulose microfibrils in the cell wall is controlled by:

a. cortical microtubules.
b. nuclear microtubules.
c. cortical actin filaments.
d. nuclear actin filaments.
e. cell wall actin filaments.

49. Cytoskeleton; p. 60; difficult; ans: b
In contrast to microtubules, actin filaments:

a. are composed of protein subunits.
b. lack distinct plus and minus ends.
c. are involved with cell wall deposition.
d. can re-form into new configurations.
e. are long, thin structures.

50. Flagella and Cilia; p. 60; moderate; ans: c
Flagella are different from cilia in that flagella are:

a. hairlike structures.
b. found on the cell surface.
c. longer.
d. found in greater numbers.
e. locomotor structures.

51. Flagella and Cilia; p. 60; moderate; ans: e
In plants, flagella are found only in:

a. leaves.
b. phloem.
c. xylem.
d. motile eggs.
e. motile sperm.

52. **Flagella and Cilia; p. 61; easy; ans: d**
    Flagella and cilia have an internal structure consisting of:
    a.   9 single microtubules.
    b.   9 pairs of microtubules.
    c.   9 triplets of microtubules.
    d.   9 pairs of microtubules surrounding 2 central microtubules.
    e.   9 triplets of microtubules surrounding 2 central microtubules.

53. **Flagella and Cilia; p. 61; easy; ans: c**
    A _____ is a cytoplasmic structure with an internal arrangement of 9 triplets
    of microtubules.
    a.   flagellum
    b.   cilium
    c.   basal body
    d.   microfilament
    e.   basal root

54. **Cell Wall; p. 61; difficult; ans: a**
    Which of the following is NOT a function of the plant cell wall?
    a.   protein synthesis
    b.   digestion
    c.   defense
    d.   absorption
    e.   secretion

55. **Cell Wall; p. 63; easy; ans: e**
    The cell wall matrix consists of all of the following molecules EXCEPT:
    a.   cellulose.
    b.   hemicelluloses.
    c.   pectins.
    d.   glycoproteins.
    e.   tubulin.

56. **Cell Wall; p. 63; moderate; ans: c**
    The principal hemicelluloses in monocots are:
    a.   glycoproteins.
    b.   pectins.
    c.   xylans.
    d.   xyloglucans.
    e.   cutins.

57. **Cell Wall; p. 63; moderate; ans: b**
    The _____ are hydrophilic polysaccharides that make the growing cell wall
    pliable.
    a.   celluloses
    b.   pectins
    c.   hemicelluloses
    d.   lignins
    e.   waxes

58. Cell Wall; p. 63; moderate; ans: d
Which of the following statements about lignin is FALSE?

a. It adds compressive strength and rigidity to the wall.
b. It is commonly found in cells that have a supporting function.
c. It replaces the water in the cell wall.
d. It is a hydrophilic compound.
e. It is first deposited in the intercellular substance at the cell corners.

59. Cell Wall; p. 63; moderate; ans: d
Which of the following substances reduce water loss from the plant?

a. waxes only
b. cutin and suberin only
c. suberin and waxes only
d. cutin, suberin, and waxes, only
e. cutin, suberin, waxes, and the cell wall only

60. Cell Wall; p. 64; difficult; ans: a
Which of the following lists the correct sequence of cell wall layers, beginning with the outermost layer and progressing inward?

a. middle lamella, primary wall, secondary wall
b. secondary wall, primary wall, middle lamella
c. secondary wall, middle lamella, primary wall
d. primary wall, middle lamella, secondary wall
e. middle lamella, secondary wall, primary wall

61. Cell Wall; p. 64; easy; ans: c
Adjacent plant cells are cemented together by the:

a. secondary wall.
b. primary wall.
c. middle lamella.
d. cutin layer.
e. waxy layer.

62. Cell Wall; p. 64; moderate; ans: e
Which of the following statements about the primary wall is FALSE?

a. It is deposited before and during cell growth.
b. It may contain lignin, suberin, or cutin.
c. Actively dividing cells have only primary walls.
d. Some mature plant cells have only a primary wall.
e. It is usually of uniform thickness.

63. Cell Wall; p. 66; easy; ans: c
An interruption in the secondary wall is called a:

a. primary pit-field.
b. pit membrane.
c. pit.
d. pit cavity.
e. pit-pair.

64. Cell Wall; p. 66; moderate; ans: e
Cellulose synthase is an enzyme situated in the:

    a. vacuole.
    b. chloroplasts.
    c. cell wall.
    d. mitochondria.
    e. plasma membrane.

65. Cell Wall; p. 66; difficult; ans: a
Which of the following cell wall components is/are NOT synthesized within the cell then carried to the wall in Golgi vesicles?

    a. cellulose
    b. pectins
    c. xylans
    d. xyloglucans
    e. glycoproteins

66. Plasmodesmata; p. 68; moderate; ans: d
Which of the following best describes a desmotubule?

    a. a component of a flagellum
    b. a microtubule associated with an actin filament
    c. a portion of cytoplasm in a nuclear pore
    d. a portion of endoplasmic reticulum that traverses a plasmodesma
    e. an actin filament associated with tubular endoplasmic reticulum

True-False Questions

1. Introduction to the Plant; p. 41; easy; ans: T
The characteristics of living organisms emerge quite suddenly at the level of the cell.

2. Nucleus; p. 47; easy; ans: T
One function of the nucleus is to determine which protein molecules are produced by the cell.

3. Nucleus; p. 47; easy; ans: F
Different types of organisms have the same number of chromosomes in their somatic cells.

4. Nucleus; p. 47; easy; ans: T
If a plant has 50 chromosomes in each of its somatic cells, then each of its gametes will have 25 chromosomes.

5. Chloroplasts and Other Plastids; p. 52; difficult; ans: F
In dividing meristematic cells, mature plastids undergo fission at a sufficient rate to keep pace with cell division.

6. Peroxisomes; pp. 53–54; moderate; ans: F
Glyoxysomes are peroxisomes involved in the conversion of sucrose to fat during seed germination.

7. **Vacuoles; p. 54; easy; ans: T**
    Most of the increase in size of a plant cell results from enlargement of the vacuole.

8. **Ribosomes; pp. 55–56; easy; ans: T**
    Polysomes are aggregations of ribosomes that are actively engaged in protein synthesis.

9. **Golgi Complex; p. 58; difficult; ans: F**
    A protein destined for secretion at the cell surface is packaged at the *trans*-Golgi network into coated vesicles rather than smooth-surfaced vesicles.

10. **Cytoskeleton; p. 58; easy; ans: T**
    The cytoskeleton of plant cells consists of microtubules, actin filaments, and, in some cases, intermediate filaments.

11. **Flagella and Cilia; p. 61; easy; ans: T**
    Flagella and cilia grow out of basal bodies.

12. **Cell Wall; p. 63; moderate; ans: F**
    The presence of extensins in the cell wall appears to make the wall more extensible.

13. **Cell Wall; p. 64; moderate; ans: T**
    A compound middle lamella may consist of the middle lamella, the two adjacent primary walls, and the first layer of the secondary wall of each cell.

14. **Cell Wall; p. 66; easy; ans: F**
    In cells that enlarge in all directions more or less uniformly, microfibrils have been deposited perpendicular to the axis of elongation.

15. **Plasmodesmata; p. 66; moderate; ans: F**
    Plasmodesmata are formed during cell division as strands of plasma membrane are trapped in the cell plate.

Essay Questions

1. **Development of the Cell Theory; p. 41; moderate**
    Explain why the cell theory is of central importance to biology.

2. **Prokaryotic Cells and Eukaryotic Cells; pp. 41, 43; easy**
    Discuss the major differences and similarities between prokaryotic cells and eukaryotic cells.

3. **Chloroplasts and Other Plastids; pp. 48–52; moderate**
    In what ways are the various types of plastids similar? In what ways are they different?

4. **Mitochondria; p. 53; easy**
    Describe the symbiotic events that probably gave rise to chloroplasts and mitochondria.

5. **Vacuoles; pp. 54–55; moderate**
   Discuss some of the roles played by vacuoles in plant cells.

6. **Golgi Complex; p. 58; moderate**
   Explain how the components of the endomembrane system are interrelated.

7. **Cytoskeleton; p. 59; easy**
   What is meant by the "dynamic instability" of microtubules?

8. **Flagella and Cilia; pp. 60–61; moderate**
   Relate the mechanism of movement of a flagellum to its structure.

9. **Cell Wall; pp. 61–62; moderate**
   Discuss the various roles of the plant cell wall.

10. **Cell Wall; pp. 62–66; difficult**
    List the major components of the primary cell wall. How do the primary and secondary walls differ in structure?

11. **Cell Wall; p. 66; moderate**
    Distinguish between (a) pit membrane and pit-pair; (b) simple and bordered pits.

12. **Cell Wall; p. 66; moderate**
    List the steps involved in the growth of the cell wall.

Chapter 4

Membrane Structure and Function

Multiple-Choice Questions

1. **Introduction; p. 74; easy; ans: d**
   Which of the following is NOT a role of cellular membranes?

   a. regulating the exchange of substances between a cell and the surrounding air, soil, and water
   b. regulating the exchange of substances among cells
   c. controlling the passage of materials among compartments within the cell
   d. preventing the establishment of differences in electrical potential between the cell and its environment
   e. permitting the establishment of difference in electrical potential between adjacent compartments of the cell.

2. **Structure of Cellular Membranes; p. 74; easy; ans: e**
   According to the fluid-mosaic model of cellular membrane structure:

   a. all membranes have a lipid-protein-lipid structure.
   b. proteins form a continuous layer on the surface of the membrane.
   c. the two dark layers visible in the electron microscope are lipid layers.
   d. lipids cannot move laterally within the bilayer.
   e. proteins form different patterns in the bilayer that vary from time to time and place to place.

3. **Structure of Cellular Membranes; p. 74; easy; ans: b**
   The portion of a transmembrane protein embedded in the bilayer is:

   a. hydrophilic.
   b. hydrophobic.
   c. a glycoprotein.
   d. a glycolipid.
   e. a lectin.

4. **Structure of Cellular Membranes; p. 74; moderate; ans: d**
   By definition, all proteins attached to protruding portions of transmembrane proteins are:

   a. transmembrane proteins.
   b. integral proteins.
   c. globular proteins.
   d. peripheral proteins.
   e. glycoproteins.

5. **Structure of Cellular Membranes; p. 74; moderate; ans: a**
   The most abundant type of lipid in the plant cell membrane is:

   a.   phospholipid.
   b.   cholesterol.
   c.   stigmasterol.
   d.   triglyceride.
   e.   glycolipid.

6. **Structure of Cellular Membranes; p. 74; difficult; ans: d**
   In the plasma membrane, carbohydrates are most likely to be found:

   a.   attached to phospholipids in the bilayer.
   b.   attached to cholesterol in the bilayer.
   c.   attached to stigmasterol in the bilayer.
   d.   on the outer membrane surface.
   e.   on the inner membrane surface.

7. **Structure of Cellular Membranes; pp. 74–75; moderate; ans: c**
   What is the hypothesized role of carbohydrates in the plasma membrane?

   a.   catalyze chemical reactions
   b.   keep the phospholipid tails aligned
   c.   recognize molecules that interact with the cell
   d.   act as a barrier to the passage of molecules
   e.   transport molecules into the cell

8. **Structure of Cellular Membranes; p. 75; difficult; ans: b**
   The portion of a transmembrane protein that is embedded in the hydrophobic
   interior of the bilayer is usually in the form of a(n):

   a.   beta pleated sheet.
   b.   alpha helix.
   c.   alpha helix and beta pleated sheet combined.
   d.   extended polypeptide without secondary structure.
   e.   folded polypeptide without secondary structure.

9. **Structure of Cellular Membranes; p. 75; moderate; ans: d**
   Which of the following is NOT a function of membrane proteins?

   a.   convert energy from one form to another
   b.   catalyze chemical reactions
   c.   transport specific molecules or ions into and out of the cell
   d.   make the membrane impermeable
   e.   act as receptors for chemical signals

10. **Movement of Water and Solutes; p. 76; easy; ans: e**
    Water potential is defined as the:

    a.   tendency of water to enter a cell.
    b.   tendency of water to leave a cell.
    c.   mechanical energy of water.
    d.   kinetic energy of water.
    e.   potential energy of water.

11. **Movement of Water and Solutes; p. 77; moderate; ans: b**
    In the absence of other factors affecting water potential, water will move
    FROM a region of _____ TO a region of _____.

    a. low water concentration; high water concentration
    b. low solute concentration; high solute concentration
    c. low water potential; high water potential
    d. low potential energy; high potential energy
    e. low pressure; high pressure

12. **Movement of Water and Solutes; p. 77; easy; ans: c**
    By convention, the water potential of pure water at atmospheric pressure and at
    sea level is set at _____ bars.

    a. −10
    b. −5
    c. 0
    d. 10
    e. 20

13. **Movement of Water and Solutes; p. 77; moderate; ans: e**
    Which of the following could be the water potential of a sucrose solution at
    atmospheric pressure and at sea level?

    a. 100 MPa
    b. 100 bars
    c. 10 bars
    d. 0 bars
    e. −10 MPa

14. **Movement of Water and Solutes; p. 77; moderate; ans: c**
    Which of the following is an example of bulk flow?

    a. perfume molecules filling a room
    b. dye molecules distributing through a tank of water
    c. sap moving in the phloem from leaves to other plant parts
    d. glucose molecules being transported across a membrane
    e. water molecules moving across the plasma membrane

15. **Movement of Water and Solutes; p. 77; moderate; ans: a**
    Which of the following statements about diffusion is FALSE?

    a. When the diffusing molecules become evenly distributed, their movement
        stops.
    b. Each of the diffusing molecules moves randomly and independently of the
        others.
    c. It is more rapid at higher than at lower temperatures.
    d. It involves the net movement of a substance.
    e. It is more rapid in gases than in liquids.

16. **Movement of Water and Solutes; p. 78; moderate; ans: b**
Suppose a drop of dye is placed in one end of a tank of water. What happens next?

 a.   Nothing happens, because the molecules are at equilibrium.
 b.   The dye molecules and the water molecules move down their respective concentration gradients.
 c.   The dye molecules move from a region of high water potential to a region of low water potential.
 d.   The dye molecules move against a concentration gradient.
 e.   The water molecules move by osmosis.

17. **Cells and Diffusion; p. 78; moderate; ans: c**
Which of the following substances is LEAST likely to diffuse across the plasma membrane?

 a.   water
 b.   a small, uncharged polar molecule
 c.   an ion
 d.   carbon dioxide
 e.   oxygen

18. **Cells and Diffusion; p. 78; difficult; ans: b**
Which of the following statements about solute movement in cells is FALSE?

 a.   Once a substance is inside a cell, it moves through the cell by diffusion.
 b.   Diffusion is an effective way to move substances between distantly separated cells.
 c.   Transport of substances within a cell may be speeded up by cytoplasmic streaming.
 d.   Metabolic activities help maintain steep concentration gradients between the inside and outside of the cell.
 e.   Within a cell, a substance may be produced in one place and used in another, thus maintaining an intracellular concentration gradient.

19. **Cells and Diffusion; p. 79; moderate; ans: c**
Osmosis is defined as the:

 a.   movement of any substance across a selectively permeable membrane.
 b.   movement of water down a solute concentration gradient.
 c.   diffusion of water across a selectively permeable membrane.
 d.   movement of water from a region of high to a region of low water potential.
 e.   diffusion of water from a region of low to a region of high water potential.

20. **Cells and Diffusion; p. 79; difficult; ans: b**
Consider two aqueous solutions, A and B. Solution A has a higher solute
concentration than solution B. Solution A is placed on one side of a selectively
permeable membrane, and solution B is placed on the other side. Which of the
following best describes this situation?

   a.   Solution A is hypotonic to solution B.
   b.   Solution B is hypotonic to solution A.
   c.   Solute will flow from solution A to solution B.
   d.   Water will flow from solution A to solution B.
   e.   There will be no net movement of either solute or water between the two
        solutions.

21. **Cells and Diffusion; p. 79; moderate; ans: e**
Osmotic pressure refers to the:

   a.   tendency of water to move across a membrane because of pressure.
   b.   tendency of water to move across a membrane because of solutes.
   c.   pressure necessary to increase the flow of water across a membrane.
   d.   pressure necessary to reverse the flow of water due to gravity.
   e.   pressure necessary to stop the flow of water across a membrane.

22. **Osmosis and Living Organisms; p. 80; moderate; ans: c**
A plant cell is placed in an aqueous solution that has a lower concentration of
dissolved solutes than the cell. Which of the following statements is correct?

   a.   Because the cell is hypotonic to the solution, water will move out of the
        cell.
   b.   Because the cell is hypotonic to the solution, water will move into the cell.
   c.   Because the cell is hypertonic to the solution, water will move into the
        cell.
   d.   Because the cell is hypertonic to the solution, water will move out of the
        cell.
   e.   Because the cell and solution are isotonic, there will be no net movement
        of water.

23. **Osmosis and Living Organisms; p. 81; difficult; ans: e**
Wall pressure:

   a.   is the same as turgor pressure.
   b.   develops within a cell and pushes outward against the wall.
   c.   develops outside the cell wall and pushes inward against the wall.
   d.   is an outwardly directed pressure of the wall.
   e.   is an inwardly directed pressure of the wall.

24. **Osmosis and Living Organisms; p. 81; moderate; ans: b**
If a plant cell is placed in an aqueous solution that is hypertonic to the cell, the
cell will:

   a.   swell and perhaps burst.
   b.   undergo plasmolysis.
   c.   build up turgor pressure.
   d.   build up wall pressure.
   e.   neither gain nor lose water.

25. **Transport of Solutes across Membranes; p. 82; difficult; ans: c**
Which of the following statements about an electrochemical gradient is FALSE?

    a.   It involves both a concentration gradient and the membrane potential.
    b.   It involves both a concentration gradient and the total electrical gradient.
    c.   It is the driving force for the movement of both charged and uncharged substances across a membrane.
    d.   Plant cells maintain an electrochemical gradient across the plasma membrane.
    e.   Plant cells maintain an electrochemical gradient across the tonoplast.

26. **Transport of Solutes across Membranes; p. 82; moderate; ans: e**
Small nonpolar molecules such as O_2 and CO_2 enter a cell by:

    a.   active transport.
    b.   osmosis.
    c.   facilitated diffusion.
    d.   vesicle-mediated transport.
    e.   simple diffusion.

27. **Transport of Solutes across Membranes; p. 83; difficult; ans: a**
If the concentration of K^+ is higher outside a plant cell than inside, K^+ will enter the cell by:

    a.   facilitated diffusion through channel proteins.
    b.   simple diffusion through channel proteins.
    c.   facilitated diffusion via carrier proteins.
    d.   active transport through channel proteins.
    e.   active transport via carrier proteins.

28. **Transport of Solutes across Membranes; p. 83; easy; ans: b**
An aquaporin is a:

    a.   hole in a water molecule.
    b.   channel protein for water.
    c.   carrier protein for water.
    d.   protein that binds water inside the cytosol.
    e.   carbohydrate that binds water to prevent bursting of the cell.

29. **Transport of Solutes across Membranes; p. 83; moderate; ans: d**
A symporter is a carrier protein in which the transport of one solute:

    a.   occurs through a channel.
    b.   occurs without the simultaneous transport of another solute.
    c.   depends on the transport of another solute in the opposite direction.
    d.   depends on the transport of another solute in the same direction.
    e.   depends on the opening of a "gate" in the carrier protein.

30. **Transport of Solutes across Membranes; p. 83; easy; ans: e**
Which of the following processes in plant cell membranes requires ATP?

    a.   simple diffusion
    b.   facilitated diffusion
    c.   osmosis
    d.   passive cotransport
    e.   active transport

31. Transport of Solutes across Membranes; p. 83; moderate; ans: b
If the concentration of glucose is higher inside a cell than outside, glucose will enter the cell only by:

    a.   simple diffusion.
    b.   active transport.
    c.   facilitated diffusion.
    d.   osmosis.
    e.   passive cotransport.

32. Transport of Solutes across Membranes; p. 84; moderate; ans: e
Which of the following statements about a proton pump is FALSE?

    a.   Proton transport requires the input of energy.
    b.   It is an active transport protein.
    c.   It is a membrane-bound ATPase.
    d.   Proton transport provides the driving force for solute uptake.
    e.   Proton transport is the "secondary active transport" process.

33. Vesicle-Mediated Transport; p. 85; easy; ans: d
Proteins and polysaccharides are most likely to enter a cell by:

    a.   active transport.
    b.   osmosis.
    c.   facilitated diffusion.
    d.   vesicle-mediated transport.
    e.   simple diffusion.

34. Vesicle-Mediated Transport; p. 85; easy; ans: b
A cell that ingests bacteria or cellular debris does so by:

    a.   exocytosis.
    b.   phagocytosis.
    c.   pinocytosis.
    d.   receptor-mediated endocytosis.
    e.   facilitated diffusion.

35. Vesicle-Mediated Transport; pp. 85–86; easy; ans: c
All eukaryotic cells ingest liquids by:

    a.   exocytosis.
    b.   phagocytosis.
    c.   pinocytosis.
    d.   receptor-mediated endocytosis.
    e.   active transport.

36. Vesicle-Mediated Transport; p. 86; easy; ans: e
In receptor-mediated endocytosis, specific receptors for the substances to be transported are localized in:

    a.   the nucleus.
    b.   Golgi bodies.
    c.   mitochondria.
    d.   the cytoplasm.
    e.   coated pits.

37. **Vesicle-Mediated Transport; p. 86; moderate; ans: b**
In receptor-mediated endocytosis, what happens immediately after the substance to be transported binds to a receptor?

   a.   The substance is broken down enzymatically.
   b.   The coated pit invaginates to form a coated vesicle.
   c.   The receptor releases the substance inside the cell.
   d.   The coated vesicle sheds its coat.
   e.   The receptor is broken down enzymatically.

38. **Cell-to-Cell Communication; p. 86; easy; ans: a**
Cells are organized into _____, which in turn are organized into _____.

   a.   tissues; organs
   b.   organs; tissues
   c.   organelles; organelle systems
   d.   specialized cells; membrane components
   e.   organelles; membranes

39. **Cell-to-Cell Communication; p. 86; difficult; ans: c**
Which of the following would NOT occur during signal recognition?

   a.   The signal molecule binds to a specific receptor on the plasma membrane.
   b.   The signal molecule is transported into the cell by endocytosis.
   c.   The signal molecule is transported out of the cell by exocytosis.
   d.   The signal molecule remains outside the cell.
   e.   The signal molecule activates a transmembrane protein.

40. **Cell-to-Cell Communication; p. 87; easy; ans: c**
Two of the most common second messengers are:

   a.   hormones and ATP.
   b.   ATP and cyclic AMP.
   c.   calcium ions and cyclic AMP.
   d.   lectins and calcium ions.
   e.   lectins and clathrin.

41. **Cell-to-Cell Communication; p. 87; moderate; ans: d**
In the signal-transduction pathway involving Ca^{2+} ions in plants, which of the following occurs during the transduction step?

   a.   The chemical signal binds to the receptor.
   b.   The Ca^{2+}-calmodulin complex activates specific enzymes.
   c.   Clathrin binds to Ca^{2+} ions.
   d.   Calcium ions are released into the cytosol from the vacuole.
   e.   Calcium ions bind to calmodulin.

42. **Cell-to-Cell Communication; p. 87; moderate; ans: e**
The symplast consists of:

   a.   the cell wall continuum.
   b.   all the plasma membranes.
   c.   all the desmotubules.
   d.   all the protoplasts but not their plasmodesmata.
   e.   all the protoplasts including their plasmodesmata.

43. Cell-to-Cell Communication; p. 87; moderate; ans: a
The cytoplasmic sleeve is the:

   a. cytoplasmic channel involved in symplastic transport through a
       plasmodesma.
   b. cytoplasm inside the lumen of a desmotubule.
   c. cytoplasm in the cortical area of the protoplast.
   d. cytoplasm inside the endoplasmic reticulum.
   e. extension of the cytoplasm into the cell wall.

44. Cell-to-Cell Communication; pp. 88–89; moderate; ans: c
"Movement proteins" aid in the transport of _____ through plasmodesmata.

   a. DNA
   b. proteins
   c. viruses
   d. cyclic AMP
   e. Ca^{2++}

True-False Questions

1. **Membrane Structure and Function; p. 74; easy; ans: F**
    In plants, all exchanges of substances among cells are regulated by the cell
    wall.

2. **Structure of Cellular Membranes; p. 74; easy; ans: T**
    All the membranes of a cell have the same basic structure.

3. **Structure of Cellular Membranes; p. 74; easy; ans: F**
    The inner and outer surfaces of a plasma membrane have the same chemical
    composition.

4. **Structure of Cellular Membranes; p. 75; moderate; ans: T**
    Lectins, proteins that bind specific carbohydrate groups, have been used to
    study the carbohydrate coat of eukaryotic cells.

5. **Movement of Water and Solutes; p. 76; easy; ans: T**
    Water moves from a region where water potential is higher to a region where
    water potential is lower.

6. **Movement of Water and Solutes; p. 77; moderate; ans: T**
    In living organisms, all movement of water and solutes is either by bulk flow
    or by diffusion.

7. **Cells and Diffusion; p. 79; moderate; ans: F**
    The rate of osmosis is influenced by the size of the solute particle.

8. **Cells and Diffusion; p. 79; moderate; ans: F**
    The osmotic pressure exerted by a solution is a measure of the tendency of
    water to move into that solution.

9. **Cells and Diffusion; p. 79; easy; ans: T**
   Osmotic potential refers to the tendency of water to move across a membrane because of the effect of solutes on water potential.

10. **Osmosis and Living Organisms; p. 81; difficult; ans: T**
    When a leaf wilts, the leaf cells most likely have already plasmolyzed.

11. **Transport of Solutes across Membranes; p. 82; moderate; ans: F**
    Unlike active transport, passive transport across cellular membranes does not require the participation of transport proteins.

12. **Transport of Solutes across Membranes; p. 84; easy; ans: F**
    The transport of solutes by carrier proteins is faster than that by channel proteins.

13. **Vesicle-Mediated Transport; p. 85; moderate; ans: F**
    Clathrin is found on the outer surface of the plasma membrane in regions called coated pits.

14. **Cell-to-Cell Communication; p. 87; moderate; ans: T**
    In plants, hormones pass through the cell wall of the target cell then bind to receptors in the plasma membrane or are taken into the cell by endocytosis.

15. **Cell-to-Cell Communication; p. 87; easy; ans: T**
    The symplast consists of all the interconnected protoplasts of the plant body together with their plasmodesmata.

Essay Questions

1. **Membrane Structure and Function; p. 74 moderate**
   Discuss the importance of membranes in the life of the cell.

2. **Structure of Cellular Membranes; p. 74; moderate**
   Which parts of a transmembrane protein are hydrophilic and which parts are hydrophobic? How does this arrangement correlate with the hydrophobic and hydrophilic portions of the phospholipid bilayer?

3. **Structure of Cellular Membranes; pp. 74–76; moderate**
   List the molecular components of the plant cell plasma membrane, and note briefly the function(s) of each.

4. **Movement of Water and Solutes; p. 78; difficult**
   Why is diffusion more rapid (a) in gases than in liquids and (b) at higher temperatures than at lower temperatures?

5. **Movement of Water and Solutes; p. 78; easy**
   When diffusing substances reach equilibrium, diffusion ceases but molecular movement continues. How is this possible?

6. **Cells and Diffusion; p. 79; moderate**
   Explain the terms hypotonic, hypertonic, and isotonic as they relate to plant cells.

7. **Osmosis and Living Organisms; pp. 80–81; moderate**
   How do nonwoody plant parts maintain their stiffness?

8. **Transport of Solutes across Membranes; pp. 82–83; moderate**
   Compare and contrast simple diffusion, facilitated diffusion, and active transport.

9. **Vesicle-Mediated Transport; p. 86; moderate**
   Explain the structural relationship between the surfaces of a coated pit, of a coated vesicle, and of the plasma membrane, and how this relates to the mechanism of receptor-mediated endocytosis.

10. **Cell-to-Cell Communication; pp. 87–89; moderate**
    Discuss the transport of materials between cells via plasmodesmata.

Chapter 5

The Flow of Energy

Multiple-Choice Questions

1. **The Flow of Energy; p. 94; easy; ans: e**
    Of the total solar energy that reaches the Earth, _____ percent is captured by
    the cells of photosynthetic organisms.

    a. about 50
    b. about 20
    c. about 10
    d. about 5
    e. less than 1

2. **The Laws of Thermodynamics; p. 95; easy; ans: c**
    The energy of a system is defined as:

    a. the amount of caloric present.
    b. the amount of ATP present.
    c. its capacity to do work.
    d. the sum total of its chemical bonds.
    e. the speed at which its atoms and molecules move.

3. **The Laws of Thermodynamics; p. 95; moderate; ans: d**
    According to the first law of thermodynamics, the total energy of _____ is

    _____.

    a. a system; constant
    b. a system; changeable
    c. the surroundings; constant
    d. the universe; constant
    e. the universe; changeable

4. **The Laws of Thermodynamics; p. 95; moderate; ans: d**
    Which of the following is NOT an example of potential energy?

    a. a charged flashlight battery
    b. a boulder at the top of a hill
    c. a tank of gasoline
    d. water at the bottom of a waterfall
    e. an apple on a branch

5. **The Laws of Thermodynamics; p. 95; difficult; ans: e**
    Which of the following is NOT an illustration of the first law of
    thermodynamics?

    a. an apple falling from a tree
    b. a nuclear reactor generating electricity
    c. a waterfall turning a waterwheel
    d. electricity causing a bulb to light up
    e. heat being dissipated into the environment

6. **The Laws of Thermodynamics; p. 95; easy; ans: a**
    The first law of thermodynamics states that the total energy of a system and its
    surroundings after an energy conversion is _____ the total energy before the
    conversion.

    a. equal to
    b. greater than or equal to
    c. greater than
    d. less than
    e. less than or equal to

7. **The Laws of Thermodynamics; pp. 96–97; moderate; ans: a**
    Which of the following statements about an exergonic reaction is FALSE?

    a. The potential energy of the final state is greater than that of the initial
        state.
    b. It takes place spontaneously.
    c. It takes place without an input of energy from the outside.
    d. It releases energy.
    e. It has a negative ΔG value.

8. **The Laws of Thermodynamics; pp. 96–97; difficult; ans: c**
    ΔH refers to the:

    a. heat released in an exergonic reaction.
    b. heat released in an endergonic reaction.
    c. change in enthalpy.
    d. change in entropy.
    e. exact change in potential energy.

9. **The Laws of Thermodynamics; p. 97; easy; ans: d**
    Entropy refers to the _____ of a system.

    a. heat content
    b. free energy
    c. potential energy
    d. disorder or randomness
    e. kinetic energy

10. **The Laws of Thermodynamics; p. 97; difficult; ans: b**
Which of the following statements about the change of state of water is
FALSE?

   a.   The change from ice to liquid water is an endothermic process.
   b.   The change from liquid water to water vapor is associated with a release
       of heat.
   c.   The change from ice to liquid water is associated with an increase in
       entropy.
   d.   The change from liquid water to water vapor is associated with an increase
       in entropy.
   e.   The change from ice to liquid water can proceed spontaneously.

11. **The Laws of Thermodynamics; p. 97; moderate; ans: b**
ΔG refers to the _____ of the system.

   a.   change in total heat content
   b.   total overall energy change
   c.   change in entropy
   d.   change in enthalpy
   e.   change in randomness

12. **The Laws of Thermodynamics; p. 97; easy; ans: d**
Which of the following equations relating enthalpy, entropy, and free energy is
correct?

   a. $\Delta S = \Delta H - T\,\Delta G$
   b. $\Delta H = \Delta G - T\,\Delta S$
   c. $\Delta S = \Delta G + T\,\Delta H$
   d. $\Delta G = \Delta H - T\,\Delta S$
   e. $\Delta G = \Delta H + T\,\Delta S$

13. **The Laws of Thermodynamics; p. 97; moderate; ans: e**
In all naturally occurring processes:

   a. ΔH is positive.
   b. ΔH is negative.
   c. ΔS is positive.
   d. ΔS is negative.
   e. ΔG is negative.

14. **The Laws of Thermodynamics; p. 97; easy; ans: c**
The second law of thermodynamics states that all naturally occurring processes:

   a.   are exothermic.
   b.   are endothermic.
   c.   are exergonic.
   d.   are endergonic.
   e.   proceed with an increase in entropy.

15. **The Laws of Thermodynamics; pp. 97–98; moderate; ans: b**
    Which of the following statements about living cells is FALSE?

    a. They expend energy to maintain order.
    b. They are at equilibrium with their surroundings.
    c. They are open systems.
    d. They transform energy.
    e. As they transform energy, they dissipate heat.

16. **Oxidation-Reduction; p. 98; easy; ans: a**
    Oxidation is defined as the:

    a. loss of electrons.
    b. loss of protons.
    c. gain of electrons.
    d. gain of protons.
    e. gain of both electrons and protons.

17. **Oxidation-Reduction; p. 99; moderate; ans: b**
    In biological reactions, when a molecule is reduced it _____ both an electron
    and a(n) _____.

    a. loses; proton
    b. gains; proton
    c. loses; oxygen atom
    d. gains; oxygen atom
    e. loses; neutron

18. **Oxidation-Reduction; p. 99; easy; ans: c**
    In the following reaction, which molecule is reduced?
    carbon dioxide + water → glucose + oxygen

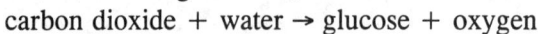

    a. glucose
    b. oxygen
    c. carbon dioxide
    d. water
    e. Both carbon dioxide and water are reduced.

19. **Oxidation-Reduction; p. 99; moderate; ans: e**
    Which of the following statements about respiration and photosynthesis is
    FALSE?

    a. Photosynthesis is an endergonic process.
    b. Respiration is an exergonic process.
    c. Both respiration and photosynthesis are redox reactions.
    d. During respiration, electrons are moved to a lower energy level.
    e. During photosynthesis, electrons are moved to a lower energy level.

20. **Oxidation-Reduction; p. 99; easy; ans: d**
    The complete oxidation of one mole of glucose releases _____ of energy.

    a. 6.86 calories
    b. 686 calories
    c. 6.86 kilocalories
    d. 686 kilocalories
    e. 686,000 kilocalories

21. **Oxidation-Reduction; p. 99; moderate; ans: b**
During the oxidation of glucose in the cell, most of the energy is released:
    a.   all at once.
    b.   in small amounts.
    c.   via mechanisms that require many kinds of molecules.
    d.   via mechanisms that result in high temperatures.
    e.   in a way that requires the input of 686 kilocalories of energy.

22. **Enzymes; pp. 99–100; easy; ans: e**
Which of the following statements about the energy of activation is FALSE?
    a.   It is the initial input of energy that gets a reaction started.
    b.   It increases the kinetic energy of the reacting molecules.
    c.   It helps overcome the repulsion between the electrons of the reacting molecules.
    d.   It helps break the existing chemical bonds in the reacting molecules.
    e.   In cells, it is usually supplied as heat.

23. **Enzymes; p. 100; easy; ans: d**
An enzyme:
    a.   is typically effective only in large amounts.
    b.   is a type of carbohydrate.
    c.   raises the activation energy of its reaction.
    d.   functions as a catalyst.
    e.   is permanently altered during the course of its reaction.

24. **Enzymes; p. 101; easy; ans: b**
Most enzymes are:
    a.   RNA molecules.
    b.   proteins.
    c.   carbohydrates.
    d.   lipids.
    e.   DNA molecules.

25. **Enzymes; p. 101; easy; ans: c**
A substrate binds to its enzyme at a location called the _____ site.
    a.   coenzyme
    b.   substrate
    c.   active
    d.   polypeptide
    e.   cofactor

26. **Enzymes; p. 101; moderate; ans: d**
A hydrophobic portion of a substrate molecule would most likely fit into a _____ region of its enzyme's active site.
    a.   positively charged
    b.   negatively charged
    c.   hydrophilic
    d.   hydrophobic
    e.   polar

27. **Enzymes; p. 101; moderate; ans: e**
    According to the induced-fit hypothesis:

    a.   a substrate binds to an active site but not to an effector site.
    b.   the shape of an enzyme is affected by temperature.
    c.   the shape of an enzyme is affected by pH.
    d.   a coenzyme may cause the active site to change shape.
    e.   a substrate may cause the active site to change shape.

28. **Cofactors in Enzyme Action; p. 102; moderate; ans: b**
    The general name for a nonprotein component required by some enzymes is a:

    a.   metal ion.
    b.   cofactor.
    c.   substrate.
    d.   prosthetic group.
    e.   coenzyme.

29. **Cofactors in Enzyme Action; pp. 102, 106; difficult; ans: b**
    Most enzymes that catalyze phosphorylation reactions require _____ as a
    cofactor.

    a. NAD^+
    b. Mg^{2+}
    c. Ca^{2+}
    d.   AMP
    e.   an iron-sulfur cluster

30. **Cofactors in Enzyme Action; pp. 102–103; moderate; ans: c**
    An example of a coenzyme that is NOT a prosthetic group is:

    a. Mg^{2+}.
    b. K^+.
    c. NAD^+.
    d.   an iron-sulfur cluster.
    e.   pyridoxal phosphate.

31. **Cofactors in Enzyme Action; p. 102; moderate; ans: e**
    The vitamin niacin is part of the _____ molecule.

    a.   ferredoxin
    b.   iron-sulfur
    c.   pyridoxal phosphate
    d.   pyrophosphate
    e. NAD^+

32. **Metabolic Pathways; p. 103; difficult; ans: c**
    Which of the following statements about metabolic pathways is FALSE?

    a.   The enzymes of some pathways are segregated inside specific organelles.
    b.   The enzymes of some pathways are embedded in specific membranes.
    c.   The enzymes of most pathways allow intermediate products to accumulate.
    d.   Exergonic reactions in the pathway will pull forward preceding reactions.
    e.   The products from exergonic reactions will push along subsequent
         reactions.

33. Metabolic Pathways; p. 103; moderate; ans: d
Isozymes are:

a.   RNA molecules that catalyze metabolic reactions.
b.   identical coenzymes that require different metal ions.
c.   identical coenzymes located in different parts of the cell.
d.   different enzymes that catalyze identical reactions.
e.   identical enzymes that catalyze different reactions.

34. Regulation of Enzyme Activity; pp. 103–104; easy; ans: b
As the temperature increases, the rate of an enzyme-catalyzed reaction usually:

a.   increases constantly.
b.   increases up to a point, then decreases.
c.   decreases constantly.
d.   decreases to a point, then increases.
e.   increases, then decreases, then increases again.

35. Regulation of Enzyme Activity; p. 104; moderate; ans: c
Denaturation of an enzyme refers to the:

a.   proper arrangement of the enzyme in a metabolic pathway.
b.   improper arrangement of the enzyme in a metabolic pathway.
c.   loss of the enzyme's proper shape.
d.   formation of the enzyme's proper shape.
e.   formation of a new isozyme for that enzyme.

36. Regulation of Enzyme Activity; p. 104; moderate; ans: b
A regulatory enzyme in a metabolic pathway is most likely to be the:

a.   last enzyme in the pathway.
b.   first enzyme in the pathway.
c.   enzyme that catalyzes the fastest reaction in the pathway.
d.   enyzme that binds its coenzyme the fastest.
e.   enzyme that lacks an effector site.

37. Regulation of Enzyme Activity; p. 104; difficult; ans: a
In an allosteric enzyme, the substrate binds at the _____ site and the regulatory substance binds at the _____ site.

a.   active; effector
b.   effector; active
c.   active; inhibitor
d.   inhibitor; active
e.   inhibitor; effector

38. The Energy Factor: ATP Revisited; p. 105; easy; ans: c
In an ATP molecule, phosphoanhydride bonds link:

a.   adenine to ribose.
b.   adenine to a phosphate group.
c.   the phosphate groups together.
d.   ribose to a phosphate group.
e.   the ribose groups together.

39. The Energy Factor: ATP Revisited; p. 105; moderate; ans: d
Which of the following statements about the energy involved in cellular reactions is FALSE?

  a. The direction in which a reaction proceeds is determined by ΔG.
  b. Enzymes reduce the energy of activation of a reaction to a level possessed by a significant proportion of the reacting molecules.
  c. ATP is the most frequently used intermediate between exergonic and endergonic reactions.
  d. In coupled reactions, an endergonic reaction provides energy to drive an exergonic reaction.
  e. When reactions are coupled, the overall process is exergonic.

40. The Energy Factor: ATP Revisited; pp. 105–106; easy; ans: b
Which type of enzyme catalyzes the following reaction?
$ATP + H_2O \rightarrow ADP + phosphate$

  a. phosphorylase
  b. ATPase
  c. ADPase
  d. kinase
  e. phosphatase

41. The Energy Factor: ATP Revisited; p. 106; moderate; ans: d
Which statement about the following reaction is FALSE?
$ATP + H_2O \rightarrow ADP + phosphate$

  a. It is catalyzed by an ATPase.
  b. It releases about 7.3 kilocalories of energy per mole
  c. It involves the breaking of phosphoanhydride bonds.
  d. The reactants are more stable than the products.
  e. It is exergonic.

42. The Energy Factor: ATP Revisited; p. 106; difficult; ans: a
Which of the following is a phosphorylation reaction catalyzed by a kinase in plant cells?

  a. $ATP + glucose \rightarrow glucose\ phosphate + ADP$
  b. $Fructose\ phosphate + ADP \rightarrow fructose + ATP$
  c. $Glucose\ phosphate + ADP \rightarrow glucose + ATP$
  d. $ATP + H_2O \rightarrow ADP + phosphate$
  e. $ADP + phosphate \rightarrow ATP + H_2O$

43. The Energy Factor: ATP Revisited; p. 106; difficult; ans: b
Which of the following statements about sucrose synthesis is FALSE?

  a. Under standard thermodynamic conditions sucrose synthesis is endergonic.
  b. In the cell, sucrose synthesis is endergonic.
  c. In the cell, sucrose synthesis involves the phosphorylation of glucose.
  d. In the cell, sucrose synthesis involves the phosphorylation of fructose.
  e. In the cell, sucrose synthesis is coupled with ATP hydrolysis.

True-False Questions

1. **The Flow of Energy; p. 94; easy; ans: F**
    A living cell is a system that prevents the change of energy from one form to another.

2. **The Laws of Thermodynamics; p. 95; moderate; ans: F**
    The heat released in the exhaust of an engine can produce work.

3. **The Laws of Thermodynamics; p. 97; easy; ans: T**
    More disorder is associated with numerous small objects than with fewer larger ones.

4. **The Laws of Thermodynamics; p. 97; moderate; ans: T**
    The *overall* charge in energy that determines the course of a chemical reaction is the free-energy charge, ΔG.

5. **The Laws of Thermodynamics; p. 97; moderate; ans: F**
    The universe is an open system: as energy leaves the universe, its potential energy declines.

6. **Oxidation-Reduction; p. 98; easy; ans: T**
    Oxidation and reduction always take place simultaneously.

7. **Oxidation-Reduction; p. 99; easy; ans: F**
    In cells, the energy released during the oxidation of glucose is released and used by the cell all at once.

8. **Enzymes; p. 100; easy; ans: T**
    A single enzyme molecule may catalyze its reaction at the rate of tens of thousands of substrate molecules per second.

9. **Enzymes; p. 101; moderate; ans: F**
    The amino acids in an enzyme's active site are always adjacent to each other on the polypeptide chain.

10. **Cofactors in Enzyme Action; p. 102; easy; ans: T**
    A metal ion can be a cofactor.

11. **Cofactors in Enzyme Action; p. 102; easy; ans: F**
    Plants, like humans, are unable to synthesize all of their required vitamins.

12. **Metabolic Pathways; p. 103; easy; ans: T**
    In a metabolic pathway, there is little accumulation of intermediate products.

13. **Regulation of Enzyme Activity; p. 103; moderate; ans: T**
    A principal factor regulating enzyme activity in cells is the amount of substrate available.

14. **Regulation of Enzyme Activity; p. 104; easy; ans: F**
    For most enzyme catalyzed reactions, the rate approximately doubles for each 5°C increase in temperature.

15. **The Energy Factor: ATP Revisited; p. 105; easy; ans: T**
    The energy currency of the cell is ATP.

16. **The Energy Factor: ATP Revisited; pp. 105, 106; moderate; ans: T**
    Kinases, enzymes that catalyze phosphorylation reactions, play an essential role in coupled reactions in cells.

Essay Questions

1. **The Laws of Thermodynamics; pp. 95–98; difficult**
   How do the first and second laws of thermodynamics relate to living organisms?

2. **The Laws of Thermodynamics; p. 97; moderate**
   State in words the meaning of the equation $\Delta G = \Delta H - T \Delta S$.

3. **The Laws of Thermodynamics; p. 98; difficult**
   Explain the meaning of the sentence, "Life can exist *because* the universe is running down."

4. **Oxidation-Reduction; p. 98; moderate**
   Explain the term "redox reactions." Why do the two types of reactions included in this term always occur simultaneously?

5. **Oxidation-Reduction; p. 99; easy**
   Why have living organisms developed mechanisms to release energy from their fuel molecules in small amounts?

6. **Enzymes; p. 100; moderate**
   How does a catalyst work? What is the role of catalysts in cells?

7. **Enzymes; p. 101; moderate**
   Explain the induced-fit hypothesis.

8. **Cofactors in Enzyme Action; pp. 102–103; moderate**
   Distinguish between the terms "cofactor," "coenzyme," and "prosthetic group."

9. **Metabolic Pathways; p. 103; easy**
   Discuss the advantages to the cell of arranging enzyme-catalyzed reactions into metabolic pathways.

10. **Regulation of Enzyme Activity; p. 104; moderate**
    How do allosteric enzymes function? What role do they play in the cell?

11. **Regulation of Enzyme Activity; p. 104; moderate**
    Describe the process of feedback inhibition. Why is this process important in the cell?

12. **The Energy Factor: ATP Revisited; p. 106; moderate**
    What is the typical role of ATP in coupled reactions in plant cells?

13. The Energy Factor: ATP Revisited; p. 106; moderate
Use the synthesis of sucrose from glucose and fructose to illustrate how an endergonic reaction can be coupled to an exergonic reaction.

Chapter 6

Respiration

Multiple-Choice Questions

1. **An Overview of Glucose Oxidation; p. 109; easy; ans: e**
    Which of the following statements about the reactions of glucose oxidation is FALSE?

    a. The glucose molecule is split apart.
    b. Hydrogen atoms are removed.
    c. Oxygen is reduced.
    d. Energy is released.
    e. Electrons go from lower to higher energy levels.

2. **An Overview of Glucose Oxidation; p. 109; moderate; ans: b**
    In respiration, electrons are extracted from glucose and ultimately accepted by:

    a. ATP.
    b. oxygen.
    c. carbon dioxide.
    d. water.
    e. pyruvate.

3. **An Overview of Glucose Oxidation; p. 109; easy; ans: d**
    Which of the following is NOT a stage of respiration?

    a. Krebs cycle
    b. electron transport chain
    c. glycolysis
    d. hydrolysis of starch to glucose
    e. oxidative phosphorylation

4. **An Overview of Glucose Oxidation; p. 109; easy; ans: a**
    The breakdown of glucose to pyruvate occurs in:

    a. glycolysis.
    b. oxidative phosphorylation.
    c. the Krebs cycle.
    d. electron transport.
    e. the hydrolysis of sucrose and starch.

5. **An Overview of Glucose Oxidation; p. 109; moderate; ans: e**
Formation of ATP from ADP and phosphate as a result of electron transport occurs in:

   a.   the Krebs cycle.
   b.   the hydrolysis of sucrose and starch.
   c.   fermentation.
   d.   glycolysis.
   e.   oxidative phosphorylation.

6. **Glycolysis; p. 109; easy; ans: c**
In glycolysis, one molecule of glucose is converted to _____ molecules of

_____.

   a.   two; fructose
   b.   three; fructose.
   c.   two; pyruvate
   d.   three; pyruvate
   e.   two; sucrose

7. **Glycolysis; p. 109; easy; ans: b**
Within the plant cell, glycolysis occurs in:

   a.   mitochondria.
   b.   the cytosol.
   c.   chloroplasts.
   d.   the endoplasmic reticulum.
   e.   the nucleus.

8. **Glycolysis; p. 109; easy; ans: e**
The cell's net energy harvest from glycolysis is represented by:

   a.   ATP only.
   b.   ADP only.
   c. NAD^+ only.
   d.   NADH only
   e.   ATP and NADH.

9. **Glycolysis; p. 111; difficult; ans: c**
As part of the first step in the first preparatory phase of glycolysis:

   a.   the glucose molecule is rearranged.
   b.   the glucose molecule is split in half.
   c.   glucose is phosphorylated.
   d. NAD^+ is reduced.
   e.   NADH is oxidized.

10. **Glycolysis; p. 111; difficult; ans: a**
As part of the cleavage step in glycolysis, glucose is:

   a.   converted to glyceraldehyde 3-phosphate and dihydroxyacetone phosphate.
   b.   phosphorylated to glucose 6-phosphate.
   c.   oxidized to 1,3-bisphosphoglycerate.
   d.   reduced to phosphoenolpyruvate.
   e.   converted by the enzyme isomerase to fructose.

11. **Glycolysis; p. 112; difficult; ans: d**
    As part of the first reaction in the payoff phase of glycolysis:

    a.   phosphoenolpyruvate is converted to pyruvate.
    b.   ATP phosphorylates a molecule of glucose.
    c.   1,3-bisphosphoglycerate phosphorylates a molecule of ADP.
    d.   glyceraldehyde 3-phosphate is oxidized.
    e.   dihydroxyacetone phosphate is oxidized.

12. **Glycolysis; pp. 112–113; easy; ans: c**
    For each molecule of glucose that completes glycolysis, how many NAD^+ molecules are reduced?

    a.   0
    b.   1
    c.   2
    d.   3
    e.   4

13. **Glycolysis; p. 113; moderate; ans: e**
    For each molecule of glucose that completes glycolysis, how many ADP undergo phosphorylation to ATP?

    a.   0
    b.   1
    c.   2
    d.   3
    e.   4

14. **Glycolysis; p. 113; moderate; ans: c**
    In glycolysis, what is the *net* energy harvest of ATP molecules per molecule of glucose?

    a.   0
    b.   1
    c.   2
    d.   3
    e.   4

15. **The Aerobic Pathway; p. 114; moderate; ans: d**
    The inner mitochondrial membrane restrains the passage of:

    a.   pyruvate.
    b.   ADP.
    c.   ATP.
    d.   protons.
    e.   all ions.

16. **The Aerobic Pathway; p. 114; moderate; ans: a**
    Before the aerobic phase of glucose breakdown begins, pyruvate is:

    a.   decarboxylated.
    b.   reduced by NADH.
    c.   phosphorylated.
    d.   joined directly to coenzyme A.
    e.   oxidized by FAD.

17. The Aerobic Pathway; p. 114; moderate; ans: b
Before entering the Krebs cycle, most of the carbon atoms originally present in pyruvate are converted to:

a.   carbon dioxide.
b.   an acetyl group.
c.   glucose.
d.   NADH.
e.   ATP.

18. The Aerobic Pathway; p. 115; moderate; ans: c
Upon entering the Krebs cycle, the acetyl group combines with _____ to produce _____.

a.   coenzyme A; pyruvate
b.   glucose; glucose 6-phosphate
c.   oxaloacetate; citrate
d.   oxaloacetate; carbon dioxide
e.   NADH; citrate

19. The Aerobic Pathway; p. 115; moderate; ans: e
After acetyl CoA enters the Krebs cycle, the coenzyme A portion of the molecule:

a.   combines with oxaloacetate.
b.   combines with citrate.
c.   is oxidized.
d.   is reduced.
e.   is released.

20. The Aerobic Pathway; pp. 115–116; moderate; ans: d
Which of the following does NOT occur during the Krebs cycle?

a.   decarboxylation
b.   substrate-level phosphorylation
c.   oxidation
d.   oxidative phosphorylation
e.   regeneration of oxaloacetate

21. The Aerobic Pathway; p. 115; easy; ans: c
In each turn of the Krebs cycle, how many molecules of carbon dioxide are produced?

a.   0
b.   1
c.   2
d.   3
e.   4

22. **The Aerobic Pathway; p. 116; moderate; ans: c**
In the Krebs cycle, how many molecules of FADH are produced per molecule of glucose?

    a. 0
    b. 1
    c. 2
    d. 3
    e. 4

23. **The Aerobic Pathway; p. 116; moderate; ans: a**
In the Krebs cycle, how many molecules of O_2 are reduced per molecule of glucose?

    a. 0
    b. 1
    c. 2
    d. 3
    e. 4

24. **The Aerobic Pathway; p. 116; moderate; ans: e**
In the Krebs cycle, how many molecules of NADH are produced per molecule of glucose?

    a. 0
    b. 1
    c. 2
    d. 4
    e. 6

25. **The Aerobic Pathway; p. 116; easy; ans: d**
Most of the carriers of the electron transport chain are:

    a. in the cytosol.
    b. in the mitochondrial matrix.
    c. contained between the outer and inner mitochondrial membranes.
    d. embedded in the inner mitochondrial membrane.
    e. embedded in the outer mitochondrial membrane.

26. **The Aerobic Pathway; p. 116; moderate; ans: b**
Which of the following components of the electron transport chain consist(s) of a protein attached to an iron-containing porphyrin ring?

    a. ubiquinone
    b. cytochromes
    c. iron-sulfur proteins
    d. coenzyme Q
    e. NAD

27. **The Aerobic Pathway; p. 116; difficult; ans: e**
Which of the following statements about iron-sulfur proteins is FALSE?

    a. They are components of the electron transport chain.
    b. Their iron is not attached to a porphyrin ring.
    c. Their iron is attached to sulfides.
    d. Their iron is attached to the sulfur of sulfur-containing amino acids.
    e. They carry both electrons and protons.

28. **The Aerobic Pathway; p. 118; moderate; ans: c**
The most abundant component of the mitochondrial electron transport chain is/are:

   a. cytochromes.
   b. iron-sulfur proteins.
   c. coenzyme Q.
   d. NAD.
   e. FAD.

29. **The Aerobic Pathway; p. 118; moderate; ans: d**
In the electron transport chain, quinone molecules shuttle protons from the _____ into the _____.

   a. cytosol; intermembrane space
   b. intermembrane space; cytosol
   c. outer membrane; inner membrane
   d. matrix; intermembrane space
   e. intermembrane space; matrix

30. **The Aerobic Pathway; p. 118; moderate; ans: e**
The energy released by the flow of electrons along the electron transport chain is used *directly* to:

   a. form ATP from ADP and phosphate.
   b. oxidize NADH.
   c. reduce FAD.
   d. decarboxylate citrate.
   e. pump protons.

31. **The Aerobic Pathway; p. 118; moderate; ans: d**
The final electron acceptor in the electron transport chain is:

   a. CoQ.
   b. a cytochrome.
   c. FMN.
   d. oxygen.
   e. carbon dioxide.

32. **The Aerobic Pathway; p. 118; easy; ans: d**
For each pair of electrons passing from NADH to oxygen, how many ATP molecules can be generated?

   a. 0
   b. 1
   c. 2
   d. 3
   e. 4

33. **The Aerobic Pathway; p. 118; difficult; ans: d**
Oxidative phosphorylation depends on a gradient of _____ across the mitochondrial membrane.

   a. ADP
   b. phosphate
   c. glucose
   d. protons
   e. electrons

34. **The Aerobic Pathway; p. 118; easy; ans: c**
The electron carriers of the electron transport chain are associated with proteins to form _____ complexes in the inner mitochondrial membrane.

   a. 1
   b. 2
   c. 3
   d. 4
   e. 5

35. **The Aerobic Pathway; p. 118; moderate; ans: a**
A primary function of the protein complexes of the electron transport chain is to:

   a. pump protons.
   b. synthesize ATP.
   c. reduce NAD^+.
   d. oxidize $FADH_2$.
   e. bind oxygen.

36. **The Aerobic Pathway; p. 119; easy; ans: e**
For each pair of electrons moving down the electron transport chain from NADH to oxygen, about _____ protons are pumped out of the matrix.

   a. 2
   b. 4
   c. 6
   d. 8
   e. 10

37. **The Aerobic Pathway; p. 119; moderate; ans: c**
The electrochemical gradient resulting from electron transport is due to differences in _____ across the inner mitochondrial membrane.

   a. electric charge only
   b. proton concentration only
   c. electric charge and proton concentration
   d. ATP concentration only
   e. ATP and NAD^+ concentrations

38. The Aerobic Pathway; pp. 119–120; moderate; ans: b
Which of the following statements about ATP synthase is FALSE?

   a.   It synthesizes ATP.
   b.   It transports electrons.
   c.   It binds ATP.
   d.   It binds ADP.
   e.   It transports hydrogen ions.

39. The Aerobic Pathway; p. 122; easy; ans: c
The net energy yield from complete oxidation of one molecule of glucose is
_____ molecules of ATP.

   a.   22
   b.   24
   c.   36
   d.   40
   e.   48

40. The Aerobic Pathway; p. 122; easy; ans: a
Most of the ATP formed in respiration is produced by reactions associated
with:

   a.   the electron transport chain.
   b.   the Krebs cycle.
   c.   glycolysis.
   d.   fermentation.
   e.   the conversion of pyruvate to acetyl CoA.

41. Other Substrates for Respiration; p. 122; moderate; ans: e
The process of β oxidation is involved in the breakdown of:

   a.   starch.
   b.   sucrose.
   c.   proteins.
   d.   nucleic acids.
   e.   triglycerides.

42. Anaerobic Pathways; pp. 122–123; difficult; ans: d
The end products of pyruvate breakdown in the presence of oxygen are _____,
but in the absence of oxygen the end products are _____.

   a.   glyceraldehyde 3-phosphate and dihydroxyacetone phosphate; citrate and
       fumarate
   b.   carbon dioxide and water; acetyl CoA and citrate
   c.   acetyl CoA and citrate; carbon dioxide and water
   d.   carbon dioxide and water; lactate or ethanol
   e.   lactate or ethanol; carbon dioxide and water

43. Anaerobic Pathways; p. 122; moderate; ans: b
In certain bacteria, fungi, protists, and animal cells under anaerobic conditions, the pyruvate formed from glycolysis is converted to:

a.   acetyl CoA.
b.   lactate.
c.   ethanol and carbon dioxide.
d.   ATP.
e.   glucose.

44. Anaerobic Pathways; p. 123; moderate; ans: d
Which of the following processes occurs in both lactate fermentation and alcohol fermentation?

a.   formation of acetyl CoA
b.   release of carbon dioxide
c.   pumping of protons
d.   oxidation of NADH
e.   activation of ATP synthase

45. Anaerobic Pathways; p. 123; easy; ans: c
In both lactate fermentation and alcohol fermentation, the net ATP production is _____ molecules of ATP per molecule of glucose.

a.   0
b.   1
c.   2
d.   3
e.   4

46. Anaerobic Pathways; p. 123; difficult; ans: c
During alcohol fermentation, the efficiency with which the cell conserves energy as ATP is about ____ percent.

a.   7
b.   14
c.   26
d.   38
e.   93

47. The Strategy of Energy Metabolism; p. 123; easy; ans: b
"Catabolism" specifically refers to the various pathways in which organisms _____ organic molecules.

a.   synthesize
b.   break down
c.   phosphorylate
d.   oxidize
e.   reduce

48. The Strategy of Energy Metabolism; p. 124; easy; ans: e
The metabolic "hub" of the cell is:
   a.   fermentation.
   b.   glycolysis.
   c.   oxidative phosphorylation.
   d.   the electron transport chain.
   e.   the Krebs cycle.

True-False Questions

1. An Overview of Glucose Oxidation; p. 109; moderate; ans: T
More energy is obtained when glucose is oxidized under aerobic conditions rather than anaerobic conditions.

2. Glycolysis: p. 109; easy; ans: F
All four stages of respiration take place in mitochondria.

3. Glycolysis; p. 112; moderate; ans: T
In substrate-level phosphorylation, an enzyme transfers a phosphate group from a metabolic intermediate to ADP.

4. Glycolysis; pp. 112–113; moderate; ans: T
In glycolysis, ATP is consumed in some reactions and produced in others.

5. Glycolysis; p. 113; moderate; ans: F
In glycolysis, NADH is consumed in some reactions and produced in others.

6. The Aerobic Pathway; p. 114; easy; ans: F
The inner mitochondrial membrane is permeable to most small molecules and ions.

7. The Aerobic Pathway; p. 114; easy; ans: T
In respiration, after pyruvate enters the mitochondrion, it is both oxidized and decarboxylated.

8. The Aerobic Pathway; p. 115; moderate; ans: T
Complete oxidation of one molecule of glucose requires two turns around the Krebs cycle.

9. The Aerobic Pathway; p. 116; moderate; ans: F
Cytochromes are unique electron carriers in that they can accept protons as well as electrons.

10. The Aerobic Pathway; p. 118; easy; ans: T
Oxidative phosphorylation requires an intact inner mitochondrial membrane.

11. The Aerobic Pathway; p. 119; moderate; ans: F
As protons flow through ATP synthase down the electrochemical gradient from the matrix into the intermembrane space, ATP is synthesized from ADP and phosphate.

12. **Other Substrates for Respiration; p. 122; moderate; ans: T**
    The catabolism of proteins and fats involves the Krebs cycle.

13. **Anaerobic Pathways; p. 122; moderate; ans: F**
    The NADH generated during fermentation donates its electrons to the
    mitochondrial electron transport chain.

14. **The Strategy of Energy Metabolism; pp. 123–124; moderate; ans: T**
    Intermediates in glycolysis and the Krebs cycle serve as precursors for anabolic
    pathways.

Essay Questions

1. **An Overview of Glucose Oxidation; p. 109; moderate**
   Distinguish between respiration and fermentation. Which process produces
   more usable energy for the cell, and why?

2. **Glycolysis; pp. 112, 118; moderate**
   What is the basic difference between substrate-level phosphorylation and
   oxidative phosphorylation? What do these processes have in common?

3. **Glycolysis; pp. 112–113; moderate**
   Summarize the reactions of glycolysis, focusing on the involvement of
   ATP/ADP and $NAD^+/NADH$.

4. **The Aerobic Pathway; pp. 115–116; moderate**
   Summarize the reactions of the Krebs cycle, focusing on the involvement of
   ATP/ADP, $NAD^+/NADH$, and $FAD/FADH_2$. In what way is this metabolic
   pathway a cycle?

5. **The Aerobic Pathway; pp. 116–118; difficult**
   Discuss the roles of the various types of electron carriers in the electron
   transport chain. Why is it important that some carriers transport only electrons
   and others transport both electrons and protons?

6. **The Aerobic Pathway; p. 118; moderate**
   Explain why electrons donated by NADH to the electron transport chain result
   in the formation of fewer ATP molecules than electrons donated by $FADH_2$.

7. **The Aerobic Pathway; pp. 118–120; moderate**
   What is meant by chemiosmotic coupling?

8. **The Aerobic Pathway; pp. 120, 122; moderate**
   Electrons carried to the electron transport chain by NADH from glycolysis
   result in formation of fewer ATP molecules than electrons carried by NADH
   from the Krebs cycle. Explain this difference.

9. **The Aerobic Pathway; p. 122; difficult**
   Account for the net gain of 36 molecules of ATP from the complete oxidation
   of one molecule of glucose.

10. **Anaerobic Pathways; pp. 122–123; moderate**
    It is said that fermentation provides a means for the cell to "recycle" NAD^+. Explain how this is true.

11. **Anaerobic Pathways; pp. 122–123; easy**
    What do lactate fermentation and alcohol fermentation have in common? How are they different?

12. **The Strategy of Energy Metabolism; pp. 123–124; moderate**
    Explain why the Krebs cycle can be regarded as the metabolic "hub" of the cell.

Chapter 7

Photosynthesis, Light, and Life

Multiple-Choice Questions

1. **Photosynthesis: A Historical Perspective; p. 127; easy; ans: a**
    Who showed that air is "restored" in sunlight only by the green parts of a plant?

    a. Jan Ingenhousz
    b. Aristotle
    c. F. F. Blackman
    d. Joseph Priestley
    e. Jan Baptista van Helmont

2. **Photosynthesis: A Historical Perspective; p. 128; moderate; ans: b**
    The O_2 evolved in photosynthesis comes from:

    a. carbon dioxide.
    b. water.
    c. glucose.
    d. (CH_2O).
    e. $(C_3H_3O_3)$.

3. **Photosynthesis: A Historical Perspective; pp. 127–128; easy; ans: c**
    Who showed that water, not carbon dioxide, is the source of the oxygen produced in photosynthesis?

    a. Jan Ingenhousz
    b. Aristotle
    c. C. B. van Niel
    d. Jan Baptista van Helmont
    e. Joseph Priestley

4. **Photosynthesis: A Historical Perspective; p. 128; difficult; ans: d**
    In reality, the end product of photosynthesis in algae and green plants is:

    a. glucose.
    b. starch.
    c. sucrose.
    d. a 3-carbon sugar.
    e. a 5-carbon sugar.

5. **Photosynthesis: A Historical Perspective; p. 128; easy; ans: c**
   F. F. Blackman showed that:

   a. air "restored" by vegetation could support the breathing of animals.
   b. air is "restored" only in the presence of light and only by the green parts
       of the plant.
   c. photosynthesis has both a light-dependent stage and a light-independent
       stage.
   d. isolated chloroplasts are able to produce O_2 in the absence of light.
   e. all the substance of a plant is provided by water and not the soil.

6. **The Nature of Light; p. 129; easy; ans: d**
   Which of the following best describes the photoelectric effect?

   a. Photons of light carry electric charges.
   b. Photons of light are electrons.
   c. Electrons are sometimes visible to the naked eye.
   d. Light dislodges electrons from a metal.
   e. Light attracts electrons to a metal.

7. **The Nature of Light; p. 130; easy; ans: e**
   Light is composed of particles called:

   a. electrons.
   b. protons.
   c. neutrons.
   d. gamma rays.
   e. photons.

8. **The Role of Pigments; p. 131; moderate; ans: d**
   Chlorophyll appears green because it:

   a. absorbs green light.
   b. both absorbs and reflects green light.
   c. both absorbs and transmits green light.
   d. both reflects and transmits green light.
   e. transmits violet light.

9. **The Role of Pigments; pp. 130–131; difficult; ans: e**
   An action spectrum is different from an absorption spectrum in that an action
   spectrum:

   a. provides evidence that a particular pigment is responsible for a particular
       process.
   b. provides information about the extent of reflectance.
   c. is the light-transmitting pattern of a pigment.
   d. is the light-absorbing pattern of a pigment.
   e. is the relative effectiveness of different wavelengths for a specific process.

10. **The Role of Pigments; p. 131; moderate; ans: c**
    Chlorophyll absorbs light principally in the _____ wavelengths.

    a. blue and green
    b. green and violet
    c. blue and violet
    d. violet and green
    e. green and red

11. **The Role of Pigments; p. 131; moderate; ans: b**
    The phenomenon in which the energy of an excited electron is converted to a combination of heat and light is called:

    a. pigment activation.
    b. fluorescence.
    c. resonance energy transfer.
    d. photosynthesis.
    e. electron transfer.

12. **The Role of Pigments; p. 133; easy; ans: a**
    Which pigment occurs in all photosynthetic eukaryotes?

    a. chlorophyll *a*
    b. chlorophyll *b*
    c. chlorophyll *c*
    d. bacteriochlorophyll
    e. chlorobium chlorophyll

13. **The Role of Pigments; p. 133; moderate; ans: c**
    Which of the following is NOT an accessory pigment in plants, green algae, and euglenoid algae?

    a. carotenes
    b. xanthophylls
    c. chlorophyll *a*
    d. chlorophyll *b*
    e. chlorophyll *c*

14. **The Role of Pigments; p. 133; moderate; ans: c**
    Which type of chlorophyll is found in the brown algae and the diatoms?

    a. chlorophyll *a*
    b. chlorophyll *b*
    c. chlorophyll *c*
    d. bacteriochlorophyll
    e. chlorobium chlorophyll

15. **The Role of Pigments; p. 133; easy; ans: b**
    The primary function of _____ is as an anti-oxidant.

    a. chlorophyll *a*
    b. carotenoids
    c. phycobilins
    d. bacteriochlorophyll
    e. chlorobium chlorophyll

16. **The Role of Pigments; p. 133; easy; ans: c**
    Xanthophylls and carotenes:

    a. are the principal photosynthetic pigments in green plants.
    b. are the principal sources of vitamin C for humans.
    c. are carotenoids.
    d. are normally present in the cytosol rather than in plastids.
    e. can substitute for chlorophylls in photosynthesis.

17. **The Reactions of Photosynthesis; p. 133; easy; ans: a**
The energy-transduction reactions of photosynthesis are also called the _____
reactions.

    a.   light
    b.   dark
    c.   light-independent
    d.   carbon-fixation
    e.   biosynthetic

18. **The Reactions of Photosynthesis; p. 134; moderate; ans: a**
$NADP^+$ is different from NAD^+ in that $NADP^+$:

    a.   contains an additional phosphate.
    b.   contains an additional ribose.
    c.   transfers its electrons to the electron transport chain.
    d.   provides energy for biosynthetic reactions.
    e.   can be reduced.

19. **The Reactions of Photosynthesis; p. 135; moderate; ans: e**
Which of the following statements about an antenna complex is FALSE?

    a.   It is part of a photosystem.
    b.   It "funnels" energy to the reaction center.
    c.   It contains chlorophyll molecules.
    d.   It contains carotenoid pigments.
    e.   It converts light energy into chemical energy.

20. **The Reactions of Photosynthesis; p. 135; moderate; ans: c**
In the antenna complex, light energy is transferred from one pigment molecule
to another by:

    a.   pigment activation.
    b.   fluorescence.
    c.   resonance energy transfer.
    d.   reduction.
    e.   oxidation.

21. **The Reactions of Photosynthesis; p. 135; difficult; ans: d**
Which of the following is characteristic of Photosystem I but NOT
Photosystem II?

    a.   It contains chlorophylls but not carotenoids.
    b.   It can function only in association with the other photosystem.
    c.   It donates electrons to an electron transport chain.
    d.   It contains P_{700} at the reaction center.
    e.   It splits water to release oxygen.

22. **The Reactions of Photosynthesis; p. 135; difficult; ans: d**
How many photons are required for photolysis to yield one molecule of O_2 gas?

    a.   1
    b.   2
    c.   3
    d.   4
    e.   5

23. The Reactions of Photosynthesis; p. 137; difficult; ans: a
Following photolysis, the resulting protons are released into the _____,
contributing to the proton gradient across the _____ membrane.

    a.   lumen of the thylakoid; thylakoid
    b.   chloroplast stroma; outer chloroplast
    c.   chloroplast stroma; thylakoid
    d.   chloroplast matrix; inner chloroplast
    e.   cytosol; inner mitochondrial

24. The Reactions of Photosynthesis; p. 137; easy; ans: c
Plastocyanin and pheophytin are components of:

    a.   Photosystem I.
    b.   Photosystem II.
    c.   the photosynthetic electron transport chain.
    d.   the photosynthetic ATP synthase.
    e.   the water-splitting enzyme.

25. The Reactions of Photosynthesis; p. 137; moderate; ans: b
In photophosphorylation, the role of the ATP synthase complex is to provide a
channel for protons to flow back into the:

    a.   lumen of the thylakoid.
    b.   chloroplast stroma.
    c.   intermembrane space of the mitochondrion.
    d.   intermembrane space of the chloroplast.
    e.   cytosol.

26. The Reactions of Photosynthesis; pp. 137–138; moderate; ans: c
Which of the following events is NOT associated with Photosystem I?

    a.   absorption of light by antenna molecules
    b.   excitation of an electron from P_{700}
    c.   transfer of electrons from cytochromes to iron-sulf proteins
    d.   reduction of $NADP^+$
    e.   reduction of A_o

27. The Reactions of Photosynthesis; pp. 137–138; easy; ans: a
The reduction of $NADP^+$ is associated with:

    a.   Photosystem I.
    b.   Photosystem II.
    c.   the photosynthetic electron transport chain.
    d.   cyclic electron flow.
    e.   the water-splitting enzyme.

28. The Reactions of Photosynthesis; p. 138; moderate; ans: e
Which of the following is/are produced during noncyclic electron flow?

    a.   ATP only
    b.   NADPH only
    c. O_2 only
    d.   ATP and O_2 only
    e.   ATP, NADPH, and O_2

29. **The Reactions of Photosynthesis; pp. 138–139; difficult; ans: d**
Which of the following is produced during BOTH noncyclic and cyclic electron flow?

   a. water
   b. NADPH
   c. sugar
   d. ATP
   e. O_2

30. **The Reactions of Photosynthesis; p. 139; easy; ans: c**
During cyclic electron flow, electrons are transferred directly from P_{700} to A_o to:

   a. P_{700}.
   b. P_{680}.
   c. the photosynthetic electron transport chain.
   d. the photosynthetic ATP synthase.
   e. the lumen of the thylakoid.

31. **The Carbon-Fixation Reactions; p. 139; moderate; ans: b**
The Calvin cycle takes place in the:

   a. lumen of the thylakoid.
   b. chloroplast stroma.
   c. thylakoid membrane.
   d. cytoplasm.
   e. mitochondrial matrix.

32. **The Carbon-Fixation Reactions; p. 140; moderate; ans: c**
Because the first detectable product of the Calvin cycle is _____, the cycle is also known as the _____ pathway.

   a. ribulose 1,5-bisphosphate; C_5
   b. oxaloacetate; C_4
   c. 3-phosphoglycerate; C_3
   d. glyceraldehyde 3-phosphate; C_3
   e. Rubisco; C_4

33. **The Carbon-Fixation Reactions; pp. 139–140; easy; ans: c**
Carbon dioxide is "fixed" by bonding to:

   a. glyceraldehyde 3-phosphate.
   b. 3-phosphoglycerate.
   c. ribulose 1,5-bisphosphate.
   d. $NADP^+$.
   e. ADP.

34. **The Carbon-Fixation Reactions; pp. 139–140; easy; ans: a**
The role of Rubisco is to catalyze the conversion of:

   a. CO_2 to an unstable six-carbon compound.
   b. CO_2 to glyceraldehyde 3-phosphate.
   c. 3-phosphoglycerate to glyceraldehyde 3-phosphate.
   d. glyceraldehyde 3-phosphate to sucrose.
   e. glyceraldehyde 3-phosphate to starch.

35. The Carbon-Fixation Reactions; p. 140; easy; ans: d
In the Calvin cycle, 3-phosphoglycerate is reduced to:

    a.   sucrose.
    b.   ribulose 1,5-bisphosphate.
    c.   Rubisco.
    d.   glyceraldehyde 3-phosphate.
    e.   an unstable six-carbon compound.

36. The Carbon-Fixation Reactions; pp. 140–141; moderate; ans: b
The conversion of 3-phosphoglycerate to glyceraldehyde 3-phosphate requires the participation of:

    a. NAD^+.
    b.   NADPH.
    c.   ATP.
    d.   ADP.
    e. CO_2.

37. The Carbon-Fixation Reactions; p. 141; easy; ans: a
How many molecules of CO_2 are fixed during each turn of the Calvin cycle?

    a.   1
    b.   2
    c.   3
    d.   4
    e.   5

38. The Carbon-Fixation Reactions; p. 141; difficult; ans: b
Which of the following does NOT occur in the Calvin cycle?

    a.   ATP is hydrolyzed.
    b.   ADP is phosphorylated to ATP.
    c.   NADPH is oxidized.
    d.   Ribulose 1,5-bisphosphate is regenerated.
    e. CO_2 is fixed.

39. The Carbon-Fixation Reactions; pp. 141–142; moderate; ans: e
Which of the following statements about the Calvin cycle is FALSE?

    a.   It requires more ATP than NADPH.
    b.   Each reaction is catalyzed by a specific enzyme.
    c.   It regenerates ribulose 1,5-bisphosphate.
    d.   It fixes CO_2.
    e.   It uses ATP from noncyclic but not cyclic photophosphorylation.

40. The Carbon-Fixation Reactions; p. 142; moderate; ans: c
The conversion of glyceraldehyde 3-phosphate to sucrose takes place in the:

    a.   lumen of the thylakoids.
    b.   chloroplast stroma.
    c.   cytosol.
    d.   mitochondria matrix.
    e.   lumen of the endoplasmic reticulum.

41. **The Carbon-Fixation Reactions; p. 142; easy; ans: d**
Most of the glyceraldehyde 3-phosphate not exported to the cytosol is converted to _____ and stored in the chloroplasts.

    a. 3-phosphoglycerate
    b. sucrose
    c. glucose
    d. starch
    e. ribulose 1,5-bisphosphate

42. **The Carbon-Fixation Reactions; p. 142; easy; ans: b**
Rubisco can use _____ or CO_2 as a substrate.

    a. 3-phosphoglycerate
    b. O_2
    c. glyceraldehyde 3-phosphate
    d. serine
    e. oxaloacetate

43. **The Carbon-Fixation Reactions; pp. 142–143; moderate; ans: a**
Which of the following statements about photorespiration is FALSE?

    a. It yields ATP but not NADPH.
    b. Phosphoglycolate is an intermediate.
    c. It consumes oxygen and releases carbon dioxide.
    d. It is a wasteful process.
    e. Three cellular organelles participate in the process.

44. **The Carbon-Fixation Reactions; pp. 143–144; difficult; ans: d**
Which of the following conditions favors photorespiration?

    a. a ratio of CO_2 to O_2 that favors CO_2
    b. conditions that cause the stomata to open
    c. plants growing far apart
    d. a hot, dry environment
    e. darkness

45. **The Carbon-Fixation Reactions; pp. 143, 144, 146; difficult; ans: d**
Which of the following statements concerning stomata is FALSE?

    a. They are found in leaves and green stems.
    b. They are specialized pores.
    c. They close when a plant is subjected to hot, dry conditions.
    d. They allow CO_2 but not O_2 to pass through.
    e. In CAM plants, they close during the day.

46. **The Carbon-Fixation Reactions; p. 144; moderate; ans: e**
In the C_4 pathway, the enzyme PEP carboxylase:

    a. uses O_2 as a substrate.
    b. uses CO_2 as a substrate.
    c. operates inefficiently when the CO_2 concentration is low.
    d. is active only in the chloroplasts of mesophyll cells.
    e. catalyzes the formation of oxaloacetate.

47. The Carbon-Fixation Reactions; p. 144; moderate; ans: c
In the C_4 pathway, what happens to the CO_2 formed when malate is decarboxylated in bundle-sheath cells?

a. It is released in the process of photorespiration.
b. It is released in the process of dark respiration.
c. It enters the Calvin cycle.
d. It reacts with serine to form phosphoglycolate.
e. It enters the vacuole.

48. The Carbon-Fixation Reactions; p. 146; difficult; ans: e
Compared with a C_3 plant, a C_4 plant:

a. carries out more photorespiration.
b. has a lower photosynthetic efficiency.
c. has more Rubisco.
d. has a higher leaf nitrogen content.
e. needs more ATP to fix CO_2.

49. The Carbon-Fixation Reactions; p. 146; difficult; ans: e
Which of the following statements about C_3 and C_4 plants is FALSE?

a. Carbon-fixation in C_4 plants has a larger energy cost than in C_3 plants.
b. CO_2-fixation in C_4 plants is more efficient than in C_3 plants.
c. The optimal temperature range for photosynthesis in C_4 plants is higher than that for C_3 plants.
d. C_4 plants have less Rubisco than C_3 plants.
e. C_4 plants have more leaf nitrogen than C_3 plants.

50. The Carbon-Fixation Reactions; p. 146; moderate; ans: d
Compared with C_3 plants, C_4 plants:

a. are well adapted to low light intensities.
b. are well adapted to low temperatures.
c. are well adapted to moist areas.
d. use nitrogen more efficiently.
e. fix CO_2 less efficiently.

51. The Carbon-Fixation Reactions; p. 147; easy; ans: a
In CAM plants, malate formed as the end product of CO_2 fixation in the dark is stored as malic acid in the:

a. vacuole.
b. chloroplast stroma.
c. thylakoid lumen.
d. cytosol.
e. nucleus.

52. The Carbon-Fixation Reactions; pp. 147, 149; moderate; ans: c
Which of the following is most likely to occur in a leaf cell of a CAM plant during the day?

a. entry of CO_2 through stomata
b. exit of water through stomata
c. decarboxylation of malic acid
d. fixation of CO_2 by PEP carboxylase
e. conversion of oxaloacetate to malate

53. The Carbon-Fixation Reactions; pp. 147, 149; difficult; ans: b
Which of the following statements about CAM plants is FALSE?

a.   Not all CAM plants are succulent.
b.   All CAM plants are flowering plants.
c.   They use both C_3 and C_4 pathways.
d.   They are dependent on nighttime accumulation of CO_2 for photosynthesis.
e.   Their water-use efficiency is higher than that of C_3 and C_4 plants.

54. The Carbon-Fixation Reactions; p. 149; difficult; ans: d
Under extremely _____ conditions, CAM plants would most likely grow _____ C_3 and C_4 plants.

a.   wet; more slowly than
b.   wet; more rapidly than
c.   wet; just as rapidly as
d.   arid; more rapidly than
e.   arid; more slowly than

True-False Questions

1.   Photosynthesis: A Historical Perspective; p. 128; moderate; ans: T
The proposal that water is the source of oxygen evolved in photosynthesis arose from studies of photosynthetic bacteria that do not evolve oxygen.

2.   The Nature of Light; p. 129; easy; ans: T
Light has the properties of both waves and particles.

3.   The Nature of Light; p. 130; easy; ans: T
The energy of a photon of light is inversely proportional to its wavelength.

4.   The Role of Pigments; p. 130; easy; ans: T
A substance that absorbs light is called a pigment.

5.   The Role of Pigments; pp. 130–131; moderate: ans: F
The relative effectiveness of different wavelengths of light for a specific light-requiring process is demonstrated by an absorption spectrum.

6.   The Role of Pigments; p. 133; moderate; ans: F
Accessory pigments are directly involved in the energy-transduction processes of photosynthesis.

7.   The Role of Pigments; p. 133; easy; ans: T
Carotenes and xanthophylls are two groups of carotenoids present in plant cell chloroplasts.

8.   The Role of Pigments; p. 133; moderate; ans: T
The phycobilins are accessory pigments found in cyanobacteria.

9.   The Reactions of Photosynthesis; pp. 133–134; easy; ans: F
The dark reactions of photosynthesis cannot occur in the light.

10. **The Reactions of Photosynthesis; p. 135; moderate; ans: T**
When a chlorophyll *a* molecule at the reaction center of a photosystem absorbs sufficient energy, it is oxidized by loss of an electron.

11. **The Reactions of Photosynthesis; p. 137; difficult; ans: F**
Magnesium is an essential cofactor for photolysis in Photosystem II.

12. **The Reactions of Photosynthesis; p. 138; easy; ans: F**
Noncyclic photosynthetic electron flow produces ATP but not NADPH.

13. **The Reactions of Photosynthesis; p. 138; moderate; ans: T**
When 6 pairs of electrons flow from H_2O to $NADP^+$ in noncyclic electron flow, the total energy harvest is 6 ATP and 6 NADPH.

14. **The Reactions of Photosynthesis; p. 139; moderate; ans: T**
Cyclic photophosphorylation is believed to be a more primitive process than noncyclic photophosphorylation.

15. **The Carbon-Fixation Reactions; p. 140; easy; ans: F**
PEP carboxylase is the world's most abundant enzyme.

16. **The Carbon-Fixation Reactions; p. 142; moderate; ans: T**
The 3-phosphoglycerate synthesized in the Calvin cycle can be converted to glucose by a reversal of some of the steps in glycolysis.

17. **The Carbon-Fixation Reactions; p. 142; moderate; ans: F**
Most of the glyceraldehyde 3-phosphate produced by carbon fixation in photosynthesizing cells is converted to free glucose.

18. **The Carbon-Fixation Reactions; p. 142; easy; ans: T**
Rubisco can use either O_2 or CO_2 as a substrate.

19. **The Carbon-Fixation Reactions; p. 143; easy; ans: F**
Photorespiration is an energetically wasteful process that seldom occurs in nature.

20. **The Carbon-Fixation Reactions; p. 144; moderate; ans: F**
The primary carboxylating enzyme in the C_4 pathway is Rubisco.

21. **The Carbon-Fixation Reactions; p. 146; moderate; ans: T**
C_4 plants need five ATP molecules to fix one CO_2 molecule, but C_3 plants need only three.

22. **The Carbon-Fixation Reactions; p. 146; moderate; ans: F**
The PEP carboxylase of C_4 plants is inhibited by O_2.

23. **The Carbon-Fixation Reactions; p. 147; moderate; ans: F**
C_3-C_4 intermediates provide evidence for the evolution of C_3 plants from C_4 ancestors.

24. **The Carbon-Fixation Reactions; p. 149; easy; ans: T**
Some CAM plants can keep their stomata closed both night and day.

25. **The Carbon-Fixation Reactions; p. 149; moderate; ans: T**
    Under extremely dry conditions, CAM plants outcompete both C_3 and C_4 plants.

Essay Questions

1. **Photosynthesis: A Historical Perspective; pp. 127–128; moderate**
    Summarize the key historical events in the elucidation of the photosynthetic process.

2. **The Nature of Light; p. 129; moderate**
    Describe the evidence that indicated the inadequacy of the wave model of light.

3. **The Role of Pigments; pp. 131–132; moderate**
    Discuss the three possible fates of an electron of a chlorophyll molecule once that electron has been raised to an excited state.

4. **The Role of Pigments; p. 133; moderate**
    List the main accessory pigments found in photosynthetic organisms. What roles do they play?

5. **The Reactions of Photosynthesis; p. 135; moderate**
    What is the role of the antenna complex? How are the reactions in the antenna complex different from those in the reaction center?

6. **The Reactions of Photosynthesis; pp. 135–137; moderate**
    Describe the events that occur during photolysis. In what way does photolysis contribute to ATP production?

7. **The Reactions of Photosynthesis, pp. 138–139; moderate**
    Explain the similarities and differences between (a) cyclic and noncyclic electron flow and (b) cyclic and noncyclic photophosphorylation.

8. **The Carbon-Fixation Reactions; pp. 139–141; moderate**
    Summarize the three main stages of the Calvin cycle.

9. **The Carbon-Fixation Reactions; p. 142; moderate**
    Under what conditions is glyceraldehyde 3-phosphate converted to starch rather than to sucrose?

10. **The Carbon-Fixation Reactions; pp. 143–144; moderate**
    Describe the main events occurring in photorespiration. Under what conditions is photorespiration enhanced?

11. **The Carbon-Fixation Reactions; p. 146; difficult**
    What factors enable C_4 plants to maintain a high $CO_2:O_2$ ratio at the site of Rubisco action?

12. **The Carbon-Fixation Reactions; p. 146; moderate**
    Even though C_4 photosynthesis requires more ATP than C_3 photosynthesis, the C_4 pathway is more efficient. Explain this apparent contradiction.

13. **The Carbon-Fixation Reactions; pp. 146–147; moderate**
    In what ways are C_4 plants better adapted to hot, dry areas than C_3 plants?

14. **The Carbon-Fixation Reactions; pp. 139–149; difficult**
    Compare and contrast C_3, C_4, and CAM photosynthesis.

Chapter 8

The Reproduction of Cells

Multiple-Choice Questions

1. **The Reproduction of Cells; p. 156; moderate; ans: c**
    Which of the following statements about cell division is FALSE?
    a. It is the mechanism by which cells reproduce.
    b. It divides the cell contents between two daughter cells.
    c. It is the means by which bacteria and many protists grow.
    d. Together with cell enlargement, it is the means by which multicellular organisms grow.
    e. It is the means by which injured or worn-out tissues are replaced.

2. **Cell Division in Eukaryotes; p. 156; moderate; ans: d**
    In contrast to a prokaryotic cell, a eukaryotic cell:
    a. divides by binary fission.
    b. contains a single, very large DNA molecule.
    c. can divide into two daughter cells.
    d. undergoes mitosis and cytokinesis.
    e. undergoes only cytoplasmic division.

3. **The Cell Cycle; p. 157; moderate; ans: b**
    The cell cycle is commonly divided into two parts:
    a. mitosis and cytokinesis.
    b. interphase and mitosis.
    c. interphase and cytokinesis.
    d. G_1 and G_2.
    e. S and G_0.

4. **The Cell Cycle; p. 157; moderate; ans: e**
    In the cell cycle, interphase consists of:
    a. mitosis and cytokinesis.
    b. mitosis and the S phase.
    c. the G_1 and G_2 phases.
    d. the G_2 and S phases.
    e. the G_1, G_2, and S phases.

5. The Cell Cycle; p. 157; easy; ans: c
The G_0 phase of the cell cycle is:

    a.   a stage of cytokinesis.
    b.   a stage of mitosis.
    c.   a resting, or dormant, state.
    d.   a stage of active growth.
    e.   characteristic of single-celled organisms.

6. The Cell Cycle; pp. 157–158; difficult; ans: e
Which of the following statements concerning checkpoints is FALSE?

    a.   Checkpoints control the progression between certain phases of the cell cycle.
    b.   Checkpoints control the rate at which cells are produced.
    c.   Checkpoints enable a cell to sense whether certain conditions have been met.
    d.   Checkpoints involve the control of DNA and protein synthesis.
    e.   Checkpoints differ significantly among eukaryotic cells.

7. The Cell Cycle; p. 158; moderate; ans: c
The checkpoint that controls the initiation of mitosis is at the end of:

    a. G_0.
    b. G_1.
    c. G_2.
    d.   S.
    e.   M.

8. Interphase; p. 158; easy; ans: d
DNA replication occurs during the _____ phase.

    a. G_0
    b. G_1
    c. G_2
    d.   S
    e.   M

9. Interphase; p. 158; moderate; ans: c
Which of the following lists the phases of the cell cycle in the correct sequence?

    a. G_1, G_2, S, M
    b. G_1, G_2, M, S
    c. G_1, S, G_2, M
    d. G_2, G_1, S, M
    e.   S, M, G_1, G_2

10. Interphase, p. 158; moderate; ans: b
If centrioles and centrosomes are present, they duplicate during the _____ phase.

    a. G_0
    b. G_1
    c. G_2
    d.   S
    e.   M

11. **Interphase; p. 158; moderate; ans: c**
During the _____ phase, the cell assembles the special structures required for nuclear and cytoplasmic division.

    a. G_0
    b. G_1
    c. G_2
    d. S
    e. M

12. **Cell Division in Plants; p. 158; easy; ans: a**
Which of the following is unique to cell division in plants?

    a. migration of the nucleus to the center of the cell
    b. migration of the mitochondria to the periphery of the cell
    c. duplication of the centrosome
    d. duplication of the mitochondria
    e. duplication of the endoplasmic reticulum

13. **Cell Division in Plants; p. 158; difficult; ans: e**
The phragmosome, a structure bisecting the cell prior to cell division:

    a. is a portion of the endoplasmic reticulum.
    b. is associated with the centrosome.
    c. is derived from the plasma membrane.
    d. is a ringlike band lying just beneath the plasma membrane.
    e. contains both microtubules and actin filaments.

14. **Cell Division in Plants; p. 158; easy; ans: d**
Which of the following is unique to cell division in plants?

    a. migration of the nucleus from the center to the periphery of the cell
    b. migration of the mitochondria to the nuclear membrane
    c. duplication of the centrosome and the centrioles
    d. formation of a band of microtubules just beneath the plasma membrane
    e. duplication of the endoplasmic reticulum

15. **Cell Division in Plants; p. 160; easy; ans: b**
Which of the following lists the stages of mitosis in the correct sequence?

    a. anaphase, metaphase, prophase, telophase
    b. prophase, metaphase, anaphase, telophase
    c. metaphase, telophase, anaphase, prophase
    d. telophase, anaphase, prophase, metaphase
    e. prophase, anaphase, metaphase, telophase

16. **Cell Division in Plants; p. 160; moderate; ans: d**
In late prophase, sister chromatids are joined by a constriction at the:

    a. phragmosome.
    b. preprophase band.
    c. centriole.
    d. centromere.
    e. centrosome.

17. Cell Division in Plants; p. 160; easy; ans: b
The _____ is the earliest manifestation of the mitotic spindle.

a.   preprophase band
b.   preprophase spindle
c.   phragmosome
d.   centrosome
e.   phragmoplast

18. Cell Division in Plants; p. 160; easy; ans: c
The nucleolus disappears during:

a.   anaphase.
b.   metaphase.
c.   prophase.
d.   interphase.
e.   telophase.

19. Cell Division in Plants; p. 161; moderate; ans: b
The preprophase band is replaced by the:

a.   phragmosome.
b.   mitotic spindle.
c.   cell plate.
d.   centrosome.
e.   phragmoplast.

20. Cell Division in Plants; p. 161; moderate; ans: e
Kinetochores are protein complexes associated with the:

a.   phragmoplast.
b.   phragmosome.
c.   centrosomes.
d.   centrioles.
e.   centromeres.

21. Cell Division in Plants; p. 163; moderate; ans: c
Sister chromatids become daughter chromosomes at the beginning of:

a.   prophase.
b.   metaphase.
c.   anaphase.
d.   telophase.
e.   interphase.

22. Cell Division in Plants; p. 163; moderate; ans: b
Which of the following events does NOT occur during anaphase?

a.   The kinetochore microtubules shorten.
b.   The spindle poles move closer together.
c.   Polar microtubules increase in length.
d.   Motor proteins pull the chromosomes along their attached microtubules.
e.   Energy is supplied by ATP.

23. **Cell Division in Plants; p. 163; difficult; ans: b**
Which of the following directly ensures that daughter cells contain identical sets of chromosomes?

a.   Sister chromatids are joined at the centromere.
b.   Sister chromatids move to opposite poles in anaphase.
c.   Nuclear envelopes form in telophase.
d.   Motor proteins pull daughter chromosomes apart.
e.   The energy of ATP is used to separate daughter chromosomes.

24. **Cell Division in Plants; p. 163; easy; ans: e**
Chromosomes lengthen and become indistinct during:

a.   anaphase.
b.   metaphase.
c.   prophase.
d.   interphase.
e.   telophase.

25. **Cell Division in Plants; p. 163; easy; ans: e**
During _____, the nuclear envelopes and nucleoli re-form.

a.   anaphase
b.   metaphase
c.   prophase
d.   interphase
e.   telophase

26. **Cell Division in Plants; p. 163; easy; ans: c**
The longest phase of mitosis is always:

a.   anaphase.
b.   metaphase.
c.   prophase.
d.   interphase.
e.   telophase.

27. **Cell Division in Plants; pp. 163, 165; moderate; ans: d**
Which of the following statements about the phragmoplast is FALSE?

a.   It forms between the two daughter nuclei.
b.   It is composed of microtubules.
c.   Its formation precedes the growth of the cell plate.
d.   It begins to form at the walls of the dividing cell and grows inward.
e.   In cells with large vacuoles, it is formed within the phragmosome.

28. **Cell Division in Plants; p. 166; difficult; ans: d**
In contrast with animals, morphogenesis in plants:

a.   involves the formation and differentiation of tissues.
b.   involves the formation and differentiation of organs.
c.   involves the migration of cells.
d.   reflects the planes of cell division and the directions of cell expansion.
e.   lacks a role for the cytoskeleton.

29. **Cell Division and the Reproduction of the Organism; p. 166; moderate; ans: b**
    Which of the following statements about cell division, reproduction, and growth is FALSE?

    a. Mitosis produces daughter cells with exactly the same set of chromosomes as the parent cell.
    b. Specialized cells usually cannot re-enter the cell cycle.
    c. Mitosis is the mechanism of plant growth and asexual reproduction.
    d. Vegetative reproduction is common in plants.
    e. Dedifferentiation followed by redifferentiation occurs during wound healing in plants.

True-False Questions

1. **Cell Division in Eukaryotes; p. 156; easy; ans: T**
    Eukaryotes typically contain much more DNA than prokaryotes.

2. **Cell Division in Eukaryotes; p. 156; easy; ans: F**
    In eukaryotes, mitosis is usually followed by binary fission.

3. **The Cell Cycle; p. 157; moderate; ans: F**
    In terms of cellular activity, interphase is best described as a period of dormancy.

4. **The Cell Cycle; p. 157; easy; ans: T**
    During winter dormancy, root initials are most likely to be in the G_0 phase of the cell cycle.

5. **Interphase; p. 158; moderate; ans: F**
    During the G_1 phase, new mitochondria and plastids are synthesized *de novo*.

6. **Cell Division in Plants; p. 158; moderate; ans: T**
    The phragmosome is a sheet of cytoplasm that anchors the nucleus in the center of the cell prior to nuclear division.

7. **Cell Division in Plants; p. 160; moderate; ans: T**
    Binding of the chromosomes to the mitotic spindle requires special DNA sequences called centromeres.

8. **Cell Division in Plants; p. 162; easy; ans: T**
    At full metaphase, all chromosomes are aligned midway between the spindle poles.

9. **Cell Division in Plants; p. 163; easy; ans: F**
    Polar microtubules are attached to kinetochores.

10. **Cell Division in Plants; p. 163; easy; ans: T**
    The most rapid phase of mitosis is anaphase.

11. **Cell Division in Plants; p. 163; moderate; ans: T**
Mitosis ensures that the hereditary blueprint for cellular protein production is transmitted to the next generation of cells.

12. **Cell Division in Plants; p. 165; easy; ans: T**
In plants, the cell plate forms within the phragmoplast.

13. **Cell Division in Plants; p. 166; moderate; ans: F**
During cytokinesis, the cell walls of the daughter cells form from the cell wall of the parent cell.

14. **Cell Division and the Reproduction of the Organism; p. 166; easy; ans: F**
All multicellular organisms produce new individuals by mitosis.

Essay Questions

1. **Cell Division in Prokaryotes; Cell Division in Eukaryotes; p. 156; easy**
In general, how does cell division in prokaryotes differ from that in eukaryotes?

2. **The Cell Cycle; pp. 157–158; moderate**
Name the phases of the cell cycle, and describe the main events that occur in each.

3. **Interphase; p. 158; moderate**
Early cell biologists called interphase the "resting stage." Is this an accurate description? Why or why not?

4. **Cell Division in Plants; p. 158; moderate**
Describe the two events in interphase that are unique to plants.

5. **Cell Division in Plants; pp. 160–163; moderate**
List the four stages of mitosis, and explain the main events that occur in each.

6. **Cell Division in Plants; p. 163; moderate**
Describe the structure and function of the mitotic spindle.

7. **Cell Division in Plants; p. 163; moderate**
What is the difference between a sister chromatid and a daughter chromosome?

8. **Cell Division in Plants; p. 166; moderate**
Discuss the mechanism by which the new cell wall forms following cell division in plant cells.

9. **Cell Division and the Reproduction of the Organism; p. 166; easy**
What is the relationship between mitosis and asexual reproduction?

Chapter 9

Meiosis and Sexual Reproduction

Multiple-Choice Questions

1. **Meiosis and Sexual Reproduction; p. 170; moderate; ans: a**
   Which of the following statements about sexual reproduction is FALSE?

   a. It is the principal mode of reproduction in eukaryotes.
   b. It involves the mixing of genomes of two individuals.
   c. It results in offspring that differ genetically from each other.
   d. It results in offspring that differ genetically from both of their parents.
   e. It involves a regular alternation between meiosis and fertilization.

2. **Haploid and Diploid; p. 170; easy; ans: c**
   In a eukaryotic organism, the gametes have _____ number of chromosomes that occur in the somatic cells.

   a. the same
   b. one-fourth the
   c. one-half the
   d. twice the
   e. three times the

3. **Haploid and Diploid; p. 170; easy; ans: d**
   The diploid chromosome number of a particular organism is 20. How many chromosomes are in its somatic cells?

   a. 5
   b. 10
   c. 15
   d. 20
   e. 40

4. **Haploid and Diploid; p. 170; easy; ans: a**
   In a plant for which $2n = 6$, how many chromosomes are in each gamete?

   a. 3
   b. 6
   c. 9
   d. 12
   e. 24

5. **Meiosis, the Life Cycle, and Diploidy; p. 171; moderate; ans: b**
Which of the following describes zygotic meiosis?

   a.   It occurs in most plants.
   b.   The zygote is the only diploid cell in the life cycle.
   c.   It results directly in gametes.
   d.   It is characteristic of organisms having an alternation of generations.
   e.   It is characteristic of organisms with isomorphic generations.

6. **Meiosis, the Life Cycle, and Diploidy; p. 172; moderate; ans: d**
Which of the following describes sporic meiosis?

   a.   It occurs in most animals.
   b.   The gametes are the only haploid cells in the life cycle.
   c.   It most likely evolved before zygotic meiosis.
   d.   It is characteristic of organisms having an alternation of generations.
   e.   The zygote is the only diploid cell in the life cycle.

7. **Meiosis, the Life Cycle, and Diploidy; p. 172; moderate; ans: d**
The gametophyte:

   a.   is the diploid generation.
   b.   is the spore-producing generation.
   c.   is the dominant generation in vascular plants.
   d.   occurs in organisms having sporic meiosis.
   e.   stores more genetic information than the sporophyte.

8. **Meiosis, the Life Cycle, and Diploidy; p. 172; easy; ans: e**
Life cycles in which the haploid and diploid forms are similar in external
appearance are said to have _____ generations.

   a.   morphospecific
   b.   isospecific
   c.   isogamous
   d.   heteromorphic
   e.   isomorphic

9. **Meiosis, the Life Cycle, and Diploidy; p. 172; easy; ans: a**
One clear evolutionary trend in the vascular plants is the increasing dominance
of:

   a.   the sporophyte.
   b.   the gametophyte.
   c.   zygotic meiosis.
   d.   gametic meiosis.
   e.   isomorphic life cycles.

10. **The Process of Meiosis; p. 172; moderate; ans: b**
Which of the following statements about meiosis is FALSE?

   a.   It consists of two successive nuclear divisions.
   b.   It produces a total of two daughter nuclei.
   c.   It produces cells with half as many chromosomes as the parent cell.
   d.   It produces cells with one homolog of each homologous pair.
   e.   It is a source of new combinations of chromosomes.

11. **The Process of Meiosis; p. 172; easy; ans: a**
Homologous chromosomes pair up during:

   a.   prophase I.
   b.   prophase II.
   c.   metaphase I.
   d.   metaphase II.
   e.   telophase I.

12. **The Process of Meiosis; p. 172; easy; ans: d**
Crossing-over takes place during:

   a.   metaphase I.
   b.   metaphase II.
   c.   telophase II.
   d.   prophase I.
   e.   prophase II.

13. **The Process of Meiosis; pp. 172–173; moderate; ans: c**
Which of the following statements about chiasmata is FALSE?

   a.   They are X-shaped configurations.
   b.   They occur during crossing-over.
   c.   They cause nondisjunction.
   d.   They help orient the homologs toward opposite poles.
   e.   They play a role analogous to that of the centromere in mitosis.

14. **The Phases of Meiosis; p. 173; easy; ans: b**
Cells in which meiosis occurs are called:

   a.   mitocytes.
   b.   meiocytes
   c.   gametes.
   d.   zygotes.
   e.   gametophytes.

15. **The Phases of Meiosis; p. 175; difficult; ans: d**
The axial core

   a.   is found in each chromosome.
   b.   consists mainly of proteins.
   c.   is attached to sister chromatids.
   d.   is part of the recombination nodule.
   e.   becomes part of the synaptonemal complex.

16. **The Phases of Meiosis; pp. 175–176; moderate; ans: a**
Which of the following statements about the synaptonemal complex is FALSE?

   a.   It mediates the crossing-over process.
   b.   It is a long zipperlike protein.
   c.   It connects the axial cores of homologs.
   d.   It is associated with synapsis.
   e.   It appears to be essential to crossing-over.

17. **The Phases of Meiosis; p. 175; moderate; ans: a**
The pairing of homologous chromosomes is called:

   a.   synapsis.
   b.   chiasma.
   c.   nondisjunction.
   d.   recombination.
   e.   bivalence.

18. **The Phases of Meiosis; p. 176; moderate; ans: e**
Recombination nodules:

   a.   are long zipperlike proteins.
   b.   occur only in haploid cells.
   c.   cause the chromosomes to shorten and thicken.
   d.   cause the chromosomes to divide.
   e.   mediate the crossing-over process.

19. **The Phases of Meiosis; p. 176; easy; ans: b**
Homologous pairs of chromosomes line up on the equatorial plane during:

   a.   prophase II.
   b.   metaphase I.
   c.   telophase I.
   d.   metaphase II.
   e.   prophase I.

20. **The Phases of Meiosis; p. 177; moderate; ans: d**
Nuclear envelopes form around double-stranded chromosomes during:

   a.   telophase II.
   b.   metaphase I.
   c.   prophase I.
   d.   telophase I.
   e.   metaphase II.

21. **The Phases of Meiosis; p. 177; moderate; ans: d**
Unpaired double-stranded chromosomes line up on the equatorial plane during:

   a.   prophase II.
   b.   metaphase I.
   c.   anaphase I.
   d.   metaphase II.
   e.   anaphase II.

22. **The Phases of Meiosis; p. 177; moderate; ans: e**
Sister chromatids separate during:

   a.   prophase II.
   b.   metaphase I.
   c.   anaphase I.
   d.   metaphase II.
   e.   anaphase II.

23. **The Phases of Meiosis; p. 177; easy; ans: d**
    If a cell has three pairs of homologous chromosomes, in how many ways could they be distributed among the haploid cells produced by meiosis?

    a. 2
    b. 3
    c. 4
    d. 8
    e. 9

24. **The Phases of Meiosis; pp. 177–179; difficult; ans: e**
    Which of the following statements about genetic variability is FALSE?

    a. One of the genetic consequences of meiosis is an increase in genetic variability.
    b. It is almost impossible for any cell produced by meiosis to be genetically identical to one of the cells that fused to produce the diploid line of cells undergoing meiosis.
    c. In meiosis, there are 2^n possible ways of distributing n pairs of chromosomes among the resulting haploid cells.
    d. Because of crossing-over, a cell having n pairs of chromosomes has the potential to produce considerably *more* than 2^n different gametes.
    e. As the number of chromosomes increases, the chance of reconstituting the same set that was present in the original diploid nucleus increases.

25. **The Phases of Meiosis; p. 179; difficult; ans: d**
    Meiosis is DIFFERENT from mitosis in that in meiosis:

    a. cytokinesis occurs only once.
    b. the DNA is replicated only once.
    c. the daughter nuclei have identical gene combinations.
    d. nuclei different from the original nucleus are produced.
    e. only one nuclear division is involved.

26. **Asexual Reproduction: An Alternative Strategy; p. 179; moderate; ans: e**
    Which of the following statements about asexual reproduction is FALSE?

    a. It is also known as vegetative reproduction.
    b. It results in offspring that are genetically identical to the parent.
    c. It is common in higher plants.
    d. Many plants reproduce both asexually and sexually.
    e. It increases the ability of a population to adapt to differing conditions.

27. **Advantages of Sexual Reproduction; pp. 180–181; moderate; ans: b**
    Which of the following statements about sexual reproduction is FALSE?

    a. It occurs only in eukaryotic organisms.
    b. It is necessary for the survival of the population.
    c. It involves a regular alternation between meiosis and fertilization.
    d. It produces and maintains an infinite array of genetic diversity in populations.
    e. It provides the basic mechanism of evolution.

True-False Questions

1. **Meiosis and Sexual Reproduction; p. 170; moderate; ans: F**
   Mitosis is believed to have evolved from meiosis.

2. **Haploid and Diploid; p. 170; easy; ans: T**
   Every organism has a chromosome number characteristic of its particular species.

3. **Haploid and Diploid; p. 170; easy: ans: F**
   A cell that has more than two sets of chromosomes is said to be diploid.

4. **Haploid and Diploid; p. 171; easy; ans: T**
   Meiosis counterbalances the effects of fertilization.

5. **Meiosis, the Life Cycle, and Diploidy; p. 171; easy; ans: T**
   The first eukaryotes were probably haploid, asexual organisms.

6. **Meiosis, the Life Cycle, and Diploidy; p. 172; easy; ans: F**
   Meiospores are haploid cells that divide by meiosis.

7. **Meiosis, the Life Cycle, and Diploidy; p. 172; moderate; ans: F**
   Heteromorphic generations are characteristic of all organisms having an alternation of generations.

8. **The Process of Meiosis; p. 172; moderate; ans: F**
   Crossing-over is the exchange of genetic material between the sister chromatids of a chromosome.

9. **The Process of Meiosis; p. 173; moderate; ans: T**
   Nondisjunction is the failure of homologous chromosomes to separate properly at anaphase I.

10. **The Phases of Meiosis; p. 175; easy; ans: T**
    The synaptonemal complex must form in order for crossing-over to occur.

11. **The Phases of Meiosis; p. 177; easy; ans: F**
    DNA is replicated during the interphase between meiosis I and II.

12. **The Phases of Meiosis; p. 178; moderate; ans: F**
    It is likely that one of the four daughter cells resulting from meiosis will be genetically identical to one of the gametes that fused to produce the diploid line of cells undergoing meiosis.

13. **Asexual Reproduction: An Alternative Strategy; p. 179; easy; ans: T**
    Plants that reproduce only asexually have evolved from ancestors that were capable of sexual reproduction.

14. **Advantages of Sexual Reproduction; p. 180; moderate; ans: T**
    In theory, sexual reproduction is unnecessary for the survival of a population of organisms in an unchanging environment.

Essay Questions

1. **Haploid and Diploid; p. 170; difficult**
   What is the relationship between haploidy/diploidy and homologous chromosomes?

2. **Meiosis, the Life Cycle, and Diploidy; pp. 171–172; moderate**
   Compare and contrast zygotic, gametic, and sporic meiosis.

3. **Meiosis, the Life Cycle, and Diploidy; pp. 171–172; moderate**
   What is the difference between a gametophyte and a sporophyte? Which is more evolutionarily advanced? Explain your answer.

4. **The Process of Meiosis; p. 172; moderate**
   What are chiasmata? What role do they play in cell division?

5. **The Phases of Meiosis; pp. 175–176; moderate**
   What is a synaptonemal complex? What role does it play in meiosis?

6. **The Phases of Meiosis; p. 176; easy**
   How is the arrangement of chromosomes in metaphase I of meiosis different from that in metaphase of mitosis?

7. **The Phases of Meiosis; pp. 177–178; moderate**
   Discuss the ways in which meiosis produces genetic variability.

8. **The Phases of Meiosis; p. 179; moderate**
   What are the differences between mitosis and meiosis? In what ways are they similar?

9. **Asexual Reproduction: An Alternative Strategy; Advantages of Sexual Reproduction; pp. 179–181; difficult**
   Compare the advantages and disadvantages of asexual and sexual reproduction.

10. **Advantages of Sexual Reproduction; p. 181; moderate**
    Explain the sentence "Another measure of the evolutionary advantage of sexual reproduction is provided by the amount of energy and other resources that it requires."

Chapter 10

Genetics and Heredity

Multiple-Choice Questions

1. **The Concept of the Gene; p. 184; easy; ans: e**
   Which of the following statements about Gregor Mendel is FALSE?

   a. He demonstrated that inherited characteristics are determined by discrete factors.
   b. He performed experiments with the garden pea.
   c. He studied the offspring of several generations.
   d. He analyzed his results mathematically.
   e. He became famous for his work during his lifetime.

2. **The Principle of Segregation; p. 185; difficult; ans: c**
   The seeds of all the progeny of a cross between yellow-seeded plants and green-seeded plants are as yellow as those of the yellow-seeded parent. From this information, it is clear that:

   a. this is an example of a dihybrid cross.
   b. the F_1 generation consists entirely of yellow-seeded plants.
   c. the F_2 generation consists entirely of yellow-seeded plants.
   d. green is a dominant characteristic.
   e. yellow is a recessive characteristic.

3. **The Principle of Segregation; p. 185; difficult; ans: d**
   When a particular trait appears in the F_2 generation but not in the F_1 generation, it is an indication that:

   a. a monohybrid cross is involved.
   b. a dihybrid cross is involved.
   c. true-breeding plants are involved.
   d. the trait is recessive.
   e. the trait is dominant.

4. **The Principle of Segregation; p. 186; moderate; ans: c**
   Which of the following statements about alleles is FALSE?

   a. They are different forms of the same gene.
   b. A diploid cell has two alleles for each gene.
   c. They occupy different sites on homologous chromosomes.
   d. The site of an allele is called its locus.
   e. Dominant alleles are represented by uppercase, recessive alleles by lowercase letters.

5. **The Principle of Segregation; p. 186; easy; ans: d**
    If a plant has a trait with the genetic constitution Bb, then:
    a. its phenotype is blue.
    b. its genotype is blue.
    c. the plant is homozygous for that trait.
    d. the plant is heterozygous for that trait.
    e. the plant has two identical alleles for that trait.

6. **The Principle of Segregation; p. 186; moderate; ans: a**
    If we say that the phenotype of an individual is purple, we are saying that:
    a. the appearance of the individual is purple.
    b. the genetic constitution of the individual is purple.
    c. the individual is homozygous.
    d. the individual is heterozygous.
    e. purple is the dominant allele.

7. **The Principle of Segregation; p. 186; moderate; ans: e**
    If the allele for red flower color is dominant over the allele for white flower color, which of the following represents a cross between a white-flowered plant and a plant heterozygous for flower color?
    a. *RR × RR*
    b. *RR × Rr*
    c. *RR × rr*
    d. *Rr × Rr*
    e. *Rr × rr*

8. **The Principle of Segregation; p. 186; easy; ans: c**
    If the allele for red flower color is dominant over the allele for white flower color, which of the following represents a cross between a plant homozygous dominant for flower color and a plant homozygous recessive for flower color?
    a. *RR × RR*
    b. *RR × Rr*
    c. *RR × rr*
    d. *Rr × Rr*
    e. *Rr × rr*

9. **The Principle of Segregation; p. 186; moderate; ans: c**
    In peas, green pod color is dominant over yellow pod color. If a plant heterozygous for pod color is crossed with a plant homozygous recessive for pod color, what phenotypes would you expect in the offspring?
    a. all with green pods
    b. all with yellow pods
    c. half with green pods and half with yellow pods
    d. 3/4 with green pods and 1/4 with yellow pods
    e. 1/4 with green pods and 3/4 with yellow pods

10. **The Principle of Segregation; p. 186; moderate; ans: c**
    In peas, tall stem length is dominant over short length. What genotypic ratio
    would you expect in the offspring of a cross between a heterozygous tall plant
    and a homozygous short plant?

    a.   3 homozygous tall: 1 homozygous short
    b.   1 homozygous tall: 1 homozygous short
    c.   1 heterozygous tall: 1 homozygous short
    d.   3 heterozygous tall: 1 homozygous short
    e.   1 homozygous tall: 2 heterozygous tall: 1 homozygous short

11. **The Principle of Segregation; p. 186; moderate; ans: b**
    If W represents the allele for purple flower color, and w represents the allele
    for white flower color, what genotypic ratio would you expect in the offspring
    of a cross between two heterozygous plants?

    a.   1 WW: 1 ww
    b.   1 WW: 2 Ww: 1 ww
    c.   3 WW: 1 ww
    d.   1 WW: 3 ww
    e.   3 WW: 2 Ww: 1 ww

12. **The Principle of Segregation; p. 187; difficult; ans: a**
    If W represents the allele for purple flower color, and w represents the allele
    for white flower color, what phenotypic ratio would you expect for the
    offspring of a testcross?

    a.   1 *WW*: 1 *ww*
    b.   1 *WW*: 2 *Ww*: 1 *ww*
    c.   3 *WW*: 1 *ww*
    d.   1 *WW*: 3 *ww*
    e.   3 *WW*: 2 *Ww*: 1 *ww*

13. **The Principle of Independent Assortment; pp. 188–189; moderate;**
    ans: e
    Consider the two traits, seed shape and seed color. *R* represents the allele for
    round seeds, *r* the allele for wrinkled seeds, *Y* the allele for yellow seeds, and
    y the allele for green seeds. Which of the following indicates a cross between
    a plant heterozygous for both traits and a plant homozygous recessive for both
    traits?

    a.   *Ry* × *ry*
    b.   *Rryy* × *rryy*
    c.   *Rryy* × *Rryy*
    d.   *RrYy* × *Rryy*
    e.   *RrYy* × *rryy*

14. **The Principle of Independent Assortment; pp. 188–189; moderate; ans: d**

Consider the two traits, seed shape and seed color. *R* represents the allele for round seeds, *r* the allele for wrinkled seeds, *Y* the allele for yellow seeds, and *y* the allele for green seeds. What phenotypic ratio would you expect in the offspring of a cross between a plant with wrinkled green seeds and a plant heterozygous for both traits?

   a.   9 round yellow: 3 round green: 3 wrinkled yellow: 1 wrinkled green
   b.   all round yellow
   c.   all wrinkled green
   d.   1 round yellow: 1 round green: 1 wrinkled yellow: 1 wrinkled green
   e.   3 round yellow: 3 round green: 1 wrinkled yellow: 1 wrinkled green

15. **The Principle of Independent Assortment; pp. 188–189; moderate; ans: b**

Consider the two traits, seed shape and seed color. *R* represents the allele for round seeds, *r* the allele for wrinkled seeds, *Y* the allele for yellow seeds, and *y* the allele for green seeds. What genotypic ratio would you expect in the offspring of a cross between a plant homozygous recessive for seed color but heterozygous for seed shape and a plant heterozygous for seed color but homozygous recessive for seed shape?

   a.   9 *RRYY*: 3 *rrYy*: 3 *RrYY*: 1 *rryy*
   b.   1 *RrYy*: 1 *rrYy*: 1 *Rryy*: 1 *rryy*
   c.   3 *RrYy*: 1 *rryy*
   d.   1 *RrYy*: 2 *RrYY*: 1 *rrYy*: 1 *rryy*
   e.   9 *RrYy*: 3 *RRyy*: 3 *rrYY*: 1 *rryy*

16. **The Principle of Independent Assortment; pp. 186, 189; difficult; ans: a**

Which of the following is true of Mendel's second law but not his first law?

   a.   The two alleles of a gene assort independently of the alleles of the other genes.
   b.   The two alleles of a gene separate from each other during the formation of gametes.
   c.   A plant heterozygous for a trait forms more gametes than a plant homozygous for that trait.
   d.   A plant homozygous for a trait forms more gametes than a plant heterozygous for that trait.
   e.   Alleles occupy the same locus on homologous chromosomes.

17. **Discovery of the Chromosomal Basis of Mendel's Laws; p. 190; difficult; ans: b**

Which of the following correctly links Mendel's first law with meiosis?

   a.   Chromosomes carry genes.
   b.   Alleles of a gene separate during anaphase I.
   c.   Homologous chromosomes pair during prophase I.
   d.   Nuclear envelopes surround double-stranded chromosomes in telophase I.
   e.   Sister chromatids separate during anaphase II.

18. **Discovery of the Chromosomal Basis of Mendel's Laws; p. 190; difficult; ans: c**
    Walter Sutton's studies of meiosis in sperm cells led him to propose that:
    a. genes are located at fixed points on chromosomes.
    b. alleles of different genes assort independently.
    c. the factors described by Mendel are located on chromosomes.
    d. homologous chromosomes segregate in anaphase I of meiosis.
    e. linked genes are located on the same chromosome.

19. **Discovery of the Chromosomal Basis of Mendel's Laws; p. 190; easy; ans: b**
    _____ first proved that genes are located at fixed positions on chromosomes.
    a. Walter Sutton
    b. A. H. Sturtevant
    c. Gregor Mendel
    d. Hugo de Vries
    e. William Bateson

20. **Linkage; p. 190; easy; ans: b**
    If two genes are linked, then by definition they:
    a. are alleles of the same gene.
    b. occur on the same chromosome.
    c. occur on different chromatids of the same chromosome.
    d. will segregate independently.
    e. will undergo independent assortment.

21. **Linkage; p. 191 and Fig. 10-9; moderate; ans: d**
    Suppose that a plant has two genes that are linked, with *Ab* on one homolog of a homologous pair of chromosomes and *aB* on the other homolog. What gametes could the plant produce if crossing-over occurred between these genes during meiosis?
    a. *Ab* and *aB* only
    b. *AB* and *ab* only
    c. *Aa* and *Bb* only
    d. *Ab*, *aB*, *AB*, and *ab* only
    e. *Ab*, *aB*, *AB*, *ab*, *Aa*, and *Bb* only

22. **Linkage; p. 191 and Fig. 10-10; difficult; ans: c**
    Suppose that genes *X*, *Y*, and *Z* are linked and that crossing-over occurs between *X* and *Y* 20 percent of the time, between *Y* and *Z* 30 percent of the time, and between *X* and *Z* 10 percent of the time. Which genes are farthest apart on the chromosome?
    a. *X* and *Y*
    b. *X* and *Z*
    c. *Y* and *Z*
    d. *X* and *Y*, *Y* and *Z* (equally far apart)
    e. *X* and *Z*, *Y* and *Z* (equally far apart)

23. **Mutations; p. 192; easy; ans: e**
Who first hypothesized the existence of mutations?

 a. Gregor Mendel
 b. Walter Sutton
 c. Barbara McClintock
 d. Erich von Tschermak
 e. Hugo de Vries

24. **Mutations; p. 192; easy; ans: d**
The substitution of one nucleotide for another in a gene is called a(n):

 a. deletion.
 b. duplication.
 c. translocation.
 d. point mutation.
 e. inversion.

25. **Mutations; p. 193; moderate; ans: e**
Which of the following statements about transposons is FALSE?

 a. They are also called "jumping genes."
 b. They are movable genetic elements.
 c. They may lead to mutation.
 d. Plasmids can function as transposons.
 e. They occur only in bacteria.

26. **Mutations; p. 193; easy; ans: b**
A(n) _____ is a mutation associated with the rotation of a chromosomal segment.

 a. deletion
 b. inversion
 c. duplication
 d. translocation
 e. point mutation

27. **Mutations; p. 193; easy; ans: d**
The exchange of segments between nonhomologous chromosomes is called:

 a. deletion.
 b. inversion.
 c. duplication.
 d. translocation.
 e. point mutation.

28. **Mutations; p. 193; moderate; ans: c**
Aneuploidy refers to the:

 a. loss of part of a chromosome.
 b. gain of part of a chromosome.
 c. gain or loss of some chromosomes.
 d. gain of a complete set of chromosomes.
 e. loss of a complete set of chromosomes

29. Mutations; p. 193; easy; ans: a
Colchicine is used in the laboratory to:

a.   increase the chromosome number of plant cells.
b.   cause the formation of transposons.
c.   stimulate the rate of point mutations.
d.   stimulate crossing-over.
e.   decrease the percentage of linked genes.

30. Mutations; pp. 193–194; difficult; ans: e
Which of the following statements concerning mutations is FALSE?

a.   Mutations may have favorable, unfavorable, or neutral effects.
b.   In haploid organisms, the phenotype associated with a mutation is immediately exposed to the environment.
c.   In diploid organisms, a mutation in one homolog may have less effect than if it affected both homologs.
d.   Mutations provide the raw materials for evolutionary change.
e.   The mutation rate in eukaryotes is about 1 mutant gene at a given locus per 2000 cell divisions.

31. Broadening the Concept of the Gene; p. 194; moderate; ans: e
When the phenotype of a heterozygote is intermediate between the phenotypes of its homozygous parents, the condition is known as:

a.   pleiotropy.
b.   aneuploidy.
c.   epistasis.
d.   polygenic inheritance.
e.   incomplete dominance.

32. Broadening the Concept of the Gene; p. 194; moderate; ans: d
In snapdragons, a cross between a red-flowered plant and a white-flowered plant produces a plant with pink flowers. What offspring would you expect from a cross between two pink-flowered plants?

a.   all pink
b.   1/2 pink and 1/2 white
c.   1/2 pink and 1/2 red
d.   1/4 red, 1/2 pink, 1/4 white
e.   1/4 pink, 1/2 red, 1/4 white

33. Broadening the Concept of the Gene; p. 194; easy; ans: b
A population of organisms that has four alleles for a given gene is said to have:

a.   self-sterility genes.
b.   multiple alleles.
c.   continuous variation.
d.   polygenic inheritance.
e.   incomplete dominance.

34. Broadening the Concept of the Gene; p. 195; moderate; ans: c
In _____, one gene interferes with the effect of another gene.

 a. pleiotropy
 b. polyploidy
 c. epistasis
 d. polygenic inheritance
 e. incomplete dominance

35. Broadening the Concept of the Gene; p. 195; moderate; ans: a
In peas, the dominant alleles A and B must both be present for purple flower pigment to be produced; otherwise the flowers are white. What offspring would you expect from a cross between plants of genotypes $Aabb$ and $aaBb$?

 a. 1 purple: 3 white
 b. 3 purple: 1 white
 c. 1 purple: 1 white
 d. 9 purple: 2 white
 e. 2 purple: 3 white

36. Broadening the Concept of the Gene; p. 195; moderate; ans: e
When a trait exhibits continuous variation in a population of organisms, it most likely is due to:

 a. one gene having multiple effects.
 b. one allele not completely masking another allele.
 c. several alleles of the same gene interacting.
 d. one gene masking the effect of another gene.
 e. several different genes interacting.

37. Broadening the Concept of the Gene; p. 195; moderate; ans: a
In _____, a single gene has many different effects on the phenotype.

 a. pleiotropy
 b. continuous variation
 c. epistasis
 d. polygenic inheritance
 e. incomplete dominance

38. Broadening the Concept of the Gene; p. 196; moderate; ans: e
Cytoplasmic inheritance in plants involves genes present in the:

 a. cytosol only.
 b. plastids only.
 c. mitochondria only.
 d. cytosol and plastids.
 e. mitochondria and plastids.

39. Broadening the Concept of the Gene; p. 196; difficult; ans: e
Which of the following is NOT a possible cause of maternal inheritance?

- a. The generative cell receives neither plastids nor mitochondria from the microspore.
- b. The generative cell receives both plastids and mitochondria but they degenerate.
- c. Plastids and mitochondria are excluded from generative and sperm cells.
- d. Plastids and mitochondria are present in sperm cells but are not transmitted into the egg.
- e. Plastids and mitochondria are present in sperm cells but only plastids are transmitted into the egg.

40. Broadening the Concept of the Gene; p. 197; easy; ans: b
An aquatic plant produces one type of leaf above water and another type of leaf under water. This is most likely an example of:

- a. one gene having multiple effects.
- b. interaction of the genotype with the environment.
- c. one allele not completely masking another allele.
- d. one gene masking the effect of another gene.
- e. several different genes interacting.

41. The Chemistry of the Gene: DNA versus Protein; p. 197; moderate; ans: c
The genetic material is composed chemically of:

- a. chromosomes.
- b. protein only.
- c. deoxyribonucleic acid only.
- d. ribonucleic acid only.
- e. protein and deoxyribonucleic acid in equal amounts.

42. The Structure of DNA; p. 198; easy; ans: c
The purine bases of DNA are:

- a. cytosine and thymine.
- b. cytosine and adenine.
- c. adenine and guanine.
- d. guanine and cytosine.
- e. guanine and thymine.

43. The Structure of DNA; p. 198; moderate; ans: b
Data obtained by Erwin Chargaff indicated that in DNA the ratio of nucleotides containing _____ to those containing _____ is approximately 1:1.

- a. adenine; cytosine
- b. adenine; thymine
- c. guanine; thymine
- d. thymine; cytosine
- e. guanine; adenine

44. The Structure of DNA; p. 198; easy; ans: b
The DNA molecule has the shape of a:

  a.   single helix.
  b.   double helix
  c.   triple helix.
  d.   step ladder.
  e.   staircase.

45. The Structure of DNA; p. 199; moderate; ans: e
In the ladder analogy of DNA structure, the rungs are composed of:

  a.   sugars.
  b.   phosphates.
  c.   alternating sugars and phosphates.
  d.   alternating sugars and nitrogenous bases.
  e.   nitrogenous bases.

46. The Structure of DNA; p. 199; moderate; ans: c
In the ladder analogy of DNA structure, the sides of the ladder are composed of:

  a.   sugars.
  b.   phosphates.
  c.   alternating sugars and phosphates.
  d.   alternating sugars and nitrogenous bases.
  e.   nitrogenous bases.

47. The Structure of DNA; p. 199; easy; ans: a
In a DNA molecule, the hydrogen bonds join:

  a.   the paired nitrogenous bases.
  b.   sugars and phosphates.
  c.   sugars and nitrogenous bases.
  d.   phosphates and nitrogenous bases.
  e.   pairs of sugars.

48. The Structure of DNA; p. 199; easy; ans: d
Which of the following are complementary bases?

  a.   adenine and cytosine
  b.   adenine and guanine
  c.   thymine and cytosine
  d.   guanine and cytosine
  e.   thymine and guanine

49. The Structure of DNA; p. 199; easy; ans: c
Adenine forms _____ hydrogen bonds with thymine.

  a.   0
  b.   1
  c.   2
  d.   3
  e.   4

50. DNA Replication; pp. 201–202; moderate; ans: e
Which of the following statements about DNA replication in eukaryotes is FALSE?

a.   It is a bidirectional process.
b.   Helicases break the hydrogen bonds at the origin of replication.
c.   DNA polymerase catalyzes the synthesis of the new strands.
d.   The two replication forks move in opposite directions away from the origin.
e.   There is only one origin of replication.

51. DNA Replication; p. 201; easy; ans: a
After a DNA molecule "unzips," the two strands are kept separated by:

a.   single-strand binding proteins.
b.   DNA polymerase.
c.   DNA ligase.
d.   helicases.
e.   telomerase.

52. DNA Replication; p. 202; moderate; ans: c
In DNA replication, the lagging strand differs from the leading strand in that the lagging strand is synthesized:

a.   in a 5′ to 3′ direction.
b.   in a 3′ to 5′ direction.
c.   in fragments.
d.   using DNA polymerase.
e.   outside the replication bubble.

53. The Problem of the Ends of Linear DNA; p. 203; easy; ans: a
Where are telomeres located?

a.   at the ends of chromosomes
b.   close to the centromeres
c.   close to the replication forks
d.   inside replication bubbles
e.   on lagging strands but not leading strands

54. The Problem of the Ends of Linear DNA; p. 203; moderate; ans: b
One function of telomeres is to:

a.   keep the "unzipped" DNA strands from joining.
b.   prevent end-to-end fusion of chromosomes.
c.   unwind the double helix.
d.   prevent chromosomes from increasing in length.
e.   catalyze the pairing of complementary bases.

55. The Problem of the Ends of Linear DNA; p. 203; easy; ans: b
When an incorrect base is inserted into a DNA molecule during replication, the error is corrected by:

a.   single-strand binding protein.
b.   DNA polymerase.
c.   DNA ligase.
d.   helicase.
e.   telomerase.

56. **The Energetics of DNA Replication; p. 204; moderate; ans: e**
For insertion of a guanine nucleotide into a DNA molecule, the guanine must be supplied as:

   a.   guanine.
   b.   GMP.
   c.   dGMP.
   d.   dGDP.
   e.   dGTP.

57. **The Energetics of DNA Replication; p. 204; moderate; ans: d**
In DNA replication, the nucleotide triphosphates dATP, dGTP, dCTP, and dGTP are sources of:

   a.   nucleotides only.
   b.   ribose sugar only.
   c.   energy only.
   d.   nucleotides and energy.
   e.   ribose sugar and energy.

True-False Questions

1. **The Concept of the Gene; p. 184; moderate; ans: T**
Mendel was the first scientist to apply quantitative methods to the study of heredity.

2. **The Principle of Segregation; p. 186; moderate; ans: F**
The term "genotype" refers to the appearance of an individual.

3. **The Principle of Segregation; p. 187; moderate; ans: T**
A testcross is used to determine whether a plant showing a dominant trait is homozygous or heterozygous for that trait.

4. **The Principle of Independent Assortment; pp. 188–189; moderate; ans: T**
Mendel's experiments using dihybrid crosses produced completely new combinations of characteristics in the F_1 generation.

5. **Discovery of the Chromosomal Basis of Mendel's Laws; p. 190; moderate; ans: T**
The behavior of chromosomes in meiosis was not observed until after Mendel had completed his work.

6. **Linkage; p. 192; easy; ans: F**
The greater the distance between two genes on a chromosome, the smaller is the chance that they will undergo crossing-over during meiosis.

7. **Mutations; p. 192; easy; ans: T**
A mutation is any change in the hereditary state of an organism.

8. **Mutations; p. 193; easy; ans: T**
Polyploidy results from the duplication of whole sets of chromosomes.

9. **Mutations; p. 193; moderate; ans: T**
A mutation in a haploid cell is more likely to have an effect on phenotype than a mutation in a diploid cell.

10. **Broadening the Concept of the Gene; p. 195; moderate; ans: F**
Epistasis is an example of interactions between the alleles of a gene.

11. **Broadening the Concept of the Gene; p. 196; moderate; ans: T**
Cytoplasmic inheritance refers to genes located in mitochondria or plastids.

12. **Broadening the Concept of the Gene; p. 197; easy; ans: F**
The genotype is the result of the phenotype interacting with the environment.

13. **The Chemical Basis of Heredity; p. 197; easy; ans: T**
The study of genes and chromosomes at the level of chemical structure is called molecular genetics.

14. **The Structure of DNA; p. 199; moderate; ans: T**
The complementarity of the two strands of DNA is a direct result of A pairing only with T and C pairing only with G.

15. **The Structure of DNA; p. 199; moderate; ans: F**
The two strands of a DNA molecule are said to be antiparallel, each strand having its 5' end at the same end of the molecule.

16. **DNA Replication; p. 201; moderate; ans: F**
The rate of DNA replication in eukaryotes is about ten times faster than in prokaryotes.

17. **DNA Replication; p. 201; easy; ans: T**
In DNA replication, each strand serves as a template for the formation of a new strand.

18. **DNA Replication; p. 202; moderate; ans: T**
DNA polymerase synthesizes new DNA strands only in the 5' to 3' direction.

19. **DNA Replication; p. 202; moderate; ans: F**
In DNA replication, the leading strand is synthesized in fragments, called Okazaki fragments, while the lagging strand is synthesized continuously.

20. **The Problem of the Ends of Linear DNA; p. 203; easy; ans: T**
Telomeres protect the rest of the chromosome from degradation.

21. **The Energetics of DNA Replication; pp. 203–204; easy; ans: F**
The nucleotides required for DNA replication are supplied as their monophosphates.

22. **DNA As a Carrier of Information; p. 204; moderate; ans: T**
The chromosomes of the cell of a lily contain about 200 billion paired bases.

Essay Questions

1. **The Concept of the Gene; pp. 184–185; moderate**
    List five reasons why Mendel's experimental approach to studying heredity was so successful.

2. **The Principle of Segregation; pp. 186, 189; moderate**
    Explain Mendel's principle of segregation. Does it deal with the alleles of one gene or more than one gene?

3. **The Principle of Segregation; p. 186; moderate**
    What is an allele? What is the relationship between alleles and homologous chromosomes?

4. **The Principle of Independent Assortment; Discovery of the Chromosomal Basis of Mendel's Laws; pp. 189–190; moderate**
    Explain Mendel's principle of independent assortment. How does this principle relate to meiosis?

5. **Linkage; pp. 190–191; moderate**
    Explain how the frequency of crossing-over can be used to determine the approximate locations of genes on a chromosome.

6. **Mutations; pp. 192–193; moderate**
    What is the difference between a deletion and a duplication in a chromosome? What do they have in common?

7. **Mutations; pp. 193–194; moderate**
    Discuss the relationship between mutations and evolutionary change.

8. **Broadening the Concept of the Gene; p. 195; difficult**
    Give some examples of polygenic inheritance and pleiotropy. How could you distinguish between these conditions?

9. **Broadening the Concept of the Gene; p. 196; moderate**
    Why are cytoplasmically inherited traits maternally, rather than paternally, inherited? Give an example of cytoplasmic inheritance involving (a) chloroplasts and (b) mitochondria.

10. **The Structure of DNA; pp. 198–199; moderate**
    Explain the correlation between Chargaff's data on nucleotide ratios and Watson and Crick's concept of complementary base-pairing.

11. **DNA Replication; pp. 201–202; moderate**
    Summarize the main steps in DNA replication. Why is the lagging strand synthesized in fragments rather than as a continuous strand?

12. **The Problem of the Ends of Linear DNA; p. 203; moderate**
    In what way do the ends of a DNA molecule pose a problem for DNA replication? How do cells solve this problem?

13. **The Energetics of DNA Replication; pp. 203–204; moderate**
In what form are nucleotides supplied for DNA replication? How is this
related to the energy required for replication?

Chapter 11

Gene Expression

Multiple-Choice Questions

1. **Gene Expression; p. 208; moderate; ans: e**
   RNA differs from DNA in that RNA:

   a. contains deoxyribose.
   b. contains thymine.
   c. forms a regular helix.
   d. contains adenine.
   e. is usually single-stranded.

2. **Gene Expression; p. 208; easy; ans: b**
   RNA contains _____ instead of the _____ found in DNA.

   a. adenine; thymine
   b. uracil; thymine
   c. adenine; uracil
   d. guanine; cytosine
   e. uracil; guanine

3. **From DNA to Protein: The Role of RNA; p. 208; easy; ans: b**
   In the process of transcription, _____ serves as a template for the synthesis of _____.

   a. DNA; DNA
   b. DNA; RNA
   c. RNA; DNA
   d. RNA; protein
   e. DNA; protein

4. **From DNA to Protein: The Role of RNA; p. 208; moderate; ans: c**
   The role of _____ is to carry amino acids to the ribosomes.

   a. DNA
   b. mRNA
   c. tRNA
   d. rRNA
   e. protein

5. **From DNA to Protein: The Role of RNA; p. 208; easy; ans: e**
   The process of translation results in the synthesis of:

   a.   DNA.
   b.   mRNA.
   c.   tRNA.
   d.   rRNA.
   e.   protein.

6. **From DNA to Protein: The Role of RNA; pp. 208–209; easy; ans: a**
   Replication is the synthesis of:

   a.   DNA.
   b.   mRNA.
   c.   tRNA.
   d.   rRNA.
   e.   protein.

7. **The Genetic Code; p. 209 and Fig. 11-4; moderate; ans: d**
   A codon consists of _____ nucleotides in a ____ molecule.

   a.   2; DNA
   b.   3; DNA
   c.   4; DNA
   d.   3; RNA
   e.   4; RNA

8. **The Genetic Code; p. 209; moderate; ans: b**
   When scientists describe the genetic code as degenerate, they mean that:

   a.   it becomes disorganized over time.
   b.   many amino acids have more than one codon.
   c.   some codons specify stop signals.
   d.   it varies with cell type.
   e.   it varies among species.

9. **Protein Synthesis; p. 210 and Fig. 11-5; moderate; ans: d**
   If one strand of a DNA molecule has the base sequence ACTTAGCAT (5′ end at the left), what is the corresponding sequence of bases in the mRNA following transcription (3′ end at the left)?

   a.   ACTTAGCAT
   b.   ACUUAGCAU
   c.   TGAATCGTA
   d.   UGAAUCGUA
   e.   GUCCGATGC

10. **Protein Synthesis; p. 210; moderate; ans: b**
    Which of the following statements about promoters is FALSE?

    a.   They are specific nucleotide sequences of DNA.
    b.   They consist of three nucleotides that bind to a codon.
    c.   They determine the position where RNA synthesis begins.
    d.   They determine which DNA strand is used as a template.
    e.   They are binding sites for RNA polymerase.

11. Protein Synthesis; p. 210; easy; ans: c
Which of the following molecules contains an anticodon?

a.   DNA
b.   mRNA
c.   tRNA
d.   rRNA
e.   protein

12. Protein Synthesis; p. 210; moderate; ans: c
Which types of molecules bind both to a nucleic acid and to an amino acid during protein synthesis?

a.   DNA
b.   mRNA
c.   tRNA
d.   rRNA
e.   ATP

13. Protein Synthesis; p. 210; easy; ans: a
A transfer RNA molecule carries a(n) _____ to the ribosomes.

a.   amino acid
b.   protein
c.   mRNA molecule
d.   ATP molecule
e.   segment of DNA nucleotides

14. Protein Synthesis; pp. 210–211; easy; ans: d
An aminoacyl-tRNA synthetase catalyzes the:

a.   attachment of a particular tRNA to mRNA.
b.   binding of RNA polymerase to the terminator.
c.   binding of RNA polymerase to the promoter.
d.   attachment of a particular amino acid to a particular tRNA.
e.   attachment of a particular tRNA to the ribosome.

15. Protein Synthesis; p. 211; easy; ans: e
A ribosome consists of _____ and _____.

a.   DNA; mRNA
b.   mRNA; protein
c.   mRNA; tRNA
d.   rRNA; mRNA
e.   rRNA; protein

16. Protein Synthesis; p. 211; moderate; ans: b
The smaller ribosomal subunit has a(n) _____ site.

a.   A (aminoacyl)
b.   mRNA-binding
c.   P (peptidyl)
d.   E (exit)
e.   promoter

17. **Protein Synthesis; pp. 211–212; moderate; ans: a**
An incoming aminoacyl-tRNA molecule binds to the ribosome at the _____ site.

   a.   A (aminoacyl)
   b.   mRNA-binding
   c.   P (peptidyl)
   d.   E (exit)
   e.   promoter

18. **Protein Synthesis; p. 212; moderate; ans: d**
During the initiation stage of translation:

   a.   the two ribosomal subunits separate.
   b.   fMet-tRNA binds at the A site.
   c.   release factors bind at the A site.
   d.   fMet-tRNA binds to the initiation codon.
   e.   a peptide bond is formed.

19. **Protein Synthesis, p. 212; difficult; ans: e**
Which of the following events is NOT part of the initiation stage of translation?

   a.   The smaller ribosomal subunit attaches to mRNA.
   b.   GTP is hydrolyzed.
   c.   The larger ribosomal subunit attaches to the smaller subunit.
   d.   fMet-tRNA binds at the P site.
   e.   The initiator tRNA is released from the E site.

20. **Protein Synthesis; p. 212; moderate; ans: e**
During the elongation stage of translation:

   a.   the two ribosomal subunits separate.
   b.   a polysome forms.
   c.   release factors bind at the A site.
   d.   fMet-tRNA binds to the initiation codon.
   e.   a peptide bond is formed.

21. **Protein Synthesis; p. 214; difficult; ans: c**
Which of the following events is NOT part of the termination stage of translation?

   a.   The ribosome encounters a stop codon.
   b.   Release factors bind at the A site.
   c.   The ribosome moves along the mRNA to the next codon.
   d.   The ribosomal subunits separate.
   e.   The polypeptide chain is released.

22. **Protein Synthesis; p. 214; moderate; ans: b**
Polypeptides destined for membranes of the Golgi complex are synthesized on ribosomes:

   a.   that are free in the cytosol.
   b.   attached to the endoplasmic reticulum.
   c.   in the nucleus.
   d.   in mitochondria.
   e.   in plastids.

23. **Protein Synthesis; p. 214; moderate; ans: d**
In the process of cotranslational import, a newly synthesized polypeptide is:

  a. taken up by the nucleus after translation is complete.
  b. taken up by the mitochondria after translation is complete.
  c. taken up by peroxisomes after translation is complete.
  d. transferred across the endoplasmic reticulum membrane during translation.
  e. transferred across the inner mitochondrial membrane during translation.

24. **Regulating Gene Expression; p. 214; difficult; ans: a**
In a cell of a multicellular organism:

  a. all enzymes are expressed continuously.
  b. some genes are expressed continuously.
  c. some genes are not expressed.
  d. some genes are switched "on" at certain times.
  e. some genes are switched "off" at certain times.

25. **The Prokaryotic Chromosome; p. 216; moderate; ans: a**
Which of the following statements concerning the prokaryotic chromosome is FALSE?

  a. It contains DNA and RNA, but not protein.
  b. It is circular.
  c. It is supercoiled.
  d. It is folded into loops.
  e. It is located in the nucleoid.

26. **The Prokaryotic Chromosome; p. 216; easy; ans: a**
The segments of DNA that code for the cell's polypeptides are called _____ genes.

  a. structural
  b. operator
  c. regulator
  d. promoter
  e. active

27. **The Prokaryotic Chromosome; p. 216; easy; ans: c**
A(n) _____ is a bacterial gene that codes for a repressor protein.

  a. operon
  b. operator
  c. regulator
  d. promoter
  e. corepressor

28. **The Prokaryotic Chromosome; p. 217; moderate; ans: b**
In the *lac* operon of bacteria, allolactose functions as a(n):

  a. repressor.
  b. inducer.
  c. corepressor.
  d. operator.
  e. promoter.

29. **The Prokaryotic Chromosome; pp. 217–218; moderate; ans: c**
After allolactose binds to the repressor of the *lac* operon, what happens next?

   a.   RNA polymerase is inactivated.
   b.   The repressor binds to the operator.
   c.   The repressor is removed from the operator.
   d.   The repressor is removed from the structural gene.
   e.   The repressor is activated.

30. **The Prokaryotic Chromosome; p. 218; moderate; ans: c**
In the *trp* operon of bacteria, tryptophan functions as a(n):

   a.   repressor.
   b.   inducer.
   c.   corepressor.
   d.   operator.
   e.   promoter.

31. **The Prokaryotic Chromosome; p. 218; moderate; ans: e**
When present in the *E. coli* growth medium, tryptophan:

   a.   binds to the promoter of the *trp* operon.
   b.   inactivates RNA polymerase.
   c.   activates RNA polymerase.
   d.   inactivates the repressor for the *trp* operon.
   e.   activates the repressor for the *trp* operon.

32. **The Prokaryotic Chromosome; p. 218; moderate; ans: d**
When tryptophan is present in the *E. coli* growth medium:

   a.   it activates RNA polymerase.
   b.   it inactivates RNA polymerase.
   c.   it binds to the regulator.
   d.   the synthesis of tryptophan stops.
   e.   the synthesis of tryptophan begins.

33. **The Eukaryotic Chromosome; p. 219; difficult; ans: e**
Chromatin consists of:

   a.   DNA only.
   b.   histones only.
   c.   DNA and histones only.
   d.   histones and other proteins only.
   e.   DNA, histones, and other proteins.

34. **The Eukaryotic Chromosome; p. 219; moderate; ans: c**
Which of the following statements about histones is FALSE?

   a.   They are the most abundant proteins in eukaryotic chromosomes.
   b.   They are present in five distinct types.
   c.   They are negatively charged.
   d.   They are primarily responsible for the folding and packaging of DNA.
   e.   They are synthesized during the S phase of the cell cycle.

35. **The Eukaryotic Chromosome; p. 219; easy; ans: a**
    A nucleosome consists of:

    a.   a core of histones wrapped with DNA.
    b.   eight histone molecules arranged in a spherical mass.
    c.   a strand of DNA that is not associated with histones.
    d.   a looped domain in the chromosome.
    e.   nonhistone chromosomal proteins.

36. **Regulation of Gene Expression in Eukaryotes; p. 220; moderate; ans: e**
    Euchromatin differs from heterochromatin in that euchromatin:

    a.   stains more strongly.
    b.   is more tightly condensed.
    c.   predominates in centromeres.
    d.   predominates in telomeres.
    e.   is more readily transcribed during interphase.

37. **Regulation of Gene Expression in Eukaryotes; p. 221; difficult; ans: b**
    Which of the following statements concerning transcription factors is FALSE?

    a.   They are proteins.
    b.   They are all histones.
    c.   They bind to specific sites on the DNA molecule.
    d.   They may directly affect the initiation of transcription.
    e.   They may indirectly affect the initiation of transcription.

38. **The DNA of the Eukaryotic Chromosome; p. 221; easy; ans: b**
    Which of the following is the best estimate for the percentage of DNA in
    eukaryotic cells that codes for proteins?

    a.   less than 1 percent
    b.   less than 10 percent
    c.   about 25 percent
    d.   about 50 percent
    e.   about 75 percent

39. **The DNA of the Eukaryotic Chromosome; pp. 221–222; difficult; ans: c**
    Which of the following statements about simple-sequence repeated DNA
    segments is FALSE?

    a.   They are tandemly repeated.
    b.   They occur in telomeres.
    c.   They are dispersed throughout the DNA.
    d.   They occur in centromeres.
    e.   They have fewer than 10 base pairs.

40. **The DNA of the Eukaryotic Chromosome; p. 222; difficult; ans: e**
    Interspersed repeated DNA units:

    a.   are arranged in tandem.
    b.   tend to be smaller than 10 base-pairs long.
    c.   constitute less than 10 percent of the DNA of most multicellular
         organisms.
    d.   are identical to one another.
    e.   are believed to have originated from transposons.

41. **The DNA of the Eukaryotic Chromosome; p. 222; difficult; ans: b**

    Single-copy DNA:

    a. constitutes 20 to 40 percent of the DNA of the organism.
    b. may function as spacer DNA.
    c. is frequently present in multiple copies.
    d. constitutes essentially all the genes that are not transcribed into protein.
    e. is almost entirely transcribed into RNA.

42. **The DNA of the Eukaryotic Chromosome; p. 222; moderate; ans: a**

    Introns are segments of a gene that:

    a. are transcribed into segments of mRNA which are snipped out before translation.
    b. occur in prokaryotic but not eukaryotic chromosomes.
    c. consist of RNA rather than DNA nucleotides.
    d. are not transcribed by RNA polymerase.
    e. are translated into protein.

43. **Transcription and Processing of mRNA in Eukaryotes; p. 223; difficult; ans: d**

    Which of the following events occurs in eukaryotic cells but NOT in prokaryotic cells?

    a. RNA polymerase attaches to a particular nucleotide sequence on the DNA.
    b. RNA is synthesized using the 3' to 5' DNA strand as a template.
    c. Two or more structural genes are transcribed onto a single RNA molecule.
    d. mRNA transcripts are extensively modified before they leave the nucleus.
    e. tRNAs, rRNAs, and mRNAs are all involved in the translation process.

44. **Recombinant DNA Technology; p. 224; moderate; ans: d**

    A restriction enzyme would most likely be used in recombinant DNA technology to:

    a. transfer a plasmid into a bacterial cell.
    b. screen for recombinant plasmids.
    c. stimulate a transformed cell to divide.
    d. cut DNA into fragments with sticky ends.
    e. join DNA fragments together.

45. **Recombinant DNA Technology; p. 224; moderate; ans: e**

    DNA ligase would most likely be used in recombinant DNA technology to:

    a. function as a vector.
    b. screen for recombinant plasmids.
    c. stimulate a transformed cell to divide.
    d. cut DNA into fragments with sticky ends.
    e. join DNA fragments together.

46. Recombinant DNA Technology; p. 225; difficult; ans: b
By definition, a transformed cell:

a. contains a plasmid.
b. contains a recombinant DNA molecule.
c. contains a virus.
d. functions as a vector.
e. divides in such a way so as to amplify the DNA fragment.

47. Recombinant DNA Technology; p. 225; difficult; ans: d
DNA cloning refers to the process by which:

a. plasmids act as vectors.
b. restriction enzymes cut up DNA.
c. recombinant plasmids are taken up by bacterial cells.
d. cells containing a recombinant plasmid divide and grow into a colony.
e. colonies are screened to identify the recombinant plasmid.

48. Recombinant DNA Technology; p. 225; difficult; ans: c
The amp^R gene is used in recombinant DNA technology to:

a. produce β-galactosidase.
b. clone a particular gene fragment.
c. screen for the presence of cells containing recombinant plasmids.
d. help the bacterium to survive on a medium containing lactose.
e. transform a bacterial cell.

49. DNA Libraries; p. 226; difficult; ans: e
Which of the following statements concerning DNA libraries is FALSE?

a. A genome library is obtained by "shotgun cloning."
b. A cDNA library is obtained by using reverse transcriptase.
c. A DNA library contains cloned fragments of most of an organism's genome.
d. A cDNA library lacks introns.
e. A genomic library lacks exons.

50. DNA Libraries; p. 226; easy; ans: a
In recombinant DNA technology, reverse transcriptase is used to:

a. synthesize cDNA.
b. synthesize mRNA.
c. cut DNA into fragments.
d. amplify a specific gene.
e. transfer a plasmid into a cell.

51. Polymerase Chain Reaction; p. 226; easy; ans: c
The *Taq* polymerase is used in recombinant DNA technology to:

a. locate specific DNA segments.
b. determine the nucleotide sequence of a specific gene.
c. make up to a millionfold copies of a DNA segment in a few hours.
d. make cDNA.
e. carry out "shotgun cloning."

52. DNA Sequencing; p. 229; difficult; ans: e
In recombinant DNA technology, cutting different samples of a single DNA molecule with different restriction enzymes allows one to:

a. amplify the DNA molecule.
b. carry out DNA hybridization.
c. locate a gene of interest.
d. make a cDNA library.
e. determine the nucleotide sequence of a gene.

53. Recombinant DNA Technology; p. 231; moderate; ans: a
The foundation of nucleic acid hybridization is the:

a. base-pairing properties of the nucleic acids.
b. antiparallel nature of the DNA strands.
c. ability of bacteria to become transformed.
d. ability of viruses to function as vectors.
e. ability of DNA to be radioactively labeled.

54. Recombinant DNA Technology; p. 231; moderate; ans: b
The purpose of using nucleic acid hybridization is to:

a. clone a particular gene.
b. locate a particular gene.
c. obtain a cDNA library.
d. obtain a genomic library.
e. determine the base sequence of a particular gene.

55. Recombinant DNA Technology; p. 231; difficult; ans: c
The technique of co-localizing a mutation and a previously cloned fragment of DNA would be used to:

a. determine the amino acid sequence of a protein.
b. prepare a cDNA library.
c. locate a gene of interest.
d. insert a plasmid into a bacterial cell.
e. hybridize a specific DNA segment.

True-False Questions

1. From DNA to Protein: The Role of RNA; p. 208; easy; ans: T
Transcription is catalyzed by the enzyme RNA polymerase.

2. The Genetic Code; p. 209; moderate; ans: T
The triplet codon was first demonstrated in experiments using bacterial extracts, mRNAs, and radioactively labeled amino acids.

3. The Genetic Code; p. 209; easy; ans: F
The genetic code in prokaryotes is different from that in eukaryotic cells.

4. Protein Synthesis; p. 210; easy; ans: F
The terminator is a section of DNA that codes for the destruction of mRNA molecules following translation.

5. **Protein Synthesis; p. 212; easy; ans: T**
   A group of ribosomes translating the same mRNA molecule is called a
   polysome.

6. **Protein Synthesis; p. 214; moderate; ans: F**
   Proteins destined for the nucleus are taken up through the nuclear envelope by
   cotranslational import.

7. **Regulating Gene Expression; pp. 214, 220; moderate; ans: T**
   In a plant, root cells have the same genetic composition as leaf cells.

8. **The Prokaryotic Chromosome; p. 216; moderate; ans: T**
   In prokaryotes, a single mRNA molecule may encode for more than one
   polypeptide.

9. **The Prokaryotic Chromosome; p. 216; moderate; ans: F**
   The sequence of nucleotides located between the promoter and the structural
   genes of an operon make up the regulator gene.

10. **The Prokaryotic Chromosome; p. 216; easy; ans: T**
    When a repressor protein binds to the operator gene, RNA polymerase binding
    is obstructed.

11. **The Eukaryotic Chromosome; p. 219; moderate; ans: F**
    Each haploid cell of a *Trillium* contains about the same amount of DNA as an
    E. coli cell.

12. **The Eukaryotic Chromosome; p. 219; easy; ans: F**
    Histones, unlike other chromosomal proteins, vary widely from one cell type to
    another.

13. **Regulation of Gene Expression in Eukaryotes; p. 220; easy; ans: T**
    In the cells of a multicellular eukaryote, some genes are active at the same time
    that others are inactive.

14. **Regulation of Gene Expression in Eukaryotes; p. 220; moderate; ans: T**
    Heterochromatin is more tightly condensed than euchromatin and in some
    regions is never transcribed.

15. **Regulation of Gene Expression in Eukaryotes; p. 221; moderate; ans: F**
    Transcription factors are RNA molecules that directly or indirectly affect the
    initiation of transcription.

16. **The DNA of the Eukaryotic Chromosome; pp. 221–222; easy; ans: T**
    The two major categories of repeated DNA are tandemly repeated DNA and
    interspersed repeated DNA.

17. **The DNA of the Eukaryotic Chromosome; p. 222; easy; ans: F**
    Most structural genes of eukaryotes are continuous, not interrupted by the
    noncoding sequences common in other types of genes.

18. **Transcription and Processing of mRNA in Eukaryotes; p. 223; moderate; ans: F**
In eukaryotes, several structural genes may be transcribed as a single RNA molecule.

19. **Transcription and Processing of mRNA in Eukaryotes; p. 223; easy; ans: T**
The 5' cap of mRNA serves to bind the mRNA molecule to the ribosome.

20. **Recombinant DNA Technology; p. 224; difficult; ans: T**
Recombinant DNA technology permits genetic recombination between species unable to hybridize by natural means.

21. **Recombinant DNA Technology; p. 224; moderate; ans: T**
DNA fragments having sticky ends can join with other fragments obtained by using the same restriction enzyme.

22. **Recombinant DNA Technology; p. 225; easy; ans: T**
Both plasmids and viruses can each be used to obtain transformed cells.

23. **Recombinant DNA Technology; p. 225; moderate; ans: F**
A bacterial cell carrying the gene amp^R is not able to grow on a medium containing ampicillin.

24. **Recombinant DNA Technology; p. 226; moderate; ans: F**
A vector carrying the *lac*Z gene would be used to obtain a DNA library.

25. **Recombinant DNA Technology; p. 226; moderate; ans: T**
A cDNA library consists of DNA without introns.

26. **Recombinant DNA Technology; p. 226; easy; ans: T**
The polymerase chain reaction is used to quickly make many copies of a DNA segment.

27. **Recombinant DNA Technology; p. 229; moderate; ans: F**
Identical DNA fragments are obtained by cutting the same DNA molecule with different restriction enzymes.

28. **Recombinant DNA Technology; p. 230; moderate; ans: T**
The extent to which segments from two DNA samples hybridize provides an estimate between the two nucleotide sequences.

29. **Recombinant DNA Technology; p. 231; moderate; ans: T**
Knowing the amino acid sequence of a protein, one can prepare a probe for the corresponding DNA sequence.

Essay Questions

1. **Gene Expression; p. 208; easy**
How do RNA and DNA differ? How are they similar?

2. **From DNA to Protein: The Role of RNA; p. 208; easy**
   Explain what is meant by the "central dogma of molecular biology."

3. **The Genetic Code; p. 209; easy**
   Explain why at least three nucleotides must specify each amino acid.

4. **Protein Synthesis; pp. 210–211; moderate**
   Why is the specificity of aminoacyl-tRNA synthetases essential for the synthesis of proteins? What would be the consequences for the cell if a particular aminoacyl-tRNA synthetase was *not* specific for a particular amino acid?

5. **Protein Synthesis; pp. 212, 214; moderate**
   Summarize the main events occurring in translation.

6. **Protein Synthesis; p. 214; moderate**
   What roles do cotranslational and posttranslational import play in the sorting of proteins in the cell?

7. **The Prokaryotic Chromosome; p. 216; easy**
   What types of genes comprise an operon? What is the function of each type?

8. **The Prokaryotic Chromosome; pp. 216–218; moderate**
   Compare and contrast enzyme induction and enzyme repression in prokaryotes, giving an example of each.

9. **The Eukaryotic Chromosome; pp. 219–220; moderate**
   Describe the levels of organization in the packaging of a eukaryotic chromosome.

10. **Regulation of Gene Expression in Eukaryotes; p. 220; moderate**
    Explain the concept of totipotency as it applies to plant cells.

11. **Regulation of Gene Expression in Eukaryotes; pp. 220–221; difficult**
    In what way is the regulation of gene expression in multicellular organisms necessarily more complex than in unicellular organisms?

12. **Regulation of Gene Expression in Eukaryotes; p. 221; difficult**
    What are some of the roles played by specific proteins in regulating the transcription of genes?

13. **The DNA of the Eukaryotic Chromosome; pp. 221–222; moderate**
    Describe the two major categories of repeated DNA in eukaryotes.

14. **Transcription and Processing of mRNA in Eukaryotes; p. 223; difficult**
    Discuss the differences between prokaryotes and eukaryotes in transcription, translation, and RNA processing.

15. **Recombinant DNA Technology; p. 224; moderate**
    In what way does recombinant DNA technology involve "recombination"?

16. Recombinant DNA Technology; p. 225; moderate

Describe two ways of screening colonies for the presence of transformed cells in recombinant DNA studies.

17. Recombinant DNA Technology; p. 226; moderate

Describe the steps of the polymerase chain reaction. What is the primary purpose of PCR?

18. Recombinant DNA Technology; p. 229; moderate

Describe how restriction enzymes can be used to obtain the nucleotide sequence of a DNA molecule.

19. Recombinant DNA Technology; p. 230; moderate

Explain how nucleic acid hybridization can be used to locate specific DNA segments. Why do investigators need to locate specific segments?

Chapter 12

The Process of Evolution

Multiple-Choice Questions

1. **The Process of Evolution; p. 236; easy; ans: c**
The process by which the vast diversity of living organisms have arisen from earlier forms is called:

    a. adaptation.
    b. special creation.
    c. evolution.
    d. artificial selection.
    e. natural selection.

2. **The Process of Evolution; p. 236; moderate; ans: e**
The process by which favorable modifications in an organism become more common with successive generations is called:

    a. evolution.
    b. special creation.
    c. descent from a common ancestor.
    d. artificial selection.
    e. natural selection.

3. **Darwin's Theory; p. 237; easy; ans: c**
_____ warned of the hazards of increased human population growth.

    a. Charles Lyell
    b. Gregor Mendel
    c. Thomas Malthus
    d. Charles Darwin
    e. Jean Baptiste de Lamarck

4. **Darwin's Theory; pp. 237–238; easy; ans: a**
Humans breed plants and animals through the process of:

    a. artificial selection.
    b. natural selection.
    c. survival of the fittest.
    d. evolution.
    e. mutation.

5. **Darwin's Theory; p. 238; moderate; ans: d**
Which of the following is NOT part of Darwin's formulation of the process of evolution?
   a. natural selection leading to changes in populations
   b. variations as the raw material of the evolutionary process
   c. formation of new species over very long periods of time
   d. the central role of artificial selection
   e. interaction between variations and the environment

6. **Darwin's Theory; p. 238; easy; ans: a**
_____ demonstrated that the Earth was older than previously thought.
   a. Charles Lyell
   b. Gregor Mendel
   c. Thomas Malthus
   d. Charles Darwin
   e. Jean Baptiste de Lamarck

7. **The Concept of the Gene Pool; p. 239; moderate; ans: b**
_____ is the sum total of all the alleles of all the genes in a population.
   a. The genotype
   b. The gene pool
   c. Evolution
   d. Population genetics
   e. A species

8. **The Concept of the Gene Pool; p. 239; easy; ans: e**
The sole criterion for the fitness of an individual is _____ relative to that of other individuals.
   a. its strength
   b. its beauty
   c. its resistance to disease
   d. the size of its gene pool
   e. the number of its surviving offspring

9. **The Behavior of Genes in Populations: The Hardy-Weinberg Law; pp. 239–240; difficult; ans: e**
Which of the following statements concerning the Hardy-Weinberg Law is FALSE?
   a. It explains why dominant alleles do not drive out recessives.
   b. It explains why both dominant and recessive alleles can remain in a population.
   c. It provides a standard for detecting evolutionary change.
   d. It provides a standard against which we can measure changes in allele frequencies.
   e. It involves conditions that are usually met in natural populations.

10. The Behavior of Genes in Populations: The Hardy-Weinberg Law;
p. 240; moderate; ans: e
Which of the following is NOT a necessary condition for a population to be in Hardy-Weinberg equilibrium?

a.   no natural selection
b.   isolation from other populations
c.   no mutations
d.   random mating
e.   small population size

11. The Behavior of Genes in Populations: The Hardy-Weinberg Law;
p. 240; moderate; ans: c
In a population that is in Hardy-Weinberg equilibrium, the frequency of a particular recessive allele is 60 percent. What proportion of the population is homozygous recessive for that allele?

a.   60 percent
b.   40 percent
c.   36 percent
d.   12 percent
e.   6 percent

12. The Behavior of Genes in Populations: The Hardy-Weinberg Law;
p. 240; moderate; ans: c
In a population that is in Hardy-Weinberg equilibrium, the frequency of a particular dominant allele is 70 percent. What percentage of the individuals in this population will be heterozygous for that trait?

a.   21 percent
b.   30 percent
c.   42 percent
d.   49 percent
e.   70 percent

13. The Behavior of Genes in Populations: The Hardy-Weinberg Law;
p. 240; easy; ans: b
In a population that is in Hardy-Weinberg equilibrium, the frequency of a particular dominant allele is 60 percent, and the frequency of the recessive allele is 40 percent. In the next generation, the frequency of the dominant allele will be _____ percent and the frequency of the recessive allele will be _____ percent.

a.   70; 30
b.   60; 40
c.   50; 50
d.   40: 60
e.   30; 70

14. The Agents of Change; p. 240; moderate; ans: c
The raw material for evolutionary change is provided by:

   a.   sexual reproduction.
   b.   asexual reproduction.
   c.   mutations.
   d.   meiosis.
   e.   mitosis.

15. The Agents of Change; p. 241; moderate; ans: c
In a population, _____ results from the immigration or emigration of individuals.

   a.   mutation
   b.   genetic drift
   c.   gene flow
   d.   natural selection
   e.   nonrandom mating

16. The Agents of Change; p. 241; difficult; ans: c
In a population of 1000 plants, the frequency of the a allele is 5 percent. Suppose a fire causes the loss of 500 individuals who are homozygous for the A allele. In this case the fire caused the frequency of a to change from _____ to _____.

   a.   5; 1
   b.   0.05; 1.0
   c.   0.05; 0.1
   d.   0.5; 1.0
   e.   0.5; 0.1

17. The Agents of Change; p. 241; moderate; ans: b
Changes in a gene pool due to chance are called:

   a.   mutation.
   b.   genetic drift.
   c.   gene flow.
   d.   natural selection.
   e.   nonrandom mating.

18. The Agents of Change; pp. 241–242; moderate; ans: d
An example of _____ is when a single plant seed initiates a new population.

   a.   nonrandom mating
   b.   natural selection
   c.   fitness
   d.   the founder effect
   e.   the bottleneck effect

19. The Agents of Change; p. 242; easy; ans: e
Inbreeding is an example of:

   a.   mutation.
   b.   genetic drift.
   c.   gene flow.
   d.   natural selection.
   e.   nonrandom mating.

20. The Agents of Change; p. 242; easy; ans: e
When plants self-pollinate, they are undergoing:

    a.   mutation.
    b.   genetic drift.
    c.   gene flow.
    d.   natural selection.
    e.   nonrandom mating.

21. The Agents of Change; p. 242; difficult; ans: c
When plants heterozygous for a trait self-pollinate there will be a(n):

    a.   decrease in the frequency of individuals homozygous recessive for that trait.
    b.   decrease in the frequency of individuals homozygous dominant for that trait.
    c.   decrease in the frequency of individuals heterozygous for that trait.
    d.   increase in the frequency of individuals heterozygous for that trait.
    e.   increase in the frequency of homozygous recessive individuals but a decrease in homozygous dominant individuals.

22. Preservation and Promotion of Variability; p. 242; moderate; ans: c
The most important method for promoting variation in eukaryotes is:

    a.   outbreeding instead of inbreeding.
    b.   inbreeding instead of outbreeding.
    c.   sexual reproduction.
    d.   diploidy rather than haploidy.
    e.   maintaining a heterozygote advantage.

23. Preservation and Promotion of Variability; pp. 242–243; moderate; ans: b
Which of the following is NOT a mechanism of promoting outbreeding in plants?

    a.   anatomical arrangements that inhibit self-pollination
    b.   increased rates of crossing-over
    c.   genes for self-sterility
    d.   pollen and stigma maturing at different times
    e.   male and female flowers on different plants

24. Preservation and Promotion of Variability; p. 243; moderate; ans: c
In heterozygote advantage, the _____ individual(s) produce(s) more offspring than the _____ individual(s).

    a.   homozygous recessive; heterozygous and homozygous dominant
    b.   homozygous dominant; heterozygous and homozygous recessive
    c.   heterozygous; homozygous recessive and homozygous dominant
    d.   homozygous recessive and homozygous dominant; heterozygous
    e.   homozygous recessive and heterozygous; homozygous dominant

25. Responses to Selection; p. 243; moderate; ans: a

In a population, natural selection:

   a.   can preserve and promote variability.

   b.   eliminates only the "less fit."

   c.   acts on the genotype.

   d.   reduces genetic variability.

   e.   reduces the potential for further evolution.

26. Responses to Selection; pp. 244–245; difficult; ans: c

As an example of rapid evolutionary change, one variety of *Agrostis* grows much faster on tailings and dumps around abandoned lead mines than does another variety. What is the explanation for this?

   a.   The high lead content of the soil causes an increase in the mutation rate.

   b.   Lead stimulates the growth of plants.

   c.   Selection has favored the development of the lead-resistant variety.

   d.   Selection has favored the development of the lead-sensitive variety.

   e.   Lead stimulates the production of poisons that protect the plants against predators.

27. The Result of Natural Selection: Adaptation; p. 245; moderate; ans: d

Which of the following statements about adaptation is FALSE?

   a.   It means a state of being adjusted to the environment.

   b.   It means a particular characteristic that aids in the adjustment of an organism to its environment.

   c.   It means an evolutionary process that produces organisms better suited to their environment.

   d.   It is correlated with selective forces exerted by the environment but not by other organisms.

   e.   It may occur over the course of many generations.

28. The Result of Natural Selection: Adaptation; p. 246; moderate; ans: b

The tendency of individuals to vary over time in response to different environmental conditions is called:

   a.   adaptation.

   b.   developmental plasticity.

   c.   heterozygote advantage.

   d.   genotypic plasticity.

   e.   hybrid vigor.

29. The Result of Natural Selection: Adaptation; p. 246; moderate; ans: a

A gradual change in the characteristics of a population from one habitat to another is called:

   a.   a cline.

   b.   an ecotype.

   c.   adaptive radiation.

   d.   gene flow.

   e.   natural variation.

30. **The Result of Natural Selection: Adaptation; pp. 246–247; difficult; ans: b**
Which of the following statements concerning ecotypes is FALSE?
   a.   They are members of a species occupying different habitats.
   b.   They exhibit a gradual change in phenotype correlated with gradual changes in the environment.
   c.   They usually differ physiologically.
   d.   They usually differ morphologically.
   e.   They usually differ genetically.

31. **The Result of Natural Selection: Adaptation; p. 246; difficult; ans: d**
The experiments with *Potentilla glandulosa* showed that this organism:
   a.   grows well only at high altitudes.
   b.   grows well only at sea level.
   c.   should be regarded as four different species.
   d.   exists as four distinct ecotypes.
   e.   exists as two ecotypes, each with the same physiological response to the environment.

32. **The Result of Natural Selection: Adaptation; p. 247; easy; ans: d**
_____ refers to the adjustments that occur when populations of different species exert a strong selective force on each other.
   a.   Genetic drift
   b.   Developmental plasticity
   c.   Heterozygote advantage
   d.   Coevolution
   e.   Speciation

33. **The Origin of Species; p. 248; moderate; ans: c**
The key criterion of the biological species concept is that the members of a species:
   a.   have a similar anatomy.
   b.   have a similar morphology.
   c.   are all genetically isolated from other species.
   d.   can hybridize with each other.
   e.   can live in the same area.

34. **The Origin of Species; p. 248; difficult; ans: a**
The existence of *Platanus* × *hybrida* is indicative of the fact that:
   a.   fertile hybrids between species can occur.
   b.   members of different species do not ordinarily interbreed in nature.
   c.   members of different species should really be considered part of the same species.
   d.   morphologically distinct species should be considered subspecies.
   e.   the London plane tree should be considered a separate species in its own right.

35. The Origin of Species; p. 248; easy; ans: c

_____ occurs when a group of individuals becomes reproductively isolated from the population and eventually becomes a new species.

a. Developmental plasticity
b. Coevolution
c. Speciation
d. Ecotypic variation
e. Clinal variation

36. How Does Speciation Occur?; p. 249; moderate; ans: a

Which of the following statements about adaptive radiation is FALSE?

a. It is a type of sympatric speciation.
b. It is a sudden diversification of organisms.
c. It is particularly striking on islands.
d. It results in the almost simultaneous formation of many new species.
e. The variety of species of Galápagos tortoises is an example.

37. How Does Speciation Occur?; p. 252; moderate; ans: b

In contrast to allopolyploidy, autopolyploidy refers to a:

a. doubling of chromosome number in a sterile hybrid.
b. doubling of chromosome numbers within individuals of a species.
c. mechanism involved in allopatric speciation.
d. condition in which mitosis occurs without cytokinesis.
e. condition in which sterile offspring are formed.

38. How Does Speciation Occur?; pp. 252–253; difficult; ans: e

The type of polyploidy known as allopolyploidy:

a. was discovered in *Oenothera lamarckiana* by Hugo de Vries.
b. involves a doubling of chromosome number within individuals of a species.
c. is involved in allopatric speciation.
d. is involved in adaptive radiation.
e. is more common than autopolyploidy.

39. How Does Speciation Occur?; p. 253; easy; ans: c

Approximately _____ percent of flowering plants are polyploid.

a. 10 to 25
b. 35 to 47
c. 47 to 70
d. 70 to 83
e. 83 to 95

40. How Does Speciation Occur?; p. 253; moderate; ans: d

Tragopogon mirus and *Tragopogon miscellus* are examples of species formed by:

a. allopatric speciation.
b. heterozygote advantage.
c. coevolution.
d. allopolyploid speciation.
e. autopolyploid speciation.

41. How Does Speciation Occur?; p. 254; moderate; ans: c
An important seed-producing polyploid that is often planted to bind mud flats is:

a. *Spartina alterniflora.*
b. *Spartina martima.*
c. *Spartina anglica.*
d. *Spartina × towsendii.*
e. *Tragopogon mirus.*

42. How Does Speciation Occur?; p. 254; easy; ans: b
One of the most important polyploid groups is the:

a. salt-marsh grasses.
b. wheats.
c. goat's beards.
d. sunflowers.
e. horsetails.

43. How Does Speciation Occur?; p. 255; moderate; ans: c
An example of a species that arose by recombination speciation is:

a. *Spartina alterniflora.*
b. *Tragopogon mirus.*
c. *Helianthus anomalus.*
d. *Triticum aestivum.*
e. *Equisetum × ferrissii.*

44. How Does Speciation Occur?; p. 255; moderate; ans: a
Which of the following statements about recombination speciation is FALSE?

a. A polyploid organism is formed.
b. Two distinct species hybridize.
c. The hybrid becomes a third species.
d. It is an example of sympatric speciation.
e. An example is *Helianthus anomalus.*

45. How Does Speciation Occur?; p. 255; easy; ans: d
Varieties of Kentucky bluegrass undergo a form of asexual reproduction known as _____ in which embryos are produced that are genetically identical to the parent.

a. hybridization
b. adaptive radiation
c. cross-pollination
d. apomixis
e. outcrossing

46. How Does Speciation Occur?; p. 255; easy; ans: b
Apomixis is a type of:

a. sexual reproduction.
b. asexual reproduction.
c. hybridization.
d. outcrossing.
e. clinal variation.

47. **How Does Speciation Occur?; p. 256; easy; ans: d**
Cross-pollination between individuals of the same species is called:

   a. polyploidization.
   b. hybridization.
   c. recombination speciation.
   d. outcrossing.
   e. apomixis.

48. **Maintaining Reproductive Isolation; p. 256; moderate; ans: a**
An example of a prezygotic isolating mechanism is:

   a. plant species that flower in different seasons.
   b. a seed carried by the wind to an island.
   c. mitosis without cytokinesis.
   d. nondisjunction during meiosis.
   e. apomixis.

49. **The Origin of Major Groups of Organisms; p. 257; easy; ans: e**
The origin of taxonomic groups above the level of the species is called:

   a. coevolution.
   b. punctuated equilibrium.
   c. gradualism.
   d. microevolution.
   e. macroevolution.

50. **The Origin of Major Groups of Organisms; p. 257; moderate; ans: c**
According to the gradualism model:

   a. the fossil record accurately reflects evolutionary history.
   b. the origin of genera requires mechanisms different from those involved in the origin of species.
   c. evolution occurs by the accumulation of many small changes in the frequencies of genes in a gene pool.
   d. species arise abruptly.
   e. new fossil forms appear rather suddenly in strata.

51. **The Origin of Major Groups of Organisms; p. 257; moderate; ans: d**
According to the theory of punctuated equilibrium:

   a. macroevolution occurs more readily than microevolution.
   b. species evolve by the gradual accumulation of many small changes.
   c. the fossil record is flawed.
   d. long periods of little phyletic change are interrupted by short periods of rapid change.
   e. new species do not "suddenly" appear.

True-False Questions

1. **Darwin's Theory; p. 238; easy; ans: T**
The entire phenotype is subject to natural selection.

2. **Darwin's Theory; p. 238; moderate; ans: F**
   Darwin's theory of evolution was finally able to explain the mechanisms of heredity.

3. **The Concept of the Gene Pool; p. 239; easy; ans: T**
   In a population an individual with a favorable combination of alleles is more likely to survive and reproduce.

4. **The Behavior of Genes in Populations: The Hardy-Weinberg Law; p. 240; easy; ans: F**
   According to the Hardy-Weinberg Law, over long periods of time a dominant allele will drive out a recessive allele from a population.

5. **The Behavior of Genes in Populations: The Hardy-Weinberg Law; p. 240; moderate; ans: T**
   The Hardy-Weinberg equation does not hold for natural populations but provides a standard that allows the detection and measurement of changes in allele frequencies.

6. **The Agents of Change; p. 241; moderate; ans: F**
   The rate of spontaneous mutations in a population is very high.

7. **The Agents of Change; p. 241; easy; ans: F**
   The overall effect of gene flow is to increase differences between populations.

8. **The Agents of Change; p. 241; easy; ans: T**
   The smaller a population size, the more likely it is that a rarely occurring allele will be lost.

9. **The Agents of Change; p. 241; easy; ans: T**
   Genetic drift tends to have a greater impact on evolution in small populations than in large populations.

10. **The Agents of Change; p. 242; moderate; ans: T**
    An example of the bottleneck effect is a volcanic eruption destroying a large proportion of a particular plant population.

11. **Preservation and Promotion of Variability; p. 242; moderate; ans: T**
    From an evolutionary point of view, the only advantage of sexual reproduction is to promote variability.

12. **Preservation and Promotion of Variability; p. 243; easy; ans: F**
    Diploidy helps preserve variability in a population by enabling variations to be stored as dominant alleles.

13. **Responses to Selection; p. 244; moderate; ans: T**
    Polygenic inheritance and gene interactions greatly affect the course of selection in a population.

14. **The Result of Natural Selection: Adaptation; p. 245; easy; ans: T**
    Adaptations result from natural selection.

15. **The Result of Natural Selection: Adaptation; p. 247; easy; ans: T**
Coevolution occurs when one species exerts a strong selective force on another.

16. **The Origin of Species; p. 248; moderate; ans: F**
The inability to form fertile hybrids is a valid basis for defining species.

17. **How Does Speciation Occur?; p. 249; moderate; ans: T**
In nature, allopatric speciation is more common than sympatric speciation.

18. **How Does Speciation Occur?; p. 249; easy; ans: T**
In sympatric speciation, new species are formed without being separated geographically.

19. **How Does Speciation Occur?; p. 249; moderate; ans: T**
Polyploidy may arise as a result of nondisjunction during mitosis or meiosis.

20. **How Does Speciation Occur?; p. 252; easy; ans: F**
Autopolyploidy is much more common than allopolyploidy.

21. **How Does Speciation Occur?; p. 255; moderate; ans: T**
Recombination speciation is a type of sympatric speciation.

22. **How Does Speciation Occur?; p. 255; easy; ans: T**
Apomixis is a type of asexual reproduction in which embryos are produced independent of fertilization.

23. **Maintaining Reproductive Isolation; p. 256; easy: ans: F**
Prezygotic isolating mechanisms limit gene exchange after a hybrid has been formed.

24. **The Origin of Major Groups of Organisms; p. 256; moderate; ans: F**
The origin of taxonomic groups above the species level requires mechanisms that are different from those responsible for the origin of species.

Essay Questions

1. **The Process of Evolution; pp. 236–237; easy**
What observations did Charles Darwin make in the Galápagos islands that were of particular importance to his formulation of an evolutionary theory?

2. **Darwin's Theory; p. 237; moderate**
How would Darwin explain the observation that the size of a population of elephants remains relatively stable over time?

3. **The Concept of the Gene Pool; p. 239; moderate**
How does the population geneticist view a population of organisms and what determines changes in the population over the generations?

4. **The Behavior of Genes in Populations: The Hardy-Weinberg Law; pp. 239–240; moderate**
   State the Hardy-Weinberg Law, and explain its usefulness for understanding evolution.

5. **The Behavior of Genes in Populations: The Hardy-Weinberg Law; p. 240; moderate**
   If the five conditions specified for a gene pool in Hardy-Weinberg equilibrium are seldom met in natural populations, how do you explain the usefulness of the Hardy-Weinberg Law?

6. **The Agents of Change; p. 241; moderate**
   Explain the concept of genetic drift. In what ways are the bottleneck effect and the founder effect examples of genetic drift?

7. **Preservation and Promotion of Variability; p. 242; easy**
   Discuss the three ways in which sexual reproduction produces new gene combinations.

8. **Preservation and Promotion of Variability; p. 243; moderate**
   Explain how self-sterility genes function. How does the self-sterility system encourage variability in populations?

9. **Responses to Selection; pp. 244–245; moderate**
   Use the example of *Agrostis* plants growing around abandoned lead mines in Wales to explain how evolutionary changes in natural populations can occur rapidly.

10. **The Result of Natural Selection: Adaptation; pp. 246–247; difficult**
    Describe how clines and ecotypes are reflections of developmental plasticity.

11. **The Result of Natural Selection: Adaptation; p. 246; difficult**
    What accounts for the differences among Scandinavian ecotypes of goldenrod?

12. **The Origin of Species; p. 248; moderate**
    Use the case of the two species of sycamores (*Platanus*) to discuss the problems associated with the biological species concept.

13. **How Does Speciation Occur?; p. 249; moderate**
    Use examples to explain the difference between allopatric and sympatric speciation.

14. **How Does Speciation Occur?; pp. 252–255; difficult**
    Discuss the types and importance of sympatric speciation in plants.

15. **The Origin of Major Groups of Organisms; pp. 256–257; moderate**
    Compare and contrast the gradualism and punctuated equilibrium models for the origin of major groups of organisms.

Chapter 13

Systematics: The Science of Biological Diversity

Multiple-Choice Questions

1. **Taxonomy and Hierarchical Classification; p. 262; easy; ans: d**
    Modern biological classification began with:

    a. Charles Darwin.
    b. Gregor Mendel.
    c. Jean Baptiste de Lamarck.
    d. Carl Linnaeus.
    e. Caspar Bauhin.

2. **Taxonomy and Hierarchical Classification; p. 262; moderate; ans: e**
    Which of the following statements about Linnaeus is FALSE?

    a. He published the book *Species Plantarum*.
    b. He described each species using a sentence of no more than twelve words.
    c. He made permanent the binomial system of nomenclature.
    d. He devised a polynomial as a proper name for each species.
    e. He developed a "shorthand" designation for each species consisting of a single word.

3. **Taxonomy and Hierarchical Classification; p. 262; moderate; ans: a**
    The binomial for poison ivy is *Toxicodendron radicans*. To what genus does this plant belong?

    a. *Toxicodendron*
    b. *radicans*
    c. *Toxicodendron radicans*
    d. poison ivy
    e. *Toxicodendron radicans poison ivy*

4. **Taxonomy and Hierarchical Classification; p. 262; moderate; ans: c**
    The binomial for the coast redwood is *Sequoia sempervirens*. What is the species name of this plant?

    a. *Sequoia*
    b. *sempervirens*
    c. *Sequoia sempervirens*
    d. coast redwood
    e. *Sequoia sempervirens coast redwood*

5. **Taxonomy and Hierarchical Classification; pp. 262–263; difficult; ans: d**
Which of the following statements about the naming of species and varieties is FALSE?

    a.   Type specimens serve as a basis for designating the species of other specimens.
    b.   Certain species consist of two or more varieties.
    c.   The varieties of a species that includes the type specimen is named by repeating the specific epithet.
    d.   When used alone, the specific epithet provides valuable taxonomic information.
    e.   Names of genera and species are printed in italic.

6. **Taxonomy and Hierarchical Classification; p. 264; moderate; ans: c**
Which of the following lists the taxonomic categories in the correct hierarchy, from most to least inclusive, under kingdom?

    a.   class, phylum, order, family, genus, species
    b.   order, class, phylum, family, genus, species
    c.   phylum, class, order, family, genus, species
    d.   phylum, order, class, family, genus, species
    e.   order, phylum, family, class, genus, species

7. **Taxonomy and Hierarchical Classification; p. 264; moderate; ans: e**
Cattleya is one genus in the *Orchidaceaea*, the orchid family. In this example:

    a.   *Cattleya* is a category.
    b.   *Orchidaceae* is a category.
    c.   *Cattleya* is a taxon but *Orchidaceae* is not.
    d.   *Orchidaceae* is a taxon but *Cattleya* is not.
    e.   both *Cattleya* and *Orchidaceae* are taxa.

8. **Taxonomy and Hierarchical Classification; p. 264; easy; ans: b**
The term "phylum" is nomenclaturally equivalent to:

    a.   class.
    b.   division.
    c.   order.
    d.   kingdom.
    e.   family.

9. **Taxonomy and Hierarchical Classification; p. 264; moderate; ans: e**
The names of almost all plant families end in:

    a.   *-ium*.
    b.   *-ica*.
    c.   *-om*.
    d.   *-ales*.
    e.   *-aceae*.

10. Classification and Phylogeny; pp. 264–265; easy; ans: d
Phylogeny refers to the:

a.   naming of organisms.
b.   placing of organisms into phyla.
c.   nomenclature of species.
d.   evolutionary history of an organism.
e.   grouping of classes of organisms.

11. Classification and Phylogeny; p. 265; easy; ans: c
When the members of a taxon are all descendents of a common single ancestral species, the taxon is said to be:

a.   phylogenetic.
b.   paraphyletic.
c.   monophyletic.
d.   polyphyletic.
e.   amphiphyletic.

12. Classification and Phylogeny; p. 266; moderate; ans: b
In contrast to a natural taxon, an artificial taxon:

a.   is monophyletic.
b.   may contain members descended from more than one ancestor.
c.   may contain members that all belong to the same species.
d.   may be polyphyletic but not paraphyletic.
e.   is not named using the binomial system.

13. Classification and Phylogeny; p. 266; easy; ans: d
Biological features that have a common origin, even if they have a different function, are said to be:

a.   artificial.
b.   natural.
c.   paraphyletic.
d.   homologous.
e.   analogous.

14. Methods of Classification; pp. 267–268; easy; ans: c
Monophyletic groups that can be defined by the possession of unique features are called:

a.   species.
b.   outgroups.
c.   clades.
d.   cladograms.
e.   character states.

15. Molecular Systematics; p. 268; easy; ans: e
Which of the following is NOT an advantage of using molecular data compared to data obtained using traditional classification methodology?

a.   They are easier to quantify.
b.   They may provide more characters for analysis.
c.   They allow comparisons of morphologically different organisms.
d.   They allow comparisons of organisms at the level of the gene.
e.   They easily allow the assessment of homologies.

16. **Molecular Systematics; p. 269; easy; ans: c**
If you analyze the cytochrome *c* molecules from two organisms and find there are few differences in their amino acid sequences, you would logically conclude that the two organisms:

    a. belong to the same species.
    b. belong to different species.
    c. are closely related.
    d. are distantly related.
    e. belong to a monophyletic taxon.

17. **Molecular Systematics; p. 270; moderate; ans: d**
Which of the following statements about nucleotide sequences is FALSE?

    a. They are technically easier to obtain than amino acid sequences.
    b. They can assess similarities and differences more sensitively than amino acid sequences.
    c. They have been used to identify domains of organisms.
    d. They have been used to show that the *Archaea* and *Bacteria* share a common lineage independent of the *Eukarya*.
    e. They have been used to study variations in the gene that codes for part of the Rubisco enzyme.

18. **Molecular Systematics; p. 270; moderate; ans: d**
Comparisons of homologous proteins indicate that prokaryotic and eukaryotic organisms last shared a common ancestor about _____ years ago.

    a. 2 million
    b. 20 million
    c. 200 million
    d. 2 billion
    e. 20 billion

19. **Molecular Systematics; p. 270; moderate; ans: c**
Sequencing of _____ from a variety of species provided the first evidence for the existence of three domains of organisms.

    a. cytochrome *c*
    b. enzymes in glycolysis
    c. ribosomal RNA
    d. the *rbc*L gene
    e. mRNA

20. **Molecular Systematics; pp. 270–271; moderate; ans: a**
The most comprehensive studies of seed plant phylogeny are based on differences among _____ sequences of _____.

    a. nucleotide; the *rbc*L gene
    b. amino acid; cytochrome *c*
    c. nucleotide; ribosomal RNA
    d. nucleotide; mitochondrial genes
    e. amino acid; Rubisco

21. Origin of the Eukaryotes; p. 272; easy; ans: a
According to the serial endosymbiotic theory, mitochondria evolved from:

   a.   bacteria.
   b.   protists.
   c.   fungi.
   d.   plasmids.
   e.   portions of the plasma membrane.

22. Origin of Eukaryotes; p. 272; easy; ans: c
By definition, an endosymbiont is an organism that:

   a.   is a parasite.
   b.   is a phagocyte.
   c.   lives within another, dissimilar organism.
   d.   lives within a nonliving substance.
   e.   forms an organelle within the cells of another organism.

23. Origin of the Eukaryotes; pp. 272, 274; difficult; ans: b
Which of the following best indicates the correct sequence in which the evolution of these organelles occurred?

   a.   mitochondrion, lysosome, chloroplast
   b.   lysosome, mitochondrion, chloroplast
   c.   lysosome, chloroplast, mitochondrion
   d.   chloroplast, mitochondrion, lysosome
   e.   chloroplast, lysosome, mitochondrion

24. Origin of the Eukaryotes; p. 274; easy; ans: d
The endomembrane system of plant cells most likely evolved from:

   a.   a chloroplast.
   b.   a mitochondrion.
   c.   a lysosome.
   d.   portions of the plasma membrane.
   e.   portions of the vacuole.

25. Origin of the Eukaryotes; p. 274; easy; ans: d
The nucleus of plant cells most likely evolved from:

   a.   a chloroplast.
   b.   a mitochondrion.
   c.   a lysosome.
   d.   portions of the plasma membrane.
   e.   portions of the vacuole.

26. Origin of the Eukaryotes; p. 274; easy; ans: e
In the *Vorticella/Chlorella* endosymbiosis:

   a.   *Vorticella* functions as the endomembrane system for *Chlorella*.
   b.   *Chlorella* provides minerals for *Vorticella*.
   c.   *Chlorella* functions as a chloroplast in *Vorticella*.
   d.   *Vorticella* functions as a mitochondrion in *Chlorella*.
   e.   *Chlorella* provides photosynthetic products for *Vorticella*.

27. Origin of the Eukaryotes; p. 274; easy; ans: c
In the course of evolution of eukaryotic cells, some mitochondrial DNA was transferred to:

a. the host cell's chloroplast.
b. a bacterial cell.
c. the host cell's nucleus.
d. the host cell's plasma membrane.
e. a plasmid.

28. The Eukaryotic Kingdoms; p. 275; easy; ans: b
Water molds and slime molds are included in the kingdom:

a. *Fungi.*
b. *Protista.*
c. *Eukarya.*
d. *Archaea.*
e. *Plantae.*

29. The Eukaryotic Kingdoms; p. 275; easy; ans: a
A unicellular, eukaryotic organism would be classified in the kingdom:

a. *Protista.*
b. *Monera.*
c. *Archaea.*
d. *Plantae.*
e. *Eukarya.*

30. The Eukaryotic Kingdoms; p. 276; easy; ans: c
A multicellular organism that ingests its food belongs to the kingdom:

a. *Protista.*
b. *Monera.*
c. *Animalia.*
d. *Plantae.*
e. *Eukarya.*

31. The Eukaryotic Kingdoms; p. 277; moderate; ans: e
A choanoflagellate-like organism is the most recent common ancestor of the:

a. red algae and brown algae.
b. plants and algae.
c. protozoa and fungi.
d. plants and fungi.
e. fungi and animals.

32. The Eukaryotic Kingdoms; pp. 277, 279; easy; ans: e
Complex multicellular eukaryotes that carry out photosynthesis belong to the kingdom:

a. *Fungi.*
b. *Protista.*
c. *Eukarya.*
d. *Archaea.*
e. *Plantae.*

True-False Questions

1. **Taxonomy and Hierarchical Classification; p. 262; easy; ans: T**
   A species name is a two-word name.

2. **Taxonomy and Hierarchical Classification; p. 264; moderate; ans: F**
   The taxon *Commelinales* is a family of plants, as indicated by the ending *-ales*.

3. **Taxonomy and Hierarchical Classification; p. 264; moderate; ans: F**
   The taxon *Poaceae* is an order of plants, as indicated by the ending *-aceae*.

4. **Taxonomy and Hierarchical Classification; p. 264; moderate; ans: T**
   In systematics, order and family are categories; *Agaricales* and *Agaricaceae* are taxa.

5. **Classification and Phylogeny; p. 266; moderate; ans: T**
   In natural taxa, all members are descended from the same ancestral species.

6. **Classification and Phylogeny; p. 266; easy; ans: T**
   Polyphyletic taxa contain members descended from more than one ancestral line.

7. **Methods of Classification; p. 267; easy; ans: F**
   Cladistics is a method of classifying organisms based on overall outward similarities called clades.

8. **Methods of Classification; p. 267; moderate; ans: T**
   A cladogram is a hypothesis about relationships between monophyletic groups, based on both unique and widespread features.

9. **Classification and Phylogeny; pp. 266–267; easy; ans: F**
   Foliage leaves and floral parts are analogous, rather than homologous, structures.

10. **Molecular systematics; p. 270; moderate; ans: F**
    Since the advent of methods of nucleic acid sequencing, protein sequencing has no longer been used in molecular systematics.

11. **The Major Groups of Organisms:** *Bacteria*, *Archaea*, **and** *Eukarya*;
    p. 271; easy; ans: T
    The prokaryotic domains are the *Bacteria* and *Archaea*.

12. **The Major Groups of Organisms:** *Bacteria*, *Archaea*, **and** *Eukarya*;
    p. 271; easy; ans: T
    The domain is the highest taxonomic category.

13. **The Eukaryotic Kingdoms; p. 275; moderate; ans: T**
    Kingdom *Protista* contains both autotrophs and heterotrophs, and both multicellular and unicellular organisms.

14. **Origin of the Eukaryotes; p. 272; moderate; ans: T**
According to the endosymbiotic theory, the evolution of the cytoskeleton and a flexible plasma membrane preceded the evolution of mitochondria and chloroplasts.

15. **The Eukaryotic Kingdoms; p. 277; moderate; ans: F**
Fungi are more closely related to plants than to animals.

Essay Questions

1. **Taxonomy and Hierarchical Classification; p. 263; easy**
What is meant by a type specimen and what is its role in classification?

2. **Classification and Phylogeny; pp. 265–266; moderate**
Distinguish between monophyletic, polyphyletic, and paraphyletic taxa.

3. **Classification and Phylogeny; pp. 266–267; moderate**
Explain the difference between homology and analogy in systematics. Why is it better to base an evolutionary classification system on homologous rather than analogous structures?

4. **Methods of Classification; pp. 267–268; moderate**
Distinguish between the traditional and cladistic methods of classification.

5. **Molecular Systematics; pp. 268–271; difficult**
What are some of the advantages of using molecular data rather than more traditional data in systematics? What are some of the problems associated with these molecular techniques?

6. **The Major Groups of Organisms:** *Bacteria*, *Archaea*, **and** *Eukarya*;
p. 271; moderate
What is a domain in systematics? Name the three domains of organisms. What is the rationale for using the domain rather than the kingdom as the highest category of classification?

7. **Origin of the Eukaryotes; pp. 272–274; moderate**
Explain how eukaryotic cells are thought to have evolved from prokaryotic cells.

8. **The Eukaryotic Kingdoms; pp. 275–279; moderate**
Describe the distinguishing features of the four eukaryotic kingdoms.

Chapter 14

Prokaryotes and Viruses

Multiple-Choice Questions

1. **Introduction; p. 282; easy; ans: d**
    The oldest known fossils of prokaryotes are found in rocks dated at about
    _____ years old.
    a. 3.5 million
    b. 35 million
    c. 350 million
    d. 3.5 billion
    e. 35 billion

2. **Characteristics of the Prokaryotic Cell; p. 282; moderate; ans: c**
    Which of the following statements about prokaryotic cells is FALSE?
    a. They lack membrane-bounded organelles.
    b. They have a single, circular chromosome in the nucleoid.
    c. They have a simple cytoskeleton.
    d. They may contain extra chromosomal DNA as plasmids.
    e. They lack a nuclear envelope.

3. **Characteristics of the Prokaryotic Cell; pp. 282–283; moderate; ans: b**
    Which of the following statements does NOT describe the plasma membrane of
    prokaryotes?
    a. Like eukaryotic plasma membranes, it is a lipid bilayer.
    b. Like eukaryotic plasma membranes, it contains proteins and sterols.
    c. In aerobic species, it is the site of the electron transport chain.
    d. In some photosynthetic species it is the site of photosynthesis.
    e. It has attachment sites for daughter chromosomes during cell division.

4. **Characteristics of the Prokaryotic Cell; pp. 282–283; easy; ans: e**
    In the photosynthetic purple and green bacteria, photosynthesis occurs in the:
    a. pili.
    b. glycocalyx.
    c. cell wall.
    d. nucleoid.
    e. plasma membrane.

5. Characteristics of the Prokaryotic Cell; p. 283; moderate; ans: c
The cell walls of *Bacteria* differ from those of *Archaea* and eukaryotes in that the cell walls of *Bacteria* contain:

    a.   cellulose.
    b.   sterols.
    c.   peptidoglycans.
    d.   phospholipids.
    e.   poly-β-hydroxybutyric acid.

6. Characteristics of the Prokaryotic Cell; p. 283; easy; ans: b
Gram-positive and gram-negative bacteria are distinguished by differences in their:

    a.   plasma membranes.
    b.   cell walls.
    c.   storage materials.
    d.   size.
    e.   shape.

7. Characteristics of the Prokaryotic Cell; p. 283; difficult; ans: b
The cell walls of gram-positive and gram-negative *Bacteria* differ in that gram-positive species have cell walls:

    a.   with less peptidoglycan.
    b.   of greater thickness.
    c.   consisting of two layers.
    d.   with a layer of lipopolysaccharides.
    e.   with a structure similar to that of the plasma membrane.

8. Characteristics of the Prokaryotic Cell; p. 284; moderate; ans: e
The glycocalyx is a:

    a.   type of bacterial plasmid.
    b.   convoluted infolding of the prokaryotic plasma membrane.
    c.   type of inclusion body in prokaryotic cells.
    d.   constituent of the cell wall of gram-positive *Bacteria*.
    e.   slimy or gummy substance coating the outer surface of the prokaryotic cell wall.

9. Characteristics of the Prokaryotic Cell; p. 284; easy; ans: d
In prokaryotes, poly-β-hydroxybutyric acid is a storage compound occurring:

    a.   in the nucleoid.
    b.   in the plasma membrane.
    c.   in the cell wall.
    d.   as inclusion bodies.
    e.   as fimbriae.

10. Characteristics of the Prokaryotic Cell; p. 284; moderate; ans: c
Prokaryotic flagella differ from eukaryotic flagella in that prokaryotic flagella:

    a.   are surrounded by a plasma membrane.
    b.   consist of microtubules.
    c.   consist of subunits of flagellin.
    d.   are long, slender appendages.
    e.   are involved in motility.

11. **Characteristics of the Prokaryotic Cell; p. 284; moderate; ans: b**
Fimbriae and pili are similar structures, except that fimbriae are:

    a. less rigid than pili.
    b. generally shorter than pili.
    c. filamentous.
    d. composed of protein subunits.
    e. involved in the process of conjugation.

12. **Diversity of Form; p. 284; moderate; ans: c**
A bacillus has a _____ shape.

    a. spherical
    b. spiral
    c. cylindrical
    d. curved
    e. triangular

13. **Diversity of Form; p. 284; moderate; ans: b**
Which of the following does NOT refer to a structure formed when prokaryotic cells remain together after cell division?

    a. colony
    b. spirillum
    c. filament
    d. cluster
    e. fruiting body

14. **Reproduction and Gene Exchange; pp. 285–286; moderate; ans: a**
Which of the following processes does NOT involve genetic recombination?

    a. binary fission
    b. transduction
    c. transformation
    d. conjugation
    e. meiosis

15. **Reproduction and Gene Exchange; p. 285; easy; ans: d**
Which of the following is the prokaryotic version of sexual reproduction?

    a. binary fission
    b. transduction
    c. transformation
    d. conjugation
    e. budding

16. **Reproduction and Gene Exchange; p. 286; moderate; ans: b**
Which of the following is the transfer of DNA from one prokaryotic cell to another by a virus?

    a. binary fission
    b. transduction
    c. transformation
    d. conjugation
    e. budding

17. **Endospores; p. 286; moderate; ans: d**
Which of the following statements about prokaryotic endospores is FALSE?

   a. They are dormant, resting cells.
   b. They form when a population of cells begins to exhaust its food supply.
   c. They greatly increase the survival capacity of the cell.
   d. They are resistant to heat but not to desiccation.
   e. They can remain viable for long periods of time.

18. **Metabolic Diversity; pp. 286–287; difficult; ans: c**
By definition, autotrophs differ from heterotrophs in that autotrophs:

   a. use inorganic compounds as an energy source.
   b. use light as an energy source.
   c. use carbon dioxide as their sole source of carbon.
   d. obtain their carbon from dead organic matter.
   e. obtain their energy from dead organic matter.

19. **Metabolic Diversity; p. 286; easy; ans: e**
Prokaryotes that obtain their carbon from dead organic matter are called:

   a. aerobes.
   b. chemosynthetic autotrophs.
   c. photosynthetic autotrophs.
   d. psychrophiles.
   e. saprobes.

20. **Metabolic Diversity; p. 287; moderate; ans: b**
Prokaryotes that use inorganic compounds rather than light as an energy source are called:

   a. heterotrophs.
   b. chemosynthetic autotrophs.
   c. photosynthetic autotrophs.
   d. psychrophiles.
   e. saprobes.

21. **Metabolic Diversity; p. 287; moderate; ans: a**
Prokaryotes that can grow in the presence of oxygen, even though they cannot use oxygen, are called:

   a. facultative anaerobes.
   b. facultative aerobes.
   c. photosynthetic autotrophs.
   d. psychrophiles.
   e. thermophiles

22. **Metabolic Diversity; p. 287; moderate; ans: d**
Prokaryotes that grow in the 140°C water near deep-sea vents are examples of:

   a. aerobes.
   b. facultative anaerobes.
   c. psychrophiles.
   d. extreme thermophiles.
   e. thermophiles.

23. **Metabolic Diversity; p. 287; difficult; ans: e**
Which of the following statements about the ecological role of prokaryotes is FALSE?

   a. Autotrophic species make major contributions to the global carbon balance.
   b. Certain species fix atmospheric nitrogen.
   c. Decomposers recycle materials from the bodies of dead organisms.
   d. Certain bacteria degrade pesticides and other synthetic substances.
   e. Certain prokaryotes provide the most widely used method for cleaning up toxic dumps.

24. **Metabolic Diversity; p. 287; moderate; ans: d**
Helicobacter pylori is most closely associated with:

   a. whooping cough.
   b. diphtheria.
   c. heart disease.
   d. stomach ulcers.
   e. cholera.

25. **Metabolic Diversity; p. 287; moderate; ans: c**
Bacterial fermentation of lactose is associated with the production of:

   a. streptomycin.
   b. tetracycline.
   c. cheese.
   d. vinegar.
   e. amino acids.

26. **Bacteria; p. 287; easy; ans: d**
Phylogenetic analysis indicates that there are _____ major lineages of *Bacteria*.

   a. 4
   b. 5
   c. 8
   d. 12
   e. 20

27. **Bacteria; p. 288; moderate; ans: d**
The _____ are prokaryotes that contain chlorophyll *a*, carotenoids, and phycobilins.

   a. mycoplasmas
   b. purple bacteria
   c. green bacteria
   d. cyanobacteria
   e. prochlorophytes

28. *Bacteria*; p. 288; moderate; ans: c
Phycobilins are types of _____ found in cyanobacteria.

   a.   carotenoids
   b.   chlorophylls
   c.   photosynthetic pigments
   d.   bacteriochlorophylls
   e.   storage compounds

29. *Bacteria*; p. 288; easy; ans: b
The _____ are believed to have given rise to the chloroplasts of photosynthetic eukaryotes.

   a.   prochlorophytes
   b.   cyanobacteria
   c.   mycoplasmas
   d.   mycoplasmalike organisms
   e.   green bacteria

30. *Bacteria*; p. 288; difficult; ans: b
Which of the following statements about the growth forms of cyanobacteria is FALSE?

   a.   Many produce a sheath.
   b.   Most form colonies.
   c.   Some form unbranched filaments.
   d.   Some are unicellular.
   e.   A few form branched filaments.

31. *Bacteria*; p. 288; moderate; ans: d
Which of the following statements about the environmental ranges of cyanobacteria is FALSE?

   a.   Most species are symbiotic.
   b.   Some species grow in hot springs.
   c.   Some species grow in antarctic lakes.
   d.   Some species grow in acidic waters.
   e.   Some species grow in the fur of polar bears.

32. *Bacteria*, p. 289; easy; ans: a
Stromatolites are produced when cyanobacteria:

   a.   bind calcium-rich sediments.
   b.   produce sheaths covered with ice crystals.
   c.   react with the minerals in hot springs.
   d.   interact with fur.
   e.   interact with the roots of higher plants.

33. *Bacteria*; p. 289; moderate; ans: b
Plankton are organisms that inhabit:

   a.   the deepest parts of a pond, lake, or ocean.
   b.   the surface layers of the water.
   c.   freshwater only.
   d.   saltwater only.
   e.   fast-moving water only.

34. *Bacteria*; p. 289; difficult; ans: c
Cyanobacteria form "blooms" when they:

    a.   secrete an abundance of toxic chemical substances.
    b.   fix too much nitrogen gas.
    c.   are unable to regulate their gas vesicles properly.
    d.   produce too much pigment.
    e.   produce too dense a sheath.

35. *Bacteria*; p. 290, moderate; ans: e
The Red Sea was given its name because of the frequent blooms of:

    a.   *Thiothrix.*
    b.   *Clostridium.*
    c.   *Azolla.*
    d.   *Prochlorococcus.*
    e.   *Trichodesmium.*

36. *Bacteria*; p. 290; difficult; ans: e
In the process of nitrogen fixation:

    a.   Photosystem II occurs but not Photosystem I.
    b.   cyclic photophosphorylation does not occur.
    c.   oxygen is produced as a waste product.
    d.   nitrogen gas is converted to amino acids.
    e.   nitrogen gas is converted to nitrates.

37. *Bacteria*; p. 290; moderate; ans: b
The heterocysts of cyanobacteria are most closely associated with:

    a.   sporulation.
    b.   nitrogen fixation.
    c.   photosynthesis.
    d.   gliding movements.
    e.   buoyancy.

38. *Bacteria*; p. 290; difficult; ans: e
Which of the following statements about nitrogenase is FALSE?

    a.   It is inhibited by oxygen.
    b.   It catalyzes the process of nitrogen fixation.
    c.   It is found in heterocysts.
    d.   It catalyzes an anaerobic process.
    e.   It is found in free-living, but not symbiotic, cyanobacteria.

39. *Bacteria*; p. 290; moderate; ans: a
Symbiotic cyanobacteria:

    a.   all lack a cell wall.
    b.   are found in association with mosses and liverworts.
    c.   are found within sponges and amoeba.
    d.   are found within vascular plants.
    e.   occur as the photosynthetic partner of many lichens.

40. *Bacteria*; **p. 290; easy; ans: d**
An akinete is a _____ of cyanobacteria.

 a. photosynthetic pigment
 b. filamentous fragment
 c. storage compound
 d. resistant spore
 e. pigment

41. *Bacteria*; **p. 291; moderate; ans: c**
The _____ are photosynthetic bacteria that do not produce oxygen.

 a. cyanobacteria
 b. mycoplasmas
 c. purple and green bacteria
 d. prochlorophytes
 e. mycoplasmalike organisms

42. *Bacteria*; **p. 291; moderate; ans: b**
Which group of photosynthetic prokaryotes have bacteriochlorophyll and only one photosystem?

 a. mycoplasmas
 b. purple and green bacteria
 c. mycoplasmalike organisms
 d. cyanobacteria
 e. prochlorophytes

43. *Bacteria*; **p. 291; moderate; ans: a**
Green sulfur bacteria differ from cyanobacteria in that green sulfur bacteria:

 a. use sulfur compounds as electron donors in photosynthesis.
 b. use water as an electron donor in photosynthesis.
 c. have two photosystems.
 d. have bacteriochlorophyll.
 e. produce hydrogen sulfide in photosynthesis.

44. *Bacteria*; **p. 291; moderate; ans: b**
Mitochondria are thought to have evolved from:

 a. mycoplasmas.
 b. purple nonsulfur bacteria.
 c. green nonsulfur bacteria.
 d. cyanobacteria.
 e. prochlorophytes.

45. *Bacteria*; **p. 292; difficult; ans: e**
The _____ are photosynthetic prokaryotes that contain chlorophylls *a* and *b* and carotenoids but lack phycobilins.

 a. mycoplasmas
 b. purple and green bacteria
 c. mycoplasmalike organisms
 d. cyanobacteria
 e. prochlorophytes

46. *Bacteria*; **p. 292; easy; ans: a**
_____ are probably the smallest organisms capable of independent growth.

   a.   Mycoplasmas
   b.   Purple bacteria
   c.   Green sulfur bacteria
   d.   Cyanobacteria
   e.   Prochlorophytes

47. *Bacteria*; **pp. 292–293; easy; ans: c**
The _____ are prokaryotes that cause highly destructive plant diseases such as X-disease of peach.

   a.   mycoplasmas
   b.   purple and green bacteria
   c.   mycoplasmalike organisms
   d.   cyanobacteria
   e.   prochlorophytes

48. *Bacteria*; **p. 293; easy; ans: b**
Which of the following statements concerning plant-pathogenic bacteria is FALSE?

   a.   Most are gram-negative.
   b.   Most are cocci.
   c.   They are parasites.
   d.   Most form spot-like symptoms.
   e.   They cause disease.

49. *Bacteria*; **p. 293; moderate; ans: a**
By definition, a parasite is:

   a.   a symbiont.
   b.   autotrophic.
   c.   heterotrophic.
   d.   a prokaryote.
   e.   a mycoplasma.

50. *Bacteria*; **p. 293; moderate; ans: d**
Fire blight in apples and pears is caused by members of the genus:

   a.   *Pseudomonas*.
   b.   *Xanthomonas*.
   c.   *Agrobacterium*.
   d.   *Erwinia*.
   e.   *Clavibacter*.

51. *Bacteria*; **pp. 293–294, easy; ans: c**
A disease in which bacteria invade the vessels of the xylem and interfere with the movement of water and minerals is a:

   a.   blight.
   b.   soft rot.
   c.   wilt.
   d.   gall.
   e.   spot.

52. *Archaea*; p. 295; easy; ans: d

The _____ are archaeans living in regions of very high salt concentration.

a.   mycoplasmas
b.   extreme thermophiles
c.   members of the genus *Thermoplasma*
d.   extreme halophiles
e.   methanogens

53. *Archaea*; p. 295; moderate; ans: d

Which archaeans synthesize ATP using light energy and bacteriorhodopsin?

a.   mycoplasmas
b.   extreme thermophiles
c.   members of the genus *Thermoplasma*
d.   extreme halophiles
e.   methanogens

54. *Archaea*; p. 295; easy; ans: e

The strict anaerobes residing in the digestive tracts of cattle are:

a.   mycoplasmas.
b.   extreme thermophiles.
c.   members of the genus *Thermoplasma*.
d.   extreme halophiles.
e.   methanogens.

55. *Archaea*; p. 296; easy; ans: b

The _____ are archaeans that grow optimally at temperatures above 80°C.

a.   mycoplasmas
b.   extreme thermophiles
c.   members of the genus *Thermoplasma*
d.   extreme halophiles
e.   methanogens

56. *Archaea*; p. 296; moderate; ans: c

Which group of archaeans lack a cell wall?

a.   mycoplasmas
b.   extreme thermophiles
c.   members of the genus *Thermoplasma*
d.   extreme halophiles
e.   methanogens

57. Viruses; p. 296; moderate; ans: e

Which of the following statements about viruses is FALSE?

a.   They can be crystallized without losing their effectiveness.
b.   They can be used as gene-transfer systems in biotechnology.
c.   They are usually associated with a specific host.
d.   They cause diseases of plants and animals.
e.   They infect all types of organisms except bacteria.

58. Viruses; p. 297; easy; ans: b
A bacteriophage is a:

a.   parasitic bacterium.
b.   virus that infects bacteria.
c.   type of mycoplasma.
d.   bacterium that digests viruses.
e.   virus that cannot be destroyed by bacteria.

59. Viruses; p. 297; moderate; ans: d
A virion consists of a:

a.   lipoprotein membrane surrounding a cytoplasm.
b.   lipoprotein membrane surrounding DNA or RNA.
c.   nucleic acid surrounded by a cell wall.
d.   nucleic acid surrounded by protein.
e.   nucleic acid surrounded by a cytoskeleton.

60. Viruses; p. 298; easy; ans: a
A viral capsid is a:

a.   protein coat.
b.   strand of DNA.
c.   strand of RNA.
d.   plasma membrane.
e.   type of organelle.

61. Viruses; pp. 298, 300; moderate; ans: e
The most common shape of a virus is a(n):

a.   coiled spring.
b.   thread.
c.   bullet.
d.   rod.
e.   icosahedron.

62. Viruses; p. 300; moderate; ans: d
Bean golden mosaic is a disease caused by a:

a.   mycoplasmalike organism.
b.   cyanobacterium.
c.   badnavirus.
d.   geminivirus.
e.   caulimovirus.

63. Viruses; pp. 300–301; moderate; ans: a
Movement proteins facilitate the movement of viruses:

a.   through plasmodesmata.
b.   through vessels.
c.   through sieve tubes.
d.   across the plasma membrane.
e.   across the nuclear envelope.

64. Viruses; p. 301; moderate; ans: e
A viral disease of plants in which the symptoms are small, light-colored flecks
intermingled with the normal green of leaves and fruit is called:

a. ring spot.
b. yellows.
c. canker.
d. leaf roll.
e. mosaic.

65. Viroids: Other Infectious Particles; p. 302; moderate; ans: e
Viroids consist only of a:

a. protein capsid.
b. linear DNA molecule.
c. circular DNA molecule.
d. linear RNA molecule.
e. circular RNA molecule.

66. The Origin of Viruses; p. 303; easy; ans: b
Viruses are thought to have originated from:

a. the ancestors of cells.
b. genomic material of host cells.
c. cyanobacterial cells.
d. fungal cells.
e. mycoplasmalike organisms.

True-False Questions

1. Introduction; p. 282; easy; ans: T
The prokaryotes are the most dominant and successful organisms on Earth.

2. Characteristics of the Prokaryotic Cell; p. 283; moderate; ans: T
Of all the prokaryotes, only the mycoplasmas lack a cell wall.

3. Characteristics of the Prokaryotic Cell; p. 283; moderate; ans: F
The cell wall of gram-positive bacteria consists of two layers, one of which
stains with the dye crystal violet.

4. Characteristics of the Prokaryotic Cell; p. 284; easy; ans: T
One function of the glycocalyx is to mediate the binding of bacteria to specific
host tissues.

5. Characteristics of the Prokaryotic Cell; p. 284; moderate; ans: F
In prokaryotes, poly-β-hydroxybutyric acid rather than glycogen is the major
storage substance.

6. Characteristics of the Prokaryotic Cell; p. 284; easy; ans: T
Pili are required in the process of conjugation between prokaryotic cells.

7. Diversity of Form; p. 284; easy; ans: F
The cells of prokaryotic organisms always separate from each other following
cell division.

8. Reproduction and Gene Exchange; p. 285; moderate; ans: T
One method by which prokaryotes reproduce asexually is binary fission.

9. Metabolic Diversity; p. 287; moderate; ans: F
Most prokaryotes are autotrophic.

10. Metabolic Diversity; p. 287; easy; ans: T
Some bacteria are capable of fixing atmospheric nitrogen to form nitrogen-containing compounds.

11. Metabolic Diversity; p. 287; easy; ans: T
Bacteria that convert lactose to lactic acid are used in cheese, yogurt, and sauerkraut production.

12. *Bacteria*; p. 288; easy; ans: T
The cyanobacteria were once called "blue-green algae," even though many species do not have a blue-green color.

13. *Bacteria*; p. 289; moderate; ans: F
The purple and green bacteria were primarily responsible for elevating the level of free oxygen in the atmosphere of the early Earth.

14. *Bacteria*; p. 290; moderate; ans: F
In filamentous cyanobacteria, nitrogen fixation often occurs within hormogonia.

15. *Bacteria*; p. 291; moderate; ans: T
The purple and green bacteria are photosynthetic bacteria that grow only under anaerobic conditions.

16. *Bacteria*; p. 293; difficult; ans: T
Most bacteria that are parasitic on plants are gram-negative bacilli.

17. *Bacteria*; p. 294; easy; ans: T
The most economically important wilt disease in plants is caused by *Pseudomonas solanacearum*.

18. *Archaea*; p. 294; easy; ans: F
The archaeans, unlike *Bacteria*, are of little importance to global ecology.

19. *Archaea*; p. 296; moderate; ans: T
Most species of extreme thermophilic archaeans metabolize sulfur.

20. Viruses; p. 297; moderate; ans: T
A bacteriophage is a virus that infects bacteria.

21. Viruses; p. 297; easy; ans: F
A virus can reproduce both inside and outside a host cell.

22. Viruses; p. 298; easy; ans: T
Some viruses have a lipid-containing envelope surrounding their capsid.

23. Viruses; p. 298; moderate; ans: F
Viruses are spread from diseased to healthy plants primarily by bacteria.

24. **Viruses; p. 300; moderate; ans: T**
    In single-stranded RNA viruses, the viral RNA acts as mRNA.

25. **Viruses; p. 300; moderate; ans: T**
    Within an infected plant, viruses move from cell to cell through plasmodesmata in a process facilitated by movement proteins.

26. **Viruses; p. 302; moderate; ans: T**
    For some crops, genetically engineered resistance to viruses is the only possible method of viral disease control.

27. **Viroids: Other Infectious Particles; p. 302; moderate; ans: F**
    Viroid RNA is replicated in the cytoplasm of the host cell then used as a template to synthesize a strand of DNA.

28. **The Origin of Viruses; p. 303; difficult; ans: F**
    Because viruses are not living cells, they are not subject to natural selection.

Essay Questions

1. **Introduction; p. 282; moderate**
   Why have prokaryotes become the dominant life forms on Earth?

2. **Characteristics of the Prokaryotic Cell; p. 282; easy**
   Discuss some of the differences between prokaryotic cells and eukaryotic cells.

3. **Characteristics of the Prokaryotic Cell; p. 283; moderate**
   Describe the differences in cell wall structure between gram-positive and gram-negative *Bacteria*.

4. **Characteristics of the Prokaryotic Cell; p. 284; moderate**
   How are prokaryotic flagella, fimbriae, and pili different? How are they similar?

5. **Diversity of Form; p. 284; moderate**
   Describe the variety of cell shapes and growth forms in prokaryotes.

6. **Reproduction and Gene Exchange; pp. 285–286; moderate**
   Compare and contrast the processes of binary fission, conjugation, transduction, and transformation.

7. **Endospores; p. 286; moderate**
   In what way does sporulation increase the survival capacity of prokaryotes?

8. **Metabolic Diversity; p. 287; moderate**
   Discuss some of the ways in which prokaryotes are (a) important to the functioning of ecosystems and (b) used commercially.

9. *Bacteria*; **pp. 288–290; moderate**
   Describe some of the environments occupied by cyanobacteria. What cellular structures facilitate adaptation?

10. *Bacteria*; **p. 290; moderate**
What is meant by "nitrogen fixation" and what is its importance? What structures are present in cyanobacteria that facilitate this process?

11. *Bacteria*; **p. 291; moderate**
What are the main differences between photosynthesis in cyanobacteria and in purple sulfur and green sulfur bacteria? In what way were these differences important for the atmosphere of the early Earth?

12. *Bacteria*; **pp. 292–294; moderate**
Discuss some of the major types of plant diseases caused by *Bacteria* species.

13. *Archaea*; **pp. 294–296; moderate**
What are the four main groups of *Archaea*? What features do they have in common? How do they differ?

14. **Viruses; p. 300; moderate**
Discuss the ways in which a virus enters a host cell and then multiplies within it. How do new viruses then move from that cell to other parts of the host plant?

Chapter 15

Fungi

Multiple-Choice Questions

1. **Fungi; p. 307; moderate; ans: c**
   Which of the following statements about fungi is FALSE?

   a. They are heterotrophic organisms.
   b. Most are multicellular.
   c. They are more closely related to plants than to animals.
   d. The largest living organism may be a fungus.
   e. Only the insects have a greater number of species.

2. **The Importance of Fungi; p. 307; moderate; ans: a**
   Which of the following statements about the role of fungi in decomposition is FALSE?

   a. Fungi are the principal decomposers on Earth.
   b. Fungi produce enzymes that degrade lignin.
   c. Fungi produce enzymes that degrade cellulose.
   d. As a group, fungi can attack virtually anything.
   e. Individual species are highly specific to particular substrates.

3. **The Importance of Fungi; p. 308; moderate; ans: d**
   Phanerochaete chrysosporium is economically important because it:

   a. attacks meat in cold storage.
   b. attacks garden plants.
   c. causes AIDS.
   d. degrades toxic organic compounds.
   e. produces the antibiotic cyclosporine.

4. **The Importance of Fungi; p. 308; easy; ans: b**
   A fungus that is important in baking, brewing, and winemaking is:

   a. *Pneumocystis carinii*.
   b. *Saccharomyces cerevisiae*.
   c. *Candida*.
   d. *Tolypocladium inflatum*.
   e. *Cladosporium herbarum*.

5. **The Importance of Fungi; p. 308; easy; ans: c**
   By definition, an endophyte:

   a.   produces a toxin.
   b.   is part of a fungus "garden."
   c.   lives within a plant.
   d.   lives in a hostile environment.
   e.   lives within roots.

6. **The Importance of Fungi; p. 309; moderate; ans: e**
   Which of the following is NOT an example of a symbiotic relationship involving a fungus?

   a.   lichens
   b.   mycorrhizae
   c.   fungus "gardens"
   d.   endophytes
   e.   chytrids

7. **Biology and Characteristics of Fungi; p. 310; easy; ans: d**
   Hyphae are types of fungal:

   a.   spores.
   b.   zygotes.
   c.   storage products.
   d.   filaments.
   e.   cell wall components.

8. **Biology and Characteristics of Fungi; p. 310; moderate; ans: a**
   Coenocytic hyphae lack:

   a.   septa.
   b.   microtubules.
   c.   flagellated cells.
   d.   chitin.
   e.   rhizoids.

9. **Biology and Characteristics of Fungi; p. 310; moderate; ans: a**
   Which of the following statements about fungal nutrition is FALSE?

   a.   Fungi are able to engulf small microorganisms.
   b.   All fungi are heterotrophic.
   c.   Yeasts are able to obtain their energy from fermentation.
   d.   Rhizoids are hyphae specialized for anchorage.
   e.   Haustoria are hyphae specialized for absorbing nutrients directly from the cells of other organisms.

10. **Biology and Characteristics of Fungi; p. 310; easy; ans: c**
    Haustoria are _____ found in _____ fungi.

    a.   spores; saprophytic
    b.   gametes; symbiotic
    c.   hyphae; parasitic
    d.   gametangia; endophytic
    e.   sporangia; pathogenic

11. **Biology and Characteristics of Fungi; p. 311; moderate; ans: b**
Which of the following statements about nuclear division in fungi is FALSE?

   a. In many species the nuclear envelope does not disintegrate.
   b. Except for the chytrids, all fungi have centrioles.
   c. In most fungi, the spindle forms within the nuclear envelope.
   d. In some fungi, spindle pole bodies function as microtubule organizing centers.
   e. In some fungi, the spindle moves into the nucleus from the cytoplasm.

12. **Biology and Characteristics of Fungi; p. 311; easy; ans: b**
Conidia are spores produced asexually:

   a. in a sporangium.
   b. from conidiogenous cells.
   c. in a gametangium.
   d. from a dikaryotic mycelium.
   e. from a coenocytic hypha.

13. **Biology and Characteristics of Fungi; p. 312; easy; ans: d**
Plasmogamy is the:

   a. separation of plasma membranes.
   b. division of chromosomes.
   c. formation of cell walls.
   d. fusion of protoplasts.
   e. fusion of nuclei.

14. **Biology and Characteristics of Fungi; p. 312; easy; ans: c**
Karyogamy is the:

   a. fusion of protoplasts.
   b. formation of a dikaryon.
   c. fusion of nuclei.
   d. formation of rhizoids.
   e. absorption of nutrients.

15. **Biology and Characteristics of Fungi; p. 312; moderate; ans: e**
Meiosis in fungi is:

   a. conidiogenic.
   b. dikaryotic.
   c. gametic.
   d. sporic.
   e. zygotic.

16. **Biology and Characteristics of Fungi; p. 312; easy; ans: c**
Two gametes would be called isogametes if they:

   a. undergo plasmogamy.
   b. undergo karyogamy.
   c. are similar in size and appearance.
   d. are different in size and appearance.
   e. form a dikaryon.

17. Evolution of the Fungi; p. 312; moderate; ans: b
The _____ are probably the most primitive fungi.

    a.   ascomycetes
    b.   chytrids
    c.   basidiomycetes
    d.   zygomycetes
    e.   deuteromycetes

18. Chytrids: Phylum *Chytridiomycota*; p. 312; moderate; ans: a
Which of the following statements about chytrids is FALSE?

    a.   They are primarily terrestrial.
    b.   They have cell walls of chitin.
    c.   Some are unicellular.
    d.   Most are coenocytic.
    e.   Most have a single, posterior, whiplash flagellum.

19. Chytrids: Phylum *Chytridiomycota*; p. 313; moderate; ans: c
Allomyces belongs to the phylum:

    a.   *Ascomycota.*
    b.   *Basidiomycota.*
    c.   *Chytridiomycota.*
    d.   *Zygomycota.*
    e.   *Deuteromycota.*

20. Phylum *Zygomycota*; p. 315; moderate; ans: d
In the life cycle of *Rhizopus stolonifer*, asexual spores are produced within a:

    a.   zygosporangium.
    b.   stolon.
    c.   rhizoid.
    d.   sporangium.
    e.   sporangiophore.

21. Phylum *Zygomycota*; p. 316; moderate; ans: c
Heterothallic species of zygomycetes require _____ for sexual reproduction.

    a.   two + strains
    b.   two − strains
    c.   + and − strains
    d.   two homothallic strains
    e.   flagellated male and female gametes

22. Phylum *Zygomycota*; p. 317; moderate; ans: a
In the life cycle of *Rhizopus stolonifer*, meiosis occurs when the _____
germinates.

    a.   zygosporangium
    b.   stolon
    c.   rhizoid
    d.   sporangium
    e.   sporangiophore

23. Phylum *Zygomycota*; p. 317; easy; ans: e

_____ are found inside insects and other arthropods.

a. *Glomus* and related genera
b. *Rhizopus* and related genera
c. *Choanephora* and related genera
d. Members of the *Entomophthoraleas*
e. The trichomycetes

24. Phylum *Ascomycota*; p. 317; moderate; ans: a

Fungi that cause powdery mildews, chestnut blight, and Dutch elm disease belong to the phylum:

a. *Ascomycota.*
b. *Basidiomycota.*
c. *Chytridiomycota.*
d. *Zygomycota.*
e. *Deuteromycota.*

25. Phylum *Ascomycota*; p. 318; easy; ans: b

In the ascomycete life cycle, meiosis takes place within a(n):

a. sporangium.
b. ascus.
c. conidium.
d. conidiophore.
e. antheridium.

26. Phylum *Ascomycota*; p. 318; easy; ans: c

Cells produced by meiosis in ascomycetes are called:

a. conidia.
b. croziers.
c. ascospores.
d. conidiogenous cells.
e. ascoma.

27. Phylum *Ascomycota*; p. 319; difficult; ans: d

An ascoma that is closed and spherical is called a(n):

a. apothecium.
b. perithecium.
c. hymenium.
d. cleistothecium.
e. ascogonium.

28. Phylum *Ascomycota*; pp. 319–320; moderate; ans: b

The layer of asci of an ascoma is called the:

a. perithecium.
b. hymenium.
c. cleistothecium.
d. apothecium.
e. ascogonium.

29. Phylum *Ascomycota*; p. 320; moderate; ans: a
In an ascomycete, the male gametangium is the:

   a.   antheridium.
   b.   sperm.
   c.   ascogonium.
   d.   ascus.
   e.   trichogyne.

30. Phylum *Ascomycota*; p. 320; moderate; ans: c
In sexual reproduction in ascomycetes, the male nuclei pass into the ascogonium via the:

   a.   ascogenous hyphae.
   b.   antheridium.
   c.   trichogyne.
   d.   crozier.
   e.   hymenium.

31. Phylum *Ascomycota*; p. 320; moderate; ans: c
In ascomycetes, plasmogamy occurs within the:

   a.   antheridium.
   b.   crozier.
   c.   ascogonium.
   d.   ascogenous hypha.
   e.   trichogyne.

32. Phylum *Basidiomycota*; p. 320; easy; ans: b
Mushrooms and toadstools belong to the phylum:

   a.   *Ascomycota*.
   b.   *Basidiomycota*.
   c.   *Chytridiomycota*.
   d.   *Zygomycota*.
   e.   *Deuteromycota*.

33. Phylum *Basidiomycota*; p. 320; moderate; ans: d
In the *Basidiomycota*, _____ are borne on a _____.

   a.   basidiospores; primary mycelium
   b.   basidiospores; parenthosome
   c.   dolipores; basidium
   d.   basidiospores; basidium
   e.   basidia; basidiospore

34. Phylum *Basidiomycota*; p. 320; easy; ans: c
A dolipore septum is characteristic of:

   a.   ascomycetes.
   b.   chytrids.
   c.   basidiomycetes.
   d.   zygomycetes.
   e.   deuteromycetes.

35. Phylum *Basidiomycota*; pp. 320, 322; difficult; ans: d
Which of the following statements about the mycelium of the *Basidiomycota* is FALSE?

  a.   The primary mycelium is monokaryotic.
  b.   The secondary mycelium is dikaryotic.
  c.   The tertiary mycelium is dikaryotic.
  d.   Clamp connections are characteristic of the monokaryotic mycelium.
  e.   The basidioma is a mycelium composed of dikaryotic hyphae.

36. Phylum *Basidiomycota*; p. 322; easy; ans: c
Masses of spores produced by the *Teliomycetes* and *Uromycetes* are called:

  a.   basidiomata.
  b.   ascomata.
  c.   sori.
  d.   hymenia.
  e.   tertiary mycelia.

37. Phylum *Basidiomycota*; p. 322; easy; ans: e
A _____ is a representative of the hymenomycetes.

  a.   stinkhorn
  b.   puffball
  c.   rust
  d.   smut
  e.   mushroom

38. Phylum *Basidiomycota*; p. 323; moderate; ans: e
Sterigmata are:

  a.   septate basidia.
  b.   aseptate basidia.
  c.   fertile layers of hyphae.
  d.   the stalks of mushrooms.
  e.   projections that bear basidiospores.

39. Phylum *Basidiomycota*; p. 324; easy; ans: b
The pileus is the _____ of a mushroom.

  a.   stalk
  b.   cap
  c.   gill
  d.   sterigma
  e.   hymenium

40. Phylum *Basidiomycota*; p. 324; difficult; ans: d
In the mushroom life cycle, karyogamy occurs in a:

  a.   stipe.
  b.   volva.
  c.   sterigma.
  d.   basidium.
  e.   basidiospore.

41. Phylum *Basidiomycota*; p. 324; difficult; ans: d
In the mushroom life cycle, plasmogamy occurs in a:

a.   stipe.
b.   volva.
c.   sterigma.
d.   basidium.
e.   basidiospore.

42. Phylum *Basidiomycota*; pp. 324–325; difficult; ans: a
_____ is a basidiomycete that forcibly discharges its basidiospores.

a.   *Agaricus campestris*
b.   A puffball
c.   *Puccinia graminis*
d.   *Ustilago maydis*
e.   A stinkhorn

43. Phylum *Basidiomycota*; p. 325; moderate; ans: c
The genus _____ includes the most poisonous of all mushrooms.

a.   *Agaricus*
b.   *Lentinula*
c.   *Amanita*
d.   *Puccinia*
e.   *Ustilago*

44. Phylum *Basidiomycota*; p. 325; moderate; ans: d
Members of the phylum *Basidiomycota* that have a peridium belong to the class:

a.   *Hymenomycetes*.
b.   *Teliomycetes*.
c.   *Ustomycetes*.
d.   *Gasteromycetes*.
e.   *Chytridiomycetes*.

45. Phylum *Basidiomycota*; p. 326; easy; ans: a
The _____ is the fertile portion of the basidioma of a stinkhorn.

a.   gleba
b.   sterigma
c.   peridium
d.   sac
e.   hymenium

46. Phylum *Basidiomycota*; p. 326; easy; ans: b
The rusts belong to the class:

a.   *Hymenomycetes*.
b.   *Teliomycetes*.
c.   *Ustomycetes*.
d.   *Gasteromycetes*.
e.   *Chytridiomycetes*.

47. Phylum *Basidiomycota*; p. 327; difficult; ans: c
In the life cycle of *Puccinia graminis*, dikaryotic hyphae extend downward
from the spermogonium and give rise to:

a. spermogonia.
b. spermatia.
c. aecia.
d. uredinia.
e. urediniospores.

48. Phylum *Basidiomycota*; p. 327; difficult; ans: e
In the life cycle of *Puccinia graminis*, which of the following are produced on
wheat?

a. spermogonia
b. aecia
c. spermatia
d. periphyses
e. urediniospores

49. Phylum *Basidiomycota*; p. 327; difficult; ans: d
In the life cycle of *Puccinia graminis*, karyogamy occurs in a(n):

a. spermogonium.
b. urediniospore.
c. aeciospore.
d. teliospore.
e. basidiospore.

50. Phylum *Basidiomycota*; pp. 327, 330; difficult; ans: b
In the *Ustilago maydis* life cycle, plasmogamy and karyogamy take place in:

a. basidia.
b. teliospores.
c. sporidia.
d. spermogonia.
e. aecia.

51. Phylum *Basidiomycota*; p. 330; easy; ans: c
The smuts belong to the class:

a. *Hymenomycetes*.
b. *Teliomycetes*.
c. *Ustomycetes*.
d. *Gasteromycetes*.
e. *Chytridiomycetes*.

52. Yeasts; p. 331; easy; ans: a
Virtually the only species of yeast now used in baking bread is:

a. *Saccharomyces cerevisiae*.
b. *Saccharomyces carlsbergensis*.
c. *Candida albicans*.
d. *Cryptococcus neoformans*.
e. *Penicillium roquefortii*.

53. Yeasts; pp. 331–332; easy; ans: d
The first eukaryote whose genome has been completed sequenced is:

a. *Penicillium.*
b. *Cryptococcus neoformans.*
c. *Candida albicans.*
d. *Saccharomyces cerevisiae.*
e. *Saccharomyces carlsbergensis.*

54. Deuteromycetes; pp. 332–333; moderate; ans: e
Which of the following statements about the deuteromycetes is FALSE?

a. In some, the sexual phase of the life cycle has been lost.
b. In some, the sexual phase of the life cycle has not been discovered.
c. Some have a sexual phase, but it is not the principal basis of classification.
d. Some have a sexual phase, but the organism is classified because of its resemblance to an organism that lacks a sexual phase.
e. Most deuteromycetes are clearly basidiomycetes.

55. Deuteromycetes; p. 333; moderate; ans: b
Heterokaryosis is the presence of _____ in a common cytoplasm.

a. more than one nucleus
b. genetically different nuclei
c. a haploid and a diploid nucleus
d. two haploid nuclei
e. two diploid nuclei

56. Deuteromycetes; p. 334; easy; ans: b
Which of the following indicates the correct sequence of events in parasexuality?

a. plasmogamy, karyogamy, meiosis
b. plasmogamy, karyogamy, haploidization
c. plasmogamy, karyogamy, diploidization
d. karyogamy, plasmogamy, diploidization
e. haploidization, karyogamy, plasmogamy

57. Symbiotic Relationships of Fungi; p. 334; easy; ans: b
In a mutualistic symbiotic relationship:

a. one species benefits and the other is harmed.
b. both species benefit.
c. both species are harmed.
d. one species benefits and the other neither benefits nor is harmed.
e. one species is harmed and the other neither benefits nor is harmed.

58. Symbiotic Relationships of Fungi; pp. 334–340; moderate; ans: c
Which of the following statements about lichens is FALSE?

a. They consist of a mycobiont and a photobiont.
b. They live in the harshest environments.
c. They are very resistant to pollutants.
d. They reproduce asexually by forming fragments, soredia, or isidia.
e. They have very slow growth rates.

59. **Symbiotic Relationships of Fungi; p. 338; easy; ans: e**
   "Fruticose" describes a lichen with a _____ appearance.

   a.   flattened
   b.   greenish-yellow
   c.   crusty
   d.   leaflike
   e.   bushy

60. **Symbiotic Relationships of Fungi; p. 338; easy; ans: b**
   The survival of lichens under severe environmental conditions is due mainly to their ability to:

   a.   produce toxic compounds for their defense.
   b.   dry out rapidly.
   c.   carry out photosynthesis under very high light intensities.
   d.   carry out respiration at very low temperatures.
   e.   parasitize plants.

61. **Symbiotic Relationships of Fungi; pp. 339–340; moderate; ans: a**
   What is the ecological role of lichen acids?

   a.   weathering rock and formation of soil
   b.   protecting trees against predators
   c.   increasing rates of photosynthesis
   d.   absorbing minerals from the soil
   e.   contributing fixed nitrogen to the soil

62. **Symbiotic Relationships of Fungi; p. 340; easy; ans: b**
   A mycorrhiza is a symbiotic association between a(n) _____ and a _____.

   a.   alga; fungus
   b.   fungus; root
   c.   lichen; root
   d.   alga; root
   e.   fungus; lichen

63. **Symbiotic Relationships of Fungi; p. 341; moderate; ans: e**
   Ectomycorrhizae differ from endomycorrhizae in that the fungal component of ectomycorrhizae:

   a.   is usually a zygomycete.
   b.   forms arbuscules and vesicles.
   c.   is not highly specific for the plant component.
   d.   penetrates the cortical cells of the root.
   e.   forms a Hartig net and mantle.

64. **Symbiotic Relationships of Fungi; p. 343; moderate; ans: b**
   The principal role of mycorrhizae in heather is to:

   a.   stimulate seed germination.
   b.   release enzymes that act on compounds in the soil.
   c.   increase the plant's absorption of water.
   d.   increase the plant's absorption of phosphorus.
   e.   protect the plant against pathogens.

True-False Questions

1. **The Importance of Fungi; p. 307; easy; ans: T**
   Fungi and bacteria are the principal decomposers of the biosphere.

2. **The Importance of Fungi; p. 307; moderate; ans: F**
   Fungi are second only to bacteria as the most important causal agents of plant disease.

3. **Biology and Characteristics of Fungi; p. 310; easy; ans: T**
   A mycelium is a mass of hyphae.

4. **Biology and Characteristics of Fungi; p. 310; easy; ans: T**
   Some fungi use glycogen as their primary storage polysaccharide.

5. **Biology and Characteristics of Fungi; p. 311; easy; ans: F**
   The most common method of asexual reproduction in all fungi except the chytrids involves flagellated spores.

6. **Biology and Characteristics of Fungi; p. 312; moderate; ans: T**
   In fungi, karyogamy may be separated from plasmogamy by several months.

7. **Chytrids: Phylum *Chytridiomycota*; p. 313; moderate; ans: T**
   Some members of the phylum *Chytridiomycota* are unique among the fungi in having an alternation of generations.

8. **Phylum *Zygomycota*; pp. 316–317; easy; ans: F**
   Homothallic zygomycetes require + and − strains for sexual reproduction.

9. **Phylum *Ascomycota*; pp. 318–319; easy; ans: T**
   Ascomycetes are also called "sac fungi."

10. **Phylum *Ascomycota*; p. 319; difficult; ans: T**
    The hymenium of ascomycotes is the layer of asci on the ascoma.

11. **Phylum *Ascomycota*; p. 320; difficult; ans: F**
    In the ascomycete life cycle, the trichogyne is dikaryotic.

12. **Phylum *Basidiomycota*; pp. 320, 322; moderate; ans: T**
    Parenthesomes are associated with the septa of basidiomycetes.

13. **Phylum *Basidiomycota*; p. 323; easy; ans: F**
    The edible mushrooms are gasteromycotes.

14. **Phylum *Basidiomycota*; p. 326; moderate; ans: F**
    Mushrooms typically possess a gleba as part of the basidioma.

15. **Phylum *Basidiomycota*; p. 327; difficult; ans: T**
    In the rust *Puccinia graminis*, meiosis occurs in a teliospore.

16. **Phylum *Basidiomycota*; p. 330; moderate; ans: F**
    Like rusts, smuts are autoecious.

17. **Yeasts; p. 331; moderate; ans: T**
    Saccharomyces cerevisiae has both haploid and diploid budding stages.

18. **Yeasts; p. 332; easy; ans: T**
    The DNA of *Saccharomyces cerevisiae* was the first eukaryotic genome to be completely sequenced.

19. **Deuteromycetes; p. 333; moderate; ans: T**
    In heterokaryosis, genetically different nuclei exist together in the same cytoplasm.

20. **Deuteromycetes; pp. 333–334; moderate; ans: F**
    In parasexuality, haploid nuclei are formed by meiosis following karyogamy.

21. **Symbiotic Relationships of Fungi; p. 335; easy; ans: T**
    In lichens, the photobionts are green algae and/or cyanobacteria.

22. **Symbiotic Relationships of Fungi; p. 340; moderate; ans: T**
    Mycorrhizae occur in the vast majority of vascular plants.

23. **Symbiotic Relationships of Fungi; p. 341; easy; ans: F**
    Ectomycorrhizae are much more common in nature than endomycorrhizae.

24. **Symbiotic Relationships of Fungi; p. 342; moderate; ans: T**
    The Hartig net is a network of fungal hyphae in the ectomycorrhizae of conifers and angiosperms.

Essay Questions

1. **The Importance of Fungi; p. 307; easy**
   Discuss the various roles—both beneficial and detrimental—of fungi as decomposers.

2. **The Importance of Fungi; pp. 307–308; easy**
   In what ways are fungi economically important?

3. **Biology and Characteristics of Fungi; p. 310; moderate**
   Explain how the structure of a fungus is well-adapted to its absorptive mode of nutrition.

4. **Biology and Characteristics of Fungi; pp. 310–311; moderate**
   Describe the unique features of mitosis and meiosis in fungi.

5. **Biology and Characteristics of Fungi; p. 312; moderate**
   Explain the relationship between plasmogamy, karyogamy, and dikaryon formation.

6. **Evolution of the Fungi; p. 312; moderate**
   What evidence suggests that *Chytridiomycota* is the most primitive phylum of fungi?

7. **Phylum *Zygomycota*; pp. 315–317; moderate**
   Describe the sexual and asexual life cycles of *Rhizopus stolonifer*.

8. **Phylum *Zygomycota*; p. 317; moderate**
   Describe the relationships between zygomycotes and insects and other arthropods.

9. **Phylum *Ascomycota*; p. 320; moderate**
   Describe the events occurring at the tip of an ascogenous hypha, eventually leading to dispersal of ascopores.

10. **Phylum *Basidiomycota*; pp. 323–324; moderate**
    Describe the life cycle of a typical mushroom.

11. **Phylum *Basidiomycota*; p. 327; difficult**
    Describe the life cycle of *Puccinia graminis*.

12. **Yeasts; pp. 330–331; moderate**
    What defines a fungus as a yeast? Describe how yeasts reproduce.

13. **Deuteromycetes; p. 332; moderate**
    What are the characteristic features of the deuteromycetes? What is the relationship of this group to the *Zygomycota*, *Ascomycota*, and *Basidiomycota*?

14. **Deuteromycetes; p. 334; difficult**
    Describe the events that take place in parasexual reproduction. What role does parasexuality play in the deuteromycete life cycle?

15. **Deuteromycetes; p. 334; moderate**
    Outline the economic importance of deuteromycetes.

16. **Symbiotic Relationships of Fungi; pp. 334, 339; moderate**
    Are lichens the result of a parasitic or mutualistic relationship? Give reasons to support your answer.

17. **Symbiotic Relationships of Fungi; pp. 339–340; moderate**
    Outline the ecological importance of lichens.

18. **Symbiotic Relationships of Fungi; pp. 340–342; moderate**
    Compare and contrast the two major types of mycorrhizae.

19. **Symbiotic Relationships of Fungi; pp. 343–344; moderate**
    What role are mycorrhizae thought to have played in the evolution of vascular plants?

Chapter 16

Protista I: Euglenoids, Slime Molds, Cryptomonads, Red Algae, Dinoflagellates, and Haptophytes

Multiple-Choice Questions

1. **Introduction; p. 348; easy; ans: b**
    Which of the following statements about protists is FALSE?

    a. Some of the oldest fossil eukaryotes resembled protists.
    b. They live mainly on land.
    c. They include autotrophic, heterotrophic, and mixotrophic algae.
    d. They include the cellular and plasmodial slime molds.
    e. Fungi, plants, and animals are derived from ancient protists.

2. **Ecology of the Algae; pp. 348–350; moderate; ans: a**
    Which of the following statements about algae is FALSE?

    a. Some are part of the zooplankton.
    b. Some are used in mariculture.
    c. Algal blooms are associated with water pollution.
    d. Some reduce the amount of CO_2 in the atmosphere.
    e. Some excrete sulfur-containing compounds that contribute to acid rain.

3. **Ecology of the Algae; p. 350; moderate; ans: c**
    Some phytoplankton reduce atmospheric levels of CO_2 by favoring formation of
    _____ during photosynthetic carbon fixation.

    a. carbohydrate
    b. protein
    c. calcium carbonate
    d. calcium nitrate
    e. sulfur-containing compounds

4. **Euglenoids: Phylum *Euglenophyta*; pp. 350–351; easy; ans: d**
    Which of the following statements about euglenoids is FALSE?

    a. All are flagellates.
    b. Some are heterotrophic.
    c. Some are autotrophic.
    d. Most have a cell wall.
    e. Most are unicellular.

5. **Euglenoids: Phylum *Euglenophyta*; p. 351; moderate; ans: e**
In euglenoids, the pellicle:

   a.   senses light.
   b.   collects excess water.
   c.   discharges excess water.
   d.   contains Rubisco.
   e.   supports the plasma membrane.

6. **Euglenoids: Phylum *Euglenophyta*; p. 352; easy; ans: d**
The storage polysaccharide in euglenoids is:

   a.   starch.
   b.   lipid.
   c.   glycogen.
   d.   paramylon.
   e.   pellicle.

7. **Euglenoids: Phylum *Euglenophyta*; p. 352; easy; ans: b**
In euglenoids, the pyrenoid is the site of:

   a.   paramylon storage.
   b.   Rubisco activity.
   c.   a red eyespot.
   d.   excess water.
   e.   mitosis.

8. **Plasmodial Slime Molds: Phylum *Myxomycota*; p. 352; moderate; ans: a**
Which of the following statements about myxomycetes is FALSE?

   a.   They are thought to be the ancestors of the fungi.
   b.   They form a protoplasmic mass called a plasmodium.
   c.   They may be the earliest protists to have acquired sexual reproduction.
   d.   Their nuclei can divide synchronously.
   e.   They undergo mitosis by a mechanism similar to that in plants.

9. **Plasmodial Slime Molds: Phylum *Myxomycota*; p. 352; moderate; ans: c**
In the life cycle of the myxomycetes, meiosis occurs within:

   a.   the plasmodiocarps.
   b.   amoebas.
   c.   spores.
   d.   sclerotia.
   e.   aethalia.

10. **Plasmodial Slime Molds: Phylum *Myxomycota*; p. 354; difficult; ans: b**
In the myxomycetes, a plasmodium rapidly produces a(n) _____ when the habitat dries out.

   a.   aethalium
   b.   sclerotium
   c.   microcyst
   d.   plasmodiocarp
   e.   sporangium

11. Plasmodial Slime Molds: Phylum *Myxomycota*; p. 354; moderate; ans: d
In the life cycle of plasmodial slime molds, when gametes fuse they give rise to a:

a.   microcyst.
b.   sporangium.
c.   spore.
d.   plasmodium.
e.   sclerotium.

12. Cellular Slime Molds: Phylum *Dictyosteliomycota*; pp. 354–355; moderate; ans: d
Which of the following statements concerning myxamoebas is FALSE?

a.   They are free-living cells.
b.   They are amoeba-like.
c.   They feed on bacteria by phagocytosis.
d.   They are found in the myxomycetes.
e.   They are found in litter-rich soils.

13. Cellular Slime Molds: Phylum *Dictyosteliomycota*; p. 355; moderate; ans: c
Members of the phylum *Dictyosteliomycota* have cell walls containing:

a.   chitin.
b.   peptidoglycan.
c.   cellulose.
d.   paramylon.
e.   laminarin.

14. Cellular Slime Molds: Phylum *Dictyosteliomycota*; p. 355; easy; ans: b
Dictyostelium is used in the study of apoptosis, which is the:

a.   aggregation of amoebas.
b.   programmed cell death.
c.   formation of resistant encysted structures.
d.   morphological differentiation of a plasmodium.
e.   loss of motility of amoebas.

15. Cellular Slime Molds: Phylum *Dictyosteliomycota*; p. 355; moderate; ans: e
In the life cycle of *Dictyostelium discoideum*, myxamoebas aggregate to form a:

a.   spore.
b.   sclerotium.
c.   sporangium.
d.   plasmodium.
e.   pseudoplasmodium.

16. Cellular Slime Molds: Phylum *Dictyosteliomycota*; p. 355; easy; ans: a
In the *Dictyostelium* life cycle, myxamoebas aggregate by migrating along a gradient of:

a.   cAMP.
b.   glucose.
c.   protein.
d.   paramylon.
e.   microtubules.

17. **Cellular Slime Molds: Phylum *Dictyosteliomycota*; p. 356; difficult; ans: d**
    In the *Dictyostelium* life cycle, the only diploid structure is the:

    a.   sclerotium.
    b.   aethalium.
    c.   myxamoeba.
    d.   zygote.
    e.   pseudoplasmodium.

18. **Cellular Slime Molds: Phylum *Dictyosteliomycota*; p. 356; difficult; ans: e**
    In *Dictyostelium discoideum*, meiosis takes place within the:

    a.   plasmodium.
    b.   stalk.
    c.   pseudoplasmodium.
    d.   sporangium.
    e.   macrocyst.

19. **Cryptomonads: Phylum *Cryptophyta*; p. 356; moderate; ans: c**
    Which of the following statements about cryptomonads is FALSE?

    a.   They are flagellates.
    b.   They are unicellular.
    c.   They live primarily in warm, surface waters.
    d.   They are rich in polyunsaturated fatty acids.
    e.   They include both autotrophs and heterotrophs.

20. **Cryptomonads: Phylum *Cryptophyta*; p. 357; moderate; ans: a**
    In cryptomonads, secondary endosymbiosis involves the fusion of one _____
    cell with one _____ cell.

    a.   heterotrophic eukaryotic; photosynthetic eukaryotic
    b.   heterotrophic eukaryotic; photosynthetic prokaryotic
    c.   heterotrophic prokaryotic; photosynthetic eukaryotic
    d.   heterotrophic prokaryotic; photosynthetic prokaryotic
    e.   heterotrophic prokaryotic; heterotrophic eukaryotic

21. **Cryptomonads: Phylum *Cryptophyta*; p. 357; easy; ans: a**
    In the cryptomonads, the nucleomorph is thought to be the:

    a.   remains of the nucleus of a red algal cell.
    b.   principal nucleus of the cryptomonad cell.
    c.   precursor of mitochondrial DNA.
    d.   precursor of DNA in the chloroplast endoplasmic reticulum.
    e.   remains of a plasmid that established a secondary endosymbiosis.

22. **Red Algae: Phylum *Rhodophyta*; pp. 357–358; moderate; ans: c**
    Which of the following statements about red algae is FALSE?

    a.   They belong to the phylum *Rhodophyta*.
    b.   Most are macroscopic seaweeds.
    c.   They are incapable of growing in deep water.
    d.   Most are filamentous.
    e.   Their red color is due to phycobilins.

23. **Red Algae: Phylum *Rhodophyta*; p. 358; moderate; ans: b**
The main food reserve in red algae is:

    a. amylopectin.
    b. floridean starch.
    c. glycogen.
    d. lipid.
    e. amylose.

24. **Red Algae: Phylum *Rhodophyta*; p. 358; moderate; ans: d**
Carrageenan is a _____ in red algae.

    a. food reserve
    b. photosynthetic pigment
    c. component of the polar rings
    d. mucilaginous layer of the cell wall
    e. a stony covering on the cell wall

25. **Red Algae: Phylum *Rhodophyta*; p. 359; easy; ans: c**
Coralline algae are tough and stony because they deposit _____ in their cell walls.

    a. carrageenan
    b. floridean starch
    c. calcium carbonate
    d. fatty acids
    e. protein

26. **Red Algae: Phylum *Rhodophyta*; p. 359; moderate; ans: c**
Which of the following statements about coralline algae is FALSE?

    a. They are common throughout oceans of the world.
    b. They grow on rocks and other stable surfaces.
    c. No fossil forms exist.
    d. Their cell walls are calcified.
    e. Some help stabilize the structure of coral reefs.

27. **Red Algae: Phylum *Rhodophyta*; pp. 359–360; moderate; ans: b**
Most red algae are:

    a. unicellular.
    b. filamentous.
    c. colonial.
    d. composed of one-layered sheets.
    e. composed of two-layered sheets.

28. **Red Algae: Phylum *Rhodophyta*; p. 360; moderate; ans: b**
Many red algae reproduce asexually by producing:

    a. carpospores.
    b. monospores.
    c. spermatia.
    d. trichogynes.
    e. carpogonia.

29. Red Algae: Phylum *Rhodophyta*; p. 360; difficult; ans: d
In red algae, the gametophyte differs from the sporophyte in that the gametophyte:

a. is multicellular.
b. produces haploid spores.
c. produces carpospores.
d. is haploid.
e. grows independently.

30. Red Algae: Phylum *Rhodophyta*; p. 360; moderate; ans: e
In red algae, the spermatium produces:

a. a trichogyne.
b. an egg.
c. carpogonia.
d. carpospores.
e. male gametes.

31. Red Algae: Phylum *Rhodophyta*; pp. 360–361; moderate; ans: d
It is thought that red algae acquired an alternation of multicellular generations in response to the:

a. presence of diploid carpospores.
b. presence of a trichogyne.
c. presence of a carpogonium.
d. lack of flagellated male gametes.
e. lack of flagellated female gametes.

32. Red Algae: Phylum *Rhodophyta*; p. 361; moderate; ans: d
In many red algae, a copy of the zygote is transferred to _____ in the gametophyte where it forms multiple copies by mitosis.

a. carpospores
b. carposporophytes
c. tetrasporophytes
d. auxiliary cells
e. trichogynes

33. Red Algae: Phylum *Rhodophyta*; p. 361; difficult; ans: e
In the following pairs, which structures are both diploid?

a. carpogonium and monospore
b. tetrasporophyte and carpogonium
c. tetrasporophyte and spermatium
d. carposporophyte and spermatangium
e. carposporophyte and tetrasporophyte

34. Red Algae: Phylum *Rhodophyta*; p. 361; easy; ans: d
The life history of most red algae consists of the following phases:

a. gametophyte and carposporophyte.
b. gametophyte and tetrasporophyte.
c. carposporophyte and tetrasporophyte.
d. gametophyte, carposporophyte, and tetrasporophyte.
e. gametophyte, sporophyte, carposporophyte, and tetrasporophyte.

35. Dinoflagellates: Phylum *Dinophyta*; pp. 361, 363; moderate; ans: a
Which of the following statements about dinoflagellates is FALSE?

a. They have three or more flagella in grooves.
b. They have permanently condensed chromosomes.
c. Many have stiff cellulosic plates.
d. They are unicellular.
e. About half are photosynthetic.

36. Dinoflagellates: Phylum *Dinophyta*; p. 363; easy; ans: d
A theca is found in members of the phylum:

a. *Euglenophyta.*
b. *Rhodophyta.*
c. *Haptophyta.*
d. *Cryptophyta.*
e. *Dinophyta.*

37. Dinoflagellates: Phylum *Dinophyta*; p. 365; moderate; ans: e
Which of the following statements about dinoflagellate nutrition is FALSE?

a. Some are photosynthetic and also ingest solid food particles.
b. Some are nonphotosynthetic and instead ingest solid food particles.
d. Some are photosynthetic and also absorb dissolved organic compounds.
c. Some are nonphotosynthetic and instead absorb dissolved organic compounds.
e. Some are photosynthetic and use a peduncle to facilitate light absorption.

38. Dinoflagellates: Phylum *Dinophyta*; p. 365; easy; ans: b
The presence of _____ in dinoflagellates supports the hypothesis that their chloroplasts were derived from ingested chrysophytes.

a. zooxanthellae
b. peridinin
c. floridean starch
d. carrageenan
e. paramylon

39. Dinoflagellates: Phylum *Dinophyta*; p. 365; difficult; ans: b
In a coral reef, zooxanthellae:

a. form a non-symbiotic association with the coral polyps.
b. produce glycerol instead of starch.
c. grow only at ocean depths greater than 60 meters.
d. do not photosynthesize.
e. are located on the outer surface of the coral polyps.

40. Dinoflagellates: Phylum *Dinophyta*; p. 365; moderate; ans: e
When nutrient levels drop, dinoflagellates typically:

a. die.
b. produce toxic compounds.
c. produce bioluminescent compounds.
d. reproduce sexually.
e. form resting cysts.

41. Dinoflagellates: Phylum *Dinophyta*; p. 365; easy; ans: c
The ecology and geography of dinoflagellate blooms can be explained by the production, movement, and germination of:

    a.   zygotes.
    b.   amoebas.
    c.   cysts.
    d.   spores.
    e.   gametes.

42. Dinoflagellates: Phylum *Dinophyta*; p. 366; moderate; ans: a
The bioluminescent compounds produced by dinoflagellates:

    a.   protect them against predators.
    b.   attract organisms on which they feed.
    c.   are toxins that paralyze fishes' respiratory systems.
    d.   provide a source of phosphorus in the water.
    e.   provide a "hit and run" feeding strategy.

43. Haptophytes: Phylum *Haptophyta*; p. 366; moderate; ans: b
Which of the following statements does NOT describe a haptonema?

    a.   It is a threadlike structure in haptophytes.
    b.   It has the typical 9-plus-2 arrangement of microtubules.
    c.   It moves by bending and coiling.
    d.   It extends from the cell along with two flagella.
    e.   In some haptophytes it has a role in predation.

44. Haptophytes: Phylum *Haptophyta*; p. 367; moderate; ans: d
A coccolith, found in haptophytes, is a(n):

    a.   membrane of the chloroplast endoplasmic reticulum.
    b.   stonelike outgrowth of the plasma membrane.
    c.   endosymbiotic bacterial cell.
    d.   calcified scale on the cell surface.
    e.   calcium-containing Golgi vesicle.

45. Haptophytes: Phylum *Haptophyta*; p. 367; moderate; ans: a
Which of the following statements about marine haptophytes is FALSE?

    a.   They can function as producers but not consumers.
    b.   They can produce sulfur oxides connected with acid rain.
    c.   They can transport organic carbon to the deep ocean.
    d.   They can transport calcium carbonate to the deep ocean.
    e.   They can use cyanobacteria as food.

True-False Questions

1. Protista I; p. 348; easy; ans: T
Members of the kingdom *Protista* range in size from single cells to organisms 30 meters long.

2. Ecology of the Algae; p. 350; moderate; ans: T
CO_2 drawdown refers to the suction effect created by the formation of calcium carbonate by phytoplankton.

 3. **Euglenoids: Phylum *Euglenophyta*; p. 351; easy; ans: F**
    In euglenoids, the contractile vacuole serves to capture water from the surroundings.

 4. **Euglenoids: Phylum *Euglenophyta*; p. 352; moderate; ans: F**
    Euglenoids reproduce both asexually. and sexually.

 5. **Plasmodial Slime Molds: Phylum *Myxomycota*; p. 352; moderate; ans: F**
    In all plasmodial slime molds, haploid spores are produced by meiosis in a sporangium.

 6. **Plasmodial Slime Molds: Phylum *Myxomycota*; p. 354; moderate; ans: T**
    In the myxomycete life cycle, the protoplast of spores forms amoeboid or flagellated cells that are interconvertible.

 7. **Cellular Slime Molds: Phylum *Dictyosteliomycota*; p. 355; easy; ans: F**
    When cells of *Dictyostelium* aggregate, they fuse to form a multinucleate pseudoplasmodium.

 8. **Cellular Slime Molds: Phylum *Dictyosteliomycota*; p. 356; moderate; ans: T**
    In cellular slime molds, sexual reproduction occurs when haploid myxamoebas fuse to form a phagocytic zygote.

 9. **Cryptomonads: Phylum *Cryptophyta*; p. 357; easy; ans: T**
    Cryptomonads arose through the fusion of a heterotrophic and a photosynthetic eukaryotic cell.

10. **Red Algae: Phylum *Rhodophyta*; p. 359; easy; ans: T**
    Coralline algae provide architectural strength to coral reefs.

11. **Red Algae: Phylum *Rhodophyta*; p. 359; moderate; ans: F**
    Many red algae form a three-dimensional body with filaments interconnected by primary pit connections.

12. **Red Algae: Phylum *Rhodophyta*; p. 361; moderate; ans: F**
    The life history of most red algae has three phases: a haploid gametophyte, a diploid zygote, and a diploid tetrasporophyte.

13. **Dinoflagellates: Phylum *Dinophyta*; p. 363; easy; ans: T**
    In some dinoflagellates, thecal plates are sail-like structures aiding in flotation.

14. **Dinoflagellates: Phylum *Dinophyta*; p. 365; easy; ans: T**
    The carbohydrate reserve in dinoflagellates is starch.

15. **Haptophytes: Phylum *Haptophyta*; p. 366; moderate; ans: F**
    All members of the phylum *Haptophyta* are motile, flagellated organisms.

16. **Haptophytes: Phylum *Haptophyta*; p. 367; easy; ans: T**
    No single photosynthetic pigment characterizes all haptophyte species.

Essay Questions

1. **Ecology of the Algae; pp. 348–350; moderate**
   Discuss some of the roles played by algae in ecosystems around the world.

2. **Euglenoids: Phylum *Euglenophyta*; pp. 350–352; moderate**
   Describe the distinctive cellular features of the euglenoids.

3. **Plasmodial Slime Molds: Phylum *Myxomycota*; Cellular Slime Molds: Phylum *Dictyosteliomycota*; pp. 352–356; difficult**
   Compare and contrast the life cycles of the plasmodial and cellular slime molds.

4. **Cryptomonads: Phylum *Cryptophyta*; pp. 356–357; moderate**
   Is a cryptomonad one organism or two? Explain your answer.

5. **Red Algae: Phylum *Rhodophyta*; pp. 357–358; easy**
   Explain how the photosynthetic pigments of red algae correlate well with their usual habitat.

6. **Red Algae: Phylum *Rhodophyta* and Dinoflagellates: Phylum *Dinophyta*; pp. 358–359, 365; difficult**
   Describe the relationships between coral reefs and both red algae and dinoflagellates.

7. **Red Algae: Phylum *Rhodophyta*; p. 360; moderate**
   Explain the rationale for the hypothesis on how red algae acquired an alternation of generations.

8. **Red Algae: Phylum *Rhodophyta*; p. 361; difficult**
   Describe the three phases in the life history of most red algae.

9. **Dinoflagellates: Phylum *Dinophyta*; pp. 365–366; moderate**
   What is the relationship between the dinoflagellate cyst and the involvement of dinoflagellates in blooms and fish kills?

10. **Haptophytes: Phylum *Haptophyta*; pp. 366–367; moderate**
    Describe the two distinctive features shared by all members of the phylum *Haptophyta*.

11. **Ecology of the Algae, Cryptomonads: Phylum *Cryptophyta*, Dinoflagellates: Phylum *Dinophyta*, Haptophytes: Phylum *Haptophyta*; pp. 348–350, 356, 361–362, 366, 367; difficult**
    Define the term phytoplankton. Which phyla discussed in this chapter contain individuals that are members of the phytoplankton?

12. **Ecology of the Algae, Dinoflagellates: Phylum *Dinophyta*, Haptophytes: Phylum *Haptophyta*; pp. 350, 365, 367; difficult**
    What are algal blooms and why are they ecologically important? Which phyla discussed in this chapter contain individuals that form blooms?

Chapter 17

Protista II: Heterokonts and Green Algae

Multiple-Choice Questions

1. **The Heterokonts; p. 371; easy; ans: c**
    Which of the following is NOT a group of heterokonts?

    a. brown algae
    b. diatoms
    c. green algae
    d. oomycetes
    e. chrysophytes

2. **Oomycetes: Phylum *Oomycota*; p. 371; moderate; ans: b**
    Which of the following statements about oomycetes is FALSE?

    a. They have cell walls of cellulose.
    b. Asexually reproduction occurs by uniflagellate zoospores.
    c. Their filaments resemble hyphae.
    d. Sexual reproduction is oogamous.
    e. Some are unicellular.

3. **Oomycetes: Phylum *Oomycota*; p. 372; moderate; ans: e**
    In the oomycetes, one or more eggs are produced in a(n):

    a. antheridium.
    b. zoospore.
    c. zygospore.
    d. oospore.
    e. oogonium.

4. **Oomycetes: Phylum *Oomycota*; p. 372; moderate; ans: d**
    In the oomycetes, the thick-walled zygote is called a(n):

    a. antheridium.
    b. zoospore.
    c. zygospore.
    d. oospore.
    e. oogonium.

5. **Oomycetes: Phylum *Oomycota*; p. 373; moderate; ans: a**
   _____ is a genus of homothallic water molds.

   a. *Saprolegnia*
   b. *Achlya*
   c. *Plasmopara*
   d. *Pythium*
   e. *Phytophthora*

6. **Oomycetes: Phylum *Oomycota*; p. 373; moderate; ans: b**
   Phytophthora infestans is responsible for:

   a. downy mildew of grapes.
   b. late blight of potatoes.
   c. damping-off diseases.
   d. rotting of avocados.
   e. parasitism of fish.

7. **Oomycetes: Phylum *Oomycota*; p. 374; moderate; ans: d**
   Damping-off diseases are caused by members of the genus:

   a. *Saprolegnia*.
   b. *Achlya*.
   c. *Plasmopara*.
   d. *Pythium*.
   e. *Phytophthora*.

8. **Diatoms: Phylum *Bacillariophyta*; p. 375; moderate; ans: c**
   As much as _____ percent of the primary production on Earth is due to marine planktonic diatoms.

   a. 5
   b. 10
   c. 25
   d. 50
   e. 75

9. **Diatoms: Phylum *Bacillariophyta*; p. 376; easy; ans: c**
   Frustules are the _____ of diatoms.

   a. pigments
   b. zygotes
   c. cell walls
   d. gametes
   e. resting cysts

10. **Diatoms: Phylum *Bacillariophyta*; p. 376; easy; ans: a**
    The cell walls of diatoms are composed of:

    a. silica.
    b. cellulose.
    c. peptidoglycan.
    d. chitin.
    e. calcium carbonate.

11. **Diatoms: Phylum *Bacillariophyta*; p. 376; difficult; ans: c**
In the Phylum *Bacillariophyta*:

    a. sexual reproduction in pennate diatoms is oogamous.
    b. sexual reproduction in centric diatoms is isogamous.
    c. centric diatoms have flagellated male gametes.
    d. pennate diatoms have flagellated male gametes.
    e. centric diatoms have flagellated female gametes.

12. **Diatoms: Phylum *Bacillariophyta*; p. 378; difficult; ans: d**
Which of the following statements about diatom blooms is FALSE?

    a. They occur in the spring and fall.
    b. In oceans, they occur when upwelling takes place.
    c. In lakes, they occur when wind-driven turnover takes place.
    d. They occur when silica becomes depleted.
    e. They may occur beneath the winter ice cover.

13. **Diatoms: Phylum *Bacillariophyta*; p. 378; easy; ans: a**
Diatomaceous earth is formed from the _____ of diatoms.

    a. frustules
    b. cytoplasm
    c. fucoxanthin
    d. chrysolaminarin
    e. plastids

14. **Diatoms: Phylum *Bacillariophyta*; p. 378; easy; ans: b**
The carbohydrate reserve in diatoms is:

    a. floridean starch.
    b. chrysolaminarin.
    c. glycogen.
    d. amylopectin.
    e. fucoxanthin.

15. **Diatoms: Phylum *Bacillariophyta*; p. 378; moderate; ans: c**
Which of the following statements about diatoms is FALSE?

    a. Most species are autotrophic.
    b. Some species lack chlorophyll.
    c. Most heterotrophic diatoms are centric.
    d. Some species are symbionts.
    e. Some species produce neurotoxins.

16. **Chrysophytes: Phylum *Chrysophyta*; p. 378; moderate; ans: d**
Algae containing a gold-colored pigment belong to the phylum:

    a. *Bacillariophyta*.
    b. *Oomycota*.
    c. *Phaeophyta*.
    d. *Chrysophyta*.
    e. *Rhodophyta*.

17. Chrysophytes: Phylum *Chrysophyta*; p. 378; moderate; ans: d
In the phylum _____, *Dinobryon* is a major ingestor of bacteria.

    a. *Bacillariophyta*
    b. *Oomycota*
    c. *Phaeophyta*
    d. *Chrysophyta*
    e. *Rhodophyta*

18. Chrysophytes: Phylum *Chrysophyta*; p. 379; moderate; ans: e
Chrysophytes differ from diatoms in that chrysophytes:

    a. have two flagella of unequal length.
    b. contain the pigment fucoxanthin.
    c. have chrysolaminarin as the carbohydrate food reserve.
    d. may be unicellular or colonial.
    e. have cell walls of cellulose or lack cell walls.

19. Brown Algae: Phylum *Phaeophyta*; pp. 379–382; moderate; ans: d
Which of the following statements about members of the phylum *Phaeophyta* is FALSE?

    a. They are almost entirely marine.
    b. They dominate rocky shores in cooler regions.
    c. Some are rockweeds and kelps.
    d. Their classification is based on thallus structure.
    e. They contain the xanthophyll fucoxanthin.

20. Brown Algae: Phylum *Phaeophyta*; p. 381; moderate; ans: a
A thallus would NOT be a(n):

    a. simple reproductive body.
    b. simple vegetative body.
    c. branched filament.
    d. aggregation of branched filaments.
    e. authentic tissue.

21. Brown Algae: Phylum *Phaeophyta*; p. 381; easy; ans: a
The characteristic brown color of *Phaeophyta* is due to the presence of:

    a. fucoxanthin.
    b. algin.
    c. chlorophyll *a*.
    d. silicon.
    e. symbiotic bacteria.

22. Brown Algae: Phylum *Phaeophyta*; p. 381; moderate; ans: c
The food reserve in brown algae is:

    a. floridean starch.
    b. chrysolaminarin.
    c. laminarin.
    d. glycogen.
    e. lipid.

23. Brown Algae: Phylum *Phaeophyta*; p. 382; easy; ans: e
Large kelps are differentiated into three regions:

a. filament, gametangia, and sporangia.
b. pseudoparenchyma, unilocular gametangia, and plurilocular gametangia.
c. gametophyte, sporophyte, and gametes.
d. blade, holdfast, and meristem.
e. blade, holdfast, and stipe.

24. Brown Algae: Phylum *Phaeophyta*; p. 382; easy; ans: a
_____ is a mucilaginous component of kelps that is used as an emulsifier for some foods.

a. Algin
b. Mannitol
c. Laminarin
d. Fucoxanthin
e. Agar

25. Brown Algae: Phylum *Phaeophyta*; p. 382; easy; ans: b
In kelps, carbohydrate is transported through food-conducting cells primarily as:

a. algin.
b. mannitol.
c. laminarin.
d. fucoxanthin.
e. sucrose.

26. Brown Algae: Phylum *Phaeophyta*; p. 383; difficult; ans: d
In the *Ectocarpus* life cycle, meiosis occurs within:

a. antheridia.
b. unilocular gametangia.
c. plurilocular gametangia.
d. unilocular sporangia.
e. plurilocular sporangia.

27. Brown Algae: Phylum *Phaeophyta*; p. 383; moderate; ans: c
Three defining characteristics of the brown algae are:

a. plurilocular sporangia, fucoxanthin, and laminarin.
b. plurilocular sporangia, unilocular sporangia, and fucoxanthin.
c. unilocular sporangia, algin, and plasmodesmata.
d. unilocular sporangia, mannitol, and plasmodesmata.
e. mannitol, fucoxanthin, and laminarin.

28. Green Algae: Phylum *Chlorophyta*; p. 383; difficult; ans: b
Green algae differ from plants in that the green algae:

a. store carbohydrate as starch.
b. include unicellular and colonial forms.
c. store their food reserves inside plastids.
d. do not have cellulose-containing cell walls.
e. have flagellated reproductive cells.

29. **Green Algae: Phylum *Chlorophyta*; p. 386; moderate; ans: e**
Cytokinesis in the *Chlorophyceae* involves a unique structure called a:

   a. phragmoplast.
   b. cleavage furrow.
   c. flagellar root.
   d. phycosome.
   e. phycoplast.

30. **Green Algae: Phylum *Chlorophyta*; pp. 388–389; easy; ans: a**
Which of the following statements about *Chlamydomonas* is FALSE?

   a. The diploid phase is the dominant phase in its life cycle.
   b. It has two flagella of equal length.
   c. It undergoes zygotic meiosis.
   d. It has no cellulose in its cell wall.
   e. Its chloroplast contains a pyrenoid.

31. **Green Algae: Phylum *Chlorophyta*; p. 389; moderate; ans: c**
_____ is a motile, colonial member of the *Chlorophyceae*.

   a. *Hydrodictyon*
   b. *Halimeda*
   c. *Volvox*
   d. *Oedogonium*
   e. *Chlorella*

32. **Green Algae: Phylum *Chlorophyta*; p. 390; difficult; ans: b**
Which of the following statements about *Volvox* is FALSE?

   a. It consists of a hollow sphere of cells.
   b. Vegetative cells are nonflagellated.
   c. Reproductive cells divide by mitosis to form juvenile spheroids.
   d. Sexual reproduction is oogamous.
   e. Male colonies may produce inducer molecules.

33. **Green Algae: Phylum *Chlorophyta*; p. 391; moderate; ans: a**
_____ is a genus of *Chlorophyceae*

   a. *Hydrodictyon*
   b. *Stigeoclonium*
   c. *Volvox*
   d. *Fritschiella*
   e. *Chlamydomonas*

34. **Green Algae; Phylum *Chlorophyta*; p. 390; moderate; ans: c**
_____ is a very common alga found in soils.

   a. *Hydrodictyon*
   b. *Chlorella*
   c. *Chlorococcum*
   d. *Oedogonium*
   e. *Volvox*

35. Green Algae: Phylum *Chlorophyta*; p. 392; moderate; ans: e
Oedogonium is best described as a(n) _____ green alga.

    a.   motile colonial
    b.   branched filamentous
    c.   parenchymatous
    d.   motile unicellular
    e.   unbranched filamentous

36. Green Algae; Phylum *Chlorophyta*; p. 392; moderate; ans: d
What is so unusual about individuals belonging to the genus *Oedogonium*?

    a.   They have flagellated sperm.
    b.   Meiosis is zygotic.
    c.   Sexual reproduction is oogamous.
    d.   Cell division results in annular scars.
    e.   They produce zoospores.

37. Green Algae: Phylum *Chlorophyta*; p. 392; moderate; ans: d
_____ is a parenchymatous member of the *Chlorophyceae*.

    a.   *Volvox*
    b.   *Cladophora*
    c.   *Ulva*
    d.   *Fritschiella*
    e.   *Acetabularia*

38. Green Algae; Phylum *Chlorophyta*; p. 392; moderate; ans: e
_____ grows on tree trunks, moist walls, and leaf surfaces.

    a.   *Volvox*
    b.   *Oedogonium*
    c.   *Stigeoclonium*
    d.   *Chlorococcum*
    e.   *Fritschiella*

39. Green Algae: Phylum *Chlorophyta*; p. 393; moderate; ans: d
Which of the following is a filamentous member of the class *Ulvophyceae* that
has multinucleate septate cells?

    a.   *Hydrodictyon*
    b.   *Halimeda*
    c.   *Volvox*
    d.   *Cladophora*
    e.   *Chara*

40. Green Algae; Phylum *Chlorophyta*; p. 394; moderate; ans: b
Siphonous marine algae are characterized by:

    a.   short, unbranched filaments.
    b.   large, branched coenocytic cells.
    c.   a parenchymatous structure.
    d.   "caps" or annual scars.
    e.   large motile colonies.

41. Green Algae: Phylum *Chlorophyta*; pp. 394–395; difficult; ans: c
Halimeda is best described as a(n) _____ green alga.

   a. motile colonial
   b. nonmotile colonial
   c. coenocytic filamentous
   d. motile unicellular
   e. unbranched filamentous

42. Green Algae; Phylum *Chlorophyta*; p. 395; moderate; ans: a
Chloroplasts of _____ line the respiratory chamber of sea slugs.

   a. *Codium*
   b. *Cladophora*
   c. *Stigeoclonium*
   d. *Chlorococcum*
   e. *Oedogonium*

43. Green Algae: Phylum *Chlorophyta*; pp. 393, 396; difficult; ans: d
Which group of green algae contains mostly marine organisms whose flagellated cells are asymmetrical with only two flagella?

   a. *Charales*
   b. *Coleochaetales*
   c. *Chlorophyceae*
   d. *Charophyceae*
   e. *Ulvophyceae*

44. Green Algae: Phylum *Chlorophyta*; p. 396; difficult; ans: e
_____ is a member of the class *Charophyceae* that has an unbranched filamentous growth form.

   a. *Halimeda*
   b. *Ulva*
   c. *Chara*
   d. *Oedogonium*
   e. *Spirogyra*

45. Green Algae; Phylum *Chlorophyta*; pp. 396–397; moderate; ans: c
Which of the following structures is NOT associated with sexual reproduction in *Spirogyra*?

   a. isogametes
   b. conjugation tube
   c. flagellated reproductive cells
   d. sporopollenin
   e. zygotic meiosis

46. Green Algae: Phylum *Chlorophyta*; p. 397; moderate; ans: c
Desmids belong to the:

   a. order *Coleochaetales*.
   b. order *Charales*.
   c. class *Charophyceae*.
   d. class *Ulvophyceae*.
   e. class *Chlorophyceae*.

47. Green Algae: Phylum *Chlorophyta*; p. 397; easy; ans: a
Which two genera of green algae contain members that most closely resemble plants?

 a. *Chara* and *Coleochaete*
 b. *Spirogyra* and *Micrasterias*
 c. *Halimeda* and *Codium*
 d. *Fritschiella* and *Stigeoclonium*
 e. *Acetabularia* and *Oedogonium*

48. Green Algae; Phylum *Chlorophyta*; p. 398; difficult; ans: c
Which of the following features is NOT shared by *Chara* and certain plants?

 a. apical growth
 b. tissue organization in the nodal regions
 c. flagellated cells other than sperm
 d. pattern of plasmodesmatal connections
 e. sporopollenin

True-False Questions

1. The Heterokonts, p. 371; moderate; ans: T
The diatoms make up one of the two lineages of pigmented heterokonts.

2. Oomycetes: Phylum *Oomycota*; p. 373; moderate; ans: F
All oomycetes are parasitic.

3. Oomycetes: Phylum *Oomycota*; p. 373; easy; ans: T
The Bordeaux mixture, used on French vineyards, was the first chemical used in the control of a plant disease.

4. Diatoms: Phylum *Bacillariophyta*; p. 375; moderate; ans: T
Approximately one-quarter of the Earth's total primary production is carried out by diatoms.

5. Diatoms: Phylum *Bacillariophyta*; p. 376; easy; ans: F
The cell walls of diatoms are composed of three siliceous valves.

6. Diatoms: Phylum *Bacillariophyta*; p. 376; easy; ans: F
Pennate diatoms are radially symmetrical.

7. Diatoms: Phylum *Bacillariophyta*; p. 378; moderate; ans: T
Phylum *Bacillariophyta* includes autotrophs and heterotrophs.

8. Chrysophytes: Phylum *Chrysophyta*; p. 379; moderate; ans: F
Most chrysophytes reproduce sexually to form silica-containing resting cysts.

9. Brown Algae: Phylum *Phaeophyta*; p. 381; moderate; ans: T
The thalli of brown algae may consist of tissuelike pseudoparenchyma.

10. Brown Algae: Phylum *Phaeophyta*; p. 382. easy; ans: T
Macrocystis is a brown alga that is harvested commercially.

11. **Brown Algae: Phylum *Phaeophyta*; p. 382; moderate; ans: F**
Some kelps have food-conducting systems consisting of phloem cells.

12. **Brown Algae: Phylum *Phaeophyta*; p. 383; moderate; ans: F**
The life cycles of most brown algae involve gametic meiosis.

13. **Brown Algae: Phylum *Phaeophyta*; p. 383; difficult; ans: T**
In the *Ectocarpus* life cycle, the sporophyte produces both plurilocular and unilocular sporangia.

14. **Brown Algae: Phylum *Phaeophyta*; p. 383; moderate; ans: F**
Fucus has a sporic life cycle with sporic meiosis.

15. **Green Algae: Phylum *Chlorophyta*; p. 386; moderate; ans: F**
In the *Chlorophyceae*, the cross-shaped pattern of microtubules at the base of a flagellum is called a phycoplast.

16. **Green Algae: Phylum *Chlorophyta*; pp. 388–389; moderate; ans: T**
In *Chlamydomonas*, sexual reproduction involves formation of a conjugation tube between different mating types.

17. **Green Algae: Phylum *Chlorophyta*; p. 392; moderate; ans: F**
Oedogonium is unique among the algae in using chemicals to attract sperm to eggs.

18. **Green Algae: Phylum *Chlorophyta*; p. 396; moderate; ans: T**
Spirogyra has ribbon-like chloroplasts arranged in a helical pattern.

19. **Green Algae: Phylum *Chlorophyta*; pp. 394–395; moderate; ans: T**
Codium, *Halimeda*, and other siphonous green algae produce cells walls only during their reproductive phase.

20. **Green Algae: Phylum *Chlorophyta*; p. 397; difficult; ans: T**
Desmids are members of the *Ulvophyceae* with cells consisting of two sections joined by an isthmus.

21. **Green Algae: Phylum *Chlorophyta*; p. 398; difficult; ans: T**
The stoneworts are green algae that produce flagellated sperm in complex multicellular antheridia.

Essay Questions

1. **Oomycetes: Phylum *Oomycota*; pp. 373–374; moderate**
Discuss the role of the oomycetes in plant diseases.

2. **Diatoms: Phylum *Bacillariophyta*; pp. 375–378; moderate**
Discuss the ecological importance of diatoms.

3. **Diatoms: Phylum *Bacillariophyta*; p. 376; moderate**
How does the unique nature of the diatom cell wall affect cell division, generation after generation?

4. **Chrysophytes: Phylum *Chrysophyta*; pp. 378–379; moderate**
Discuss some of the feeding habits of chrysophytes.

5. **Brown Algae: Phylum *Phaeophyta*; pp. 380, 382–383; moderate**
Discuss some of the commercial uses of the brown algae.

6. **Brown Algae: Phylum *Phaeophyta*; p. 383; difficult**
Describe the formation of the male and female gametangia of *Ectocarpus*.

7. **Green Algae: Phylum *Chlorophyta*; pp. 386–387; moderate**
Describe the events that occur during cytokinesis in the *Chlorophyceae*.

8. **Green Algae: Phylum *Chlorophyta*; pp. 388–389; moderate**
Describe the sexual and asexual life cycles of *Chlamydomonas*.

9. **Green Algae: Phylum *Chlorophyta*; pp. 389–391; moderate**
Using *Volvox* and *Hydrodictyon* as examples, describe the colonial *Chlorophyceae*.

10. **Green Algae: Phylum *Chlorophyta*; pp. 393–395; moderate**
Using *Chladophora*, *Ulva*, and *Codium* as examples, describe the different growth habits of ulvophytes.

11. **Green Algae: Phylum *Chlorophyta*; pp. 397–398; moderate**
Describe the two orders of *Charophyceae* that most closely resemble plants.

Chapter 18

Bryophytes

Multiple-Choice Questions

1. **The Relationships of Bryophytes to Other Groups; p. 401;**
    moderate; ans: c
    Bryophytes are a group of organisms at the transition between:
    a. brown algae and green algae.
    b. fungi and plants.
    c. green algae and vascular plants.
    d. nonvascular and vascular plants.
    e. aquatic and terrestrial plants.

2. **The Relationships of Bryophytes to Other Groups; p. 402;**
    moderate; ans: d
    Which of the following is NOT a characteristic shared by bryophytes and vascular plants?
    a. multicellular sporangia
    b. sporopollenin in spore walls
    c. cuticle on aerial parts
    d. retention of female gametophyte within sporophyte
    e. antheridia and archegonia with sterile jacket layers

3. **Comparative Structure and Reproduction of Bryophytes; p. 403;**
    moderate; ans: d
    Which of the following statements about bryophytes is FALSE?
    a. They may be thalloid or leafy.
    b. Some species contain strands with conducting functions.
    c. They do not have true stems and leaves.
    d. They have rhizoids that absorb water and nutrients.
    e. Some species have plasmodesmata with a desmotubule.

4. **Comparative Structure and Reproduction of Bryophytes; p. 404; difficult; ans: a**
    Which of the following is found in many bryophytes, vascular plants, and charophytes?

    a. a single large plastid in some cells
    b. many small, disk-shaped plastids per cell
    c. many large, disk-shaped plastids per cell
    d. preprophase bands
    e. desmotubules

5. **Comparative Structure and Reproduction of Bryophytes; p. 404; easy; ans: e**
    Gemmae are multicellular structures involved in:

    a. protecting the young embryo.
    b. anchoring the plant to the soil.
    c. water and nutrient conduction.
    d. sexual reproduction.
    e. asexual reproduction.

6. **Comparative Structure and Reproduction of Bryophytes; pp. 404, 406; difficult; ans: d**
    Which of the following statements about sexual reproduction in bryophytes is FALSE?

    a. Sperm are the only flagellated cells produced by bryophytes.
    b. The zygote is matrotrophic.
    c. Sex is governed by the distribution of sex chromosomes.
    d. In the antheridium, the spermatogenous cells are "sterile."
    e. Chemicals released from the archegonium attract sperm.

7. **Comparative Structure and Reproduction of Bryophytes; p. 404; easy; ans: c**
    Which of the following is a flagellated cell in bryophytes?

    a. zoospore
    b. egg
    c. sperm
    d. gemma
    e. zygote

8. **Comparative Structure and Reproduction of Bryophytes; p. 404; moderate; ans: d**
    The base of the bryophyte archegonium, called the _____, contains _____.

    a. placenta; a single egg
    b. capsule; spermatogenous cells
    c. calyptra; several eggs
    d. venter; a single egg
    e. seta; spermatogenous cells

9. **Comparative Structure and Reproduction of Bryophytes; p. 404; difficult; ans: b**
   Matrotrophy refers to the:
   a. attraction of sperm by the egg.
   b. nourishment of the zygote by the archegonium.
   c. transport of sugars through the placenta.
   d. division of the zygote within the venter.
   e. movement of sugars through plasmodesmata.

10. **Comparative Structure and Reproduction of Bryophytes; p. 404; easy; ans: e**
    In bryophytes, fertilization takes places in the:
    a. open water.
    b. capsule.
    c. seta.
    d. antheridium.
    e. archegonium.

11. **Comparative Structure and Reproduction of Bryophytes; p. 406; moderate; ans: c**
    Embryophytes are characterized by having:
    a. few mitotic divisions between fertilization and meiosis.
    b. a multicellular embryo that has stomata.
    c. a multicellular, matrotrophic embryo.
    d. a dominant gametophyte generation.
    e. vascular tissues.

12. **Comparative Structure and Reproduction of Bryophytes; p. 406; easy; ans: d**
    At maturity, the sporophyte of most bryophytes consists of the following structures:
    a. foot and seta only.
    b. seta and capsule only.
    c. capsule and calyptra only.
    d. foot, seta, and capsule.
    e. seta, capsule, and calyptra.

13. **Comparative Structure and Reproduction of Bryophytes; p. 406; moderate; ans: d**
    Which of the following statements about stomata is FALSE?
    a. Each stoma is bordered by two guard cells.
    b. They aid in the uptake of CO_2.
    c. They generate a flow of water between sporophyte and gametophyte.
    d. They occur in all bryophytes.
    e. They are sites of loss of water vapor.

14. Comparative Structure and Reproduction of Bryophytes; pp. 406–407; easy; ans: b

A major difference between the spore walls of bryophytes and charophytes is that the bryophyte spore walls contain:

a. lignin.
b. sporopollenin.
c. cellulose.
d. phenolic materials.
e. stomata.

15. Comparative Structure and Reproduction of Bryophytes; p. 407; moderate; ans: d

Which of the following bryophyte structures is haploid?

a. foot
b. seta
c. embryo
d. protonema
e. capsule

16. Comparative Structure and Reproduction of Bryophytes; p. 407; moderate; ans: a

Takakia is now classified as a:

a. moss.
b. hornwort.
c. leafy liverwort.
d. thalloid liverwort.
e. vascular plant.

17. Liverworts: Phylum *Hepatophyta*; p. 408; moderate; ans: b

_____ is a liverwort that carries its gametangia on gametophores.

a. *Frullania*
b. *Marchantia*
c. *Anthoceros*
d. *Riccia*
e. *Ricciocarpus*

18. Liverworts: Phylum *Hepatophyta*; p. 408; moderate; ans: d

In *Marchantia*, the mature capsule contains:

a. spores only.
b. elaters only.
c. gemmae only.
d. spores and elaters only.
e. spores, elaters, and gemmae.

19. Liverworts: Phylum *Hepatophyta*; p. 408; moderate; ans: a

The function of elaters is to:

a. help disperse spores.
b. help disperse gemmae.
c. strengthen the spore wall.
d. conduct water and nutrients through the plant body.
e. support the archegoniophores.

20. Liverworts: Phylum *Hepatophyta*; p. 412; difficult; ans: e
Leaves of leafy liverworts differ from those of mosses in that liverwort leaves:

   a.   are of equal size.
   b.   are spirally arranged.
   c.   have a thickened midrib.
   d.   are entire rather than lobed or dissected.
   e.   are arranged in two rows with a third row of smaller leaves.

21. Liverworts: Phylum *Hepatophyta*; p. 412; moderate; ans: a
An example of a leafy liverwort is:

   a.   *Frullania*.
   b.   *Marchantia*.
   c.   *Anthoceros*.
   d.   *Riccia*.
   e.   *Ricciocarpus*.

22. Liverworts: Phylum *Hepatophyta*; p. 412; moderate; ans: c
In the liverworts, an androecium is a:

   a.   structure producing sperm.
   b.   group of water-conducting cells.
   c.   short side branch bearing antheridia.
   d.   tubular sheath surrounding the archegonium.
   e.   structure of the gemma cup.

23. Liverworts: Phylum *Hepatophyta*; p. 412; moderate; ans: d
A perianth is characteristically found in the:

   a.   thalloid liverworts.
   b.   hornworts.
   c.   mosses.
   d.   leafy liverworts.
   e.   genus *Takakia*.

24. Hornworts: Phylum *Anthocerophyta*; p. 412; easy; ans: b
Anthoceros is an example of a:

   a.   moss.
   b.   hornwort.
   c.   leafy liverwort.
   d.   thalloid liverwort.
   e.   vascular plant.

25. Mosses: Phylum *Bryophyta*; p. 412; easy; ans: c
Which of the following is NOT a group of mosses of phylum *Bryophyta*?

   a.   peat mosses
   b.   granite mosses
   c.   club mosses
   d.   *Bryidae*
   e.   *Andreaeidae*

26. **Mosses: Phylum *Bryophyta*; p. 414; easy; ans: a**
Peat mosses belong to the phylum _____, class _____.

    a.   *Bryophyta*; *Sphagnidae*
    b.   *Bryophyta*; *Andreaeidae*
    c.   *Bryophyta*; *Bryidae*
    d.   *Hepatophyta*; *Sphagnidae*
    e.   *Hepatophyta*; *Bryidae*

27. **Mosses: Phylum *Bryophyta*; p. 414; easy; ans: b**
The genus _____ is characterized by capsules raised on a pseudopodium.

    a.   *Andreaea*
    b.   *Sphagnum*
    c.   *Polytrichum*
    d.   *Anthoceros*
    e.   *Ricciocarpus*

28. **Mosses: Phylum *Bryophyta*; p. 415; moderate; ans: d**
The protonema of *Sphagnum* consists of _____ of cells.

    a.   a single row
    b.   two or more rows
    c.   a two-layer-thick plate
    d.   a one-layer-thick plate
    e.   two overlapping plates

29. **Mosses: Phylum *Bryophyta*; p. 416; easy; ans: b**
The granite mosses belong to the phylum _____, class _____.

    a.   *Bryophyta*; *Sphagnidae*
    b.   *Bryophyta*; *Andreaeidae*
    c.   *Bryophyta*; *Bryidae*
    d.   *Hepatophyta*; *Sphagnidae*
    e.   *Hepatophyta*; *Bryidae*

30. **Mosses: Phylum *Bryophyta*; p. 416; moderate; ans: b**
The protonema of *Andreaea* consists of _____ of cells.

    a.   a single row
    b.   two or more rows
    c.   a two-layer-thick plate
    d.   a one-layer-thick plate
    e.   two overlapping plates

31. **Mosses: Phylum *Bryophyta*; p. 416; easy; ans: c**
The "true mosses" belong to the phylum _____, class _____.

    a.   *Bryophyta*; *Sphagnidae*
    b.   *Bryophyta*; *Andreaeidae*
    c.   *Bryophyta*; *Bryidae*
    d.   *Hepatophyta*; *Sphagnidae*
    e.   *Hepatophyta*; *Bryidae*

32. Mosses: Phylum *Bryophyta*; pp. 416, 420; moderate; ans: c
Which of the following genera has hadrom and leptom?

   a. *Andreaea*
   b. *Sphagnum*
   c. *Polytrichum*
   d. *Anthoceros*
   e. *Ricciocarpus*

33. Mosses: Phylum *Bryophyta*; p. 416; easy; ans: c
Hadrom contains:

   a. gametes.
   b. food-conducting cells.
   c. water-conducting cells.
   d. spores.
   e. degenerate nuclei.

34. Mosses: Phylum *Bryophyta*; p. 416; easy; ans: d
What is the function of hydroids?

   a. anchoring the gametophyte
   b. photosynthesis
   c. aiding in spore dispersal
   d. conducting water
   e. conducting food

35. Mosses: Phylum *Bryophyta*; p. 421; moderate; ans: c
A peristome is a capsular structure characteristic of members of the phylum
_____, class _____.

   a. *Bryophyta*; *Sphagnidae*
   b. *Bryophyta*; *Andreaeidae*
   c. *Bryophyta*; *Bryidae*
   d. *Hepatophyta*; *Sphagnidae*
   e. *Hepatophyta*; *Bryidae*

True-False Questions

1. The Relationships of Bryophytes to Other Groups; p. 402; easy; ans: T
Bryophytes were the first extant plant group to diverge from a monophyletic plant lineage.

2. Comparative Structure and Reproduction of Bryophytes; p. 404; easy; ans: F
All bryophytes contain only a single, large plastid in each cell.

3. Comparative Structure and Reproduction of Bryophytes; p. 404; moderate; ans: T
Sperm are the only flagellated cells in bryophytes.

4. Comparative Structure and Reproduction of Bryophytes; p. 404; easy; ans: T
In bryophytes, sperm must swim through water to reach the egg.

5. **Comparative Structure and Reproduction of Bryophytes; p. 406; moderate; ans: F**
The venter of a bryophyte develops into a capsule.

6. **Comparative Structure and Reproduction of Bryophytes; p. 406; moderate; ans: T**
A moss is an embryophyte.

7. **Liverworts: Phylum *Hepatophyta*; p. 408; moderate; ans: F**
Antheridiophores and archegoniophores are distinctive features of *Riccia* and *Ricciocarpus*.

8. **Liverworts: Phylum Hepatophyta; p. 408; moderate; ans: T**
Elaters are structures that aid in spore dispersal.

9. **Liverworts: Phylum *Hepatophyta*; p. 412; moderate; ans: T**
In liverworts, a perianth is a tubular sheath that surrounds the developing sporophyte.

10. **Hornworts: Phylum *Anthocerophyta*; p. 412; moderate; ans: T**
Despite a superficial resemblance between their gametophytes, hornworts and thallose liverworts are only distantly related.

11. **Hornworts: Phylum *Anthocerophyta*; p. 412; moderate; ans: T**
Some hornwort species have a symbiotic relationship with nitrogen-fixing bacteria.

12. **Mosses: Phylum *Bryophyta*; p. 415; easy; ans: T**
Sphagnum has antiseptic properties.

13. **Mosses: Phylum *Bryophyta*; p. 415; easy; ans: F**
Peat mosses normally release hydroxide ions, thereby raising the pH of their environment.

14. **Mosses: Phylum *Bryophyta*; p. 416; moderate; ans: T**
Leptoids are somewhat similar to a type of food-conducting cells in vascular plants.

15. **Mosses: Phylum *Bryophyta*; p. 421; moderate; ans: F**
In the true mosses, the gametophyte carries out photosynthesis but the sporophyte does not.

16. **Mosses: Phylum *Bryophyta*; p. 421; moderate; ans: F**
The "cushiony" growth habit is characteristic of epiphytic mosses.

Essay Questions

1. **The Relationships of Bryophytes to Other Groups; pp. 401–402; moderate**
What features do bryophytes share with the green algae? What features do they share with vascular plants?

2. **Comparative Structure and Reproduction of Bryophytes; p. 404; moderate**
   Describe the sexual reproductive structures of bryophytes.

3. **Comparative Structure and Reproduction of Bryophytes; pp. 404, 406; difficult**
   What is a placenta? What is its role in plant development? What are the advantages of matrotrophy in the evolution of plants?

4. **Comparative Structure and Reproduction of Bryophytes; pp. 406–407; moderate**
   Discuss the importance of sporopollenin for bryophytes.

5. **Liverworts: Phylum *Hepatophyta*; pp. 407–412; moderate**
   Discuss some of the features that distinguish phylum *Anthocerophyta* from phylum *Hepatophyta*.

6. **Liverworts: Phylum *Hepatophyta*; Hornworts: Phylum *Anthocerophyta*; pp. 407–412; moderate**
   Compare and contrast the complex thalloid liverworts, the leafy liverworts, and the hornworts.

7. **Liverworts: Phylum *Hepatophyta*; p. 412; moderate**
   How would you distinguish a leafy liverwort from a moss?

8. **Mosses: Phylum *Bryophyta*; p. 415; moderate**
   In what ways is *Sphagnum* ecologically important?

9. **Mosses: Phylum *Bryophyta*; pp. 420–421; moderate**
   Describe the life cycle of a true moss.

10. **Mosses: Phylum *Bryophyta*; p. 421; moderate**
    Explain the differences between the "cushiony" and "feathery" growth patterns of mosses.

Chapter 19

Seedless Vascular Plants

Multiple-Choice Questions

1. **Evolution of Vascular Plants; p. 425; moderate; ans: c**
   Which of the following is NOT a characteristic shared by bryophytes and vascular plants?

   a. multicellular embryos
   b. a *Coleochaete*-like ancestor
   c. a monophyletic lineage
   d. dominant gametophytes
   e. an alternation of heteromorphic generations

2. **Evolution of Vascular Plants; p. 425; difficult; ans: d**
   In the evolution of vascular plants, there is a trend toward the:

   a. above-ground parts becoming structurally similar to the below-ground parts.
   b. progressive reduction of the sporophyte.
   c. sporophyte becoming nutritionally dependent on the gametophyte.
   d. increased protection of the gametophyte by the sporophyte.
   e. production of seeds in all lineages.

3. **Organization of the Vascular Plant Body; p. 426; easy; ans: d**
   The main tissue systems of the vascular plant are the _____ systems.

   a. root and shoot
   b. root, shoot, and reproductive
   c. root, stem, and leaf
   d. dermal, vascular, and ground
   e. xylem, phloem, and ground

4. **Organization of the Vascular Plant Body; pp. 426–427; moderate; ans: c**
   Which of the following statements about primary growth is FALSE?

   a. It occurs close to the tips of stems and roots.
   b. It is initiated by the apical meristems.
   c. It primarily leads to thickening of the plant body.
   d. It gives rise to primary tissues.
   e. It gives rise to the primary plant body.

5. **Organization of the Vascular Plant Body; p. 427; moderate; ans: c**
   Which of the following statements about lateral meristems is FALSE?

   a.   Their activity leads to thickening of the stem and root.
   b.   Their activity leads to secondary growth.
   c.   An example is the periderm.
   d.   An example is the vascular cambium.
   e.   They give rise to the secondary plant body.

6. **Organization of the Vascular Plant Body; p. 427; easy; ans: d**
   The conducting cells of the phloem are called:

   a.   tracheids.
   b.   vessel elements.
   c.   leaf traces.
   d.   sieve elements.
   e.   hydroids.

7. **Organization of the Vascular Plant Body; p. 428; moderate; ans: b**
   Tracheids differ from vessel elements in that tracheids:

   a.   have lignified thickenings.
   b.   are less specialized cells.
   c.   are a type of tracheary element.
   d.   conduct water and minerals.
   e.   provide support.

8. **Organization of the Vascular Plant Body; p. 428; moderate; ans: a**
   What do ALL steles have in common?

   a.   primary xylem and primary phloem
   b.   a pith
   c.   leaf gaps
   d.   dermal tissue
   e.   vascular and cork cambia

9. **Organization of the Vascular Plant Body; p. 429; moderate; ans: e**
   The siphonostele of ferns:

   a.   is the most primitive type of stele.
   b.   consists of a solid core of vascular tissues.
   c.   occurs only in the roots.
   d.   has a series of discrete strands around a central pith.
   e.   has leaf gaps.

10. **Organization of the Vascular Plant Body; p. 429; moderate; ans: c**
    Microphylls differ from megaphylls in that microphylls:

    a.   occur in most vascular plants.
    b.   are associated with leaf gaps.
    c.   are associated with protosteles.
    d.   have branched veins.
    e.   evolved from branch systems.

11. **Organization of the Vascular Plant Body; p. 429; moderate; ans: d**
Megaphylls are leaves that:

    a.   contain a single strand of vascular tissue.
    b.   evolved from enations.
    c.   are not associated with leaf gaps.
    d.   are associated with siphonosteles and eusteles.
    e.   characterize the lycophytes.

12. **Organization of the Vascular Plant Body; p. 430; moderate; ans: b**
Which sequence of events most likely describes the evolution of megaphylls?

    a.   overtopping, dichotomous branching, planation, webbing
    b.   dichotomous branching, overtopping, planation, webbing
    c.   planation, dichotomous branching, overtopping, webbing
    d.   webbing, dichotomous branching, overtopping, planation
    e.   overtopping, planation, webbing, dichotomous branching

13. **Reproductive Systems; p. 430; moderate; ans: b**
Which of the following statements about reproduction in vascular plants is FALSE?

    a.   The eggs are nonmotile.
    b.   The gametophyte is structurally more complex than the sporophyte.
    c.   All vascular plants are oogamous.
    d.   There is an alternation of heteromorphic generations.
    e.   The sporophyte is the dominant phase of the life cycle.

14. **Reproductive Systems; p. 430; moderate; ans: e**
Homosporous plants:

    a.   produce two kinds of spores.
    b.   produce two kinds of gametophytes.
    c.   are evolutionarily more advanced than heterosporous plants.
    d.   produce antheridia but not archegonia.
    e.   can produce bisexual gametophytes.

15. **Reproductive Systems; p. 431; moderate; ans: a**
Heterospory differs from homospory in that heterospory involves:

    a.   gametophytes with endosporic development.
    b.   larger gametophytes.
    c.   bisexual gametophytes.
    d.   spores differentiated on the basis of size not function.
    e.   different types of spores produced in the same sporangium.

16. **Reproductive Systems; p. 431; moderate; ans: d**
Which of the following is NOT an evolutionary trend in the vascular plants?

    a.   nutritional dependency of the gametophyte on the sporophyte
    b.   reduction in size of the gametophyte
    c.   reduction in complexity of the gametophyte
    d.   increased prominence of antheridia and archegonia
    e.   decreased reliance on water for transferring sperm to egg

17. **The Phyla of Seedless Vascular Plants; pp. 431–432; moderate; ans: c**
The four major groups of vascular plants are as follows:
I.   ferns, lycophytes, sphenophytes, and progymnosperms
II.   flowering plants
III. rhyniophytes, zosterophyllophytes, and trimerophytes
IV.   gymnosperms
Which of the following is the correct sequence—from earliest to most recent—of the time period in which they were dominant on Earth?

  a.   I, II, III, IV
  b.   I, III, IV, II
  c.   III, I, IV, II
  d.   III, IV, II, I
  e.   IV, III, I, II

18. **Phylum *Rhyniophyta*; p. 434; difficult; ans: d**
Which of the following statements about protracheophytes is FALSE?

  a.   An example is *Aglaophyton major*.
  b.   They have multiple sporangia.
  c.   They have branched axes.
  d.   They have tracheids.
  e.   They represent an intermediate stage in the evolution of vascular plants.

19. **Phylum *Rhyniophyta*; p. 434; moderate; ans: b**
Cooksonia, a member of the phylum _____, is the oldest known _____.

  a.   *Psilotophyta*; fern
  b.   *Rhyniophyta*; vascular plant
  c.   *Rhyniophyta*; protracheophyte
  d.   *Zosterophyllophyta*; vascular plant
  e.   *Zosterophyllophyta*; protracheophyte

20. **Phylum *Zosterophyllophyta*; p. 435; difficult; ans: e**
The *Zosterophyllophyta* differ from the *Rhyniophyta* in that the *Zosterophyllophyta*:

  a.   were homosporous.
  b.   were not differentiated into roots, stems, and leaves.
  c.   had a protostele.
  d.   became extinct.
  e.   produced lateral sporangia.

21. **Phylum *Lycophyta*; p. 435; moderate; ans: c**
Microphylls are highly characteristic of the phylum:

  a.   *Trimerophytophyta*.
  b.   *Zosterophyllophyta*.
  c.   *Lycophyta*.
  d.   *Rhyniophyta*.
  e.   *Psilotophyta*.

22. **Phylum *Lycophyta*; p. 438; moderate; ans: e**
Which of the following statements about sporophylls in the *Lycopodiaceae* is FALSE?

    a. They are modified leaves.
    b. They are borne on the gametophyte.
    c. They bear sporangia.
    d. They may be interspersed among sterile microphylls.
    e. They may be grouped into strobili.

23. **Phylum *Lycophyta*; p. 438; moderate; ans: b**
In the club moss life cycle:

    a. the gametophytes are unisexual.
    b. a gametophyte may produce a series of sporophytes.
    c. water is not required for fertilization.
    d. microphylls but not strobili are formed.
    e. the sporophyte does not usually become independent of the gametophyte.

24. **Phylum *Lycophyta*; pp. 442, 443; moderate; ans: d**
Lycopodium differs from *Selaginella* and *Isoetes* in that *Lycopodium*:

    a. is differentiated into roots, stems, and leaves.
    b. has sporophylls.
    c. is heterosporous.
    d. lacks ligules.
    e. has a protostele.

25. **Phylum *Lycophyta*; p. 442; moderate; ans: b**
In the *Selaginella* life cycle, the archegonia:

    a. protrude through the megasporophyll.
    b. protrude through a rupture in the megaspore wall.
    c. develop on the microsporophyll.
    d. develop from the suspensor.
    e. cause the microsporangial wall to rupture.

26. **Phylum *Lycophyta*; pp. 442, 443; moderate; ans: b**
In contrast to the *Lycopodiceae*, both the *Selaginellaceae* and the *Isoetaceae*:

    a. lack ligules.
    b. are heterosporous.
    c. have megaphylls.
    d. are vascular plants.
    e. produce bisexual gametophytes.

27. **Phylum *Lycophyta*; pp. 442–443; moderate; ans: a**
Which of the following statements about quillworts is FALSE?

    a. They have a short, fleshy aboveground stem.
    b. They have a specialized cambium.
    c. Some obtain carbon for photosynthesis from sediments.
    d. Some have CAM photosynthesis.
    e. Each leaf is a potential sporophyll.

28. Phylum *Trimerophytophyta*; p. 443; moderate; ans: c

Members of the phylum _____ probably evolved directly from the rhyniophytes and most likely are the ancestors of the ferns and progymnosperms.

a. *Psilotophyta*
b. *Lycophyta*
c. *Trimerophytophyta*
d. *Zosterophyllophyta*
e. *Sphenophyta*

29. Phylum *Psilotophyta*; pp. 443–445; easy; ans: a

Living members of the phylum _____ are homosporous species that lack roots and leaves.

a. *Psilotophyta*
b. *Rhyniophyta*
c. *Trimerophytophyta*
d. *Zosterophyllophyta*
e. *Sphenophyta*

30. Phylum *Psilotophyta*; pp. 443–445; moderate; ans: d

Psilotum is unique among living vascular plants in that it lacks:

a. spores.
b. sperm.
c. multicellular gametophytes.
d. roots and leaves.
e. mycorrhizal fungi.

31. Phylum *Sphenophyta*; p. 445; easy; ans: d

Carinal canals and conspicuously jointed stems are characteristic of the:

a. club mosses.
b. quillworts.
c. whisk ferns.
d. horsetails.
e. resurrection plants.

32. Phylum *Sphenophyta*; p. 445; moderate; ans: b

In *Equisetum*, carinal canals are located:

a. in the center of the pith.
b. surrounding the pith.
c. within the rhizome.
d. within the leaves.
e. within the sporangia.

33. Phylum *Sphenophyta*; pp. 448–449; moderate; ans: b

Which of the following statements about the *Equisetum* life cycle is FALSE?

a. Sporangia are borne along the margins of sporangiophores.
b. Spore dispersal is facilitated by elaters similar to those of *Marchantia*.
c. Gametophytes may be bisexual.
d. Sperm are multiflagellated.
e. Gametophytes are green and free-living.

34. Phylum *Pterophyta*; pp. 453–454; difficult; ans: a
Which of the following statements about the sporangia of vascular plants is
FALSE?

a.   A eusporangium typically contains fewer spores than a leptosporangium.
b.   A eusporangium is characteristic of the sphenophytes, lycophytes, and
     psilotophytes.
c.   The initial cell of a leptosporangium divides transversely or obliquely.
d.   The spores in a leptosporangium are discharged in a catapultlike manner.
e.   The initials of a eusporangium divide by forming walls parallel to the
     surface.

35. Phylum *Pterophyta*, p. 454; easy; ans: b
The nutritive tissue of a leptosporangium is called the:

a.   rachis.
b.   tapetum.
c.   annulus.
d.   lip cell.
e.   sporocarp.

36. Phylum *Pterophyta*; p. 454; easy; ans: c
The function of the annulus in a leptosporangium is to:

a.   aid in formation of the lip cells.
b.   give rise to the stalk.
c.   aid in dispersal of the spores.
d.   nourish the developing spores.
e.   attach the spores to the sporangial wall.

37. Phylum *Pterophyta*; p. 454; moderate; ans: c
_____ is a genus of phylum *Pterophyta*, order *Ophioglossales*.

a.   *Azolla*
b.   *Psaronius*
c.   *Botrychium*
d.   *Salvinia*
e.   *Trichomanes*

38. Phylum *Pterophyta*; p. 454; moderate; ans: d
A eusporangiate fern with a leaf having two parts—a vegetative portion and a
fertile portion—belongs to the:

a.   genus *Marsilea*.
b.   order *Filicales*.
c.   genus *Azolla*.
d.   genus *Ophioglossum*.
e.   order *Marattiales*.

39. Phylum *Pterophyta*; p. 455; moderate; ans: e
The _____ are the only order of homosporous leptosporangiate ferns.

a.   *Ophioglossales*
b.   *Salviniales*
c.   *Marattiales*
d.   *Marsileales*
e.   *Filicales*

40. **Phylum *Pterophyta*; p. 458; difficult; ans: a**
    In a typical member of the *Filicales*, the:

    a. antheridia and archegonia form on the prothallus.
    b. gametophyte persists long after the sporophyte has become independent.
    c. roots develop by circinate vernation.
    d. sporangia are produced on the upper surface of the leaves.
    e. rhizomes are the most conspicuous part of the sporophyte.

41. **Phylum *Pterophyta*; p. 458; easy; ans: b**
    The indusium of *Filicales* is a:

    a. cluster of sporangia.
    b. leaf outgrowth covering a sorus.
    c. leaf outgrowth covering a prothallus.
    d. megaphyll.
    e. "fiddlehead."

42. **Phylum *Pterophyta*; p. 458; easy; ans: b**
    The prothallus of *Filicales* is a(n):

    a. cluster of sporangia.
    b. heart-shaped gametophyte.
    c. outgrowth of a leaf.
    d. type of pinna.
    e. portion of the rachis.

43. **Phylum *Pterophyta*; p. 462; moderate; ans: a**
    Trichomanes speciosum is an example of a fern:

    a. lacking a sporophyte stage.
    b. lacking a gametophyte stage.
    c. lacking gemmae.
    d. producing thalloid sporophytes.
    e. producing filamentous sporophytes.

44. **Phylum *Pterophyta*; p. 462; moderate; ans: c**
    An example of a water fern that produces bean-shaped sporocarps is:

    a. *Azolla.*
    b. *Salvinia.*
    c. *Marsilea.*
    d. *Botrychium.*
    e. *Ophioglossum.*

45. **Phylum *Pterophyta*; p. 462; easy; ans: b**
    Azolla is a water fern of the order:

    a. *Ophioglossales.*
    b. *Salviniales.*
    c. *Marattiales.*
    d. *Marsileales.*
    e. *Filicales.*

46. Phylum *Pterophyta*; p. 463; easy; ans: b
        _____ is a water fern that bears sporangia on submerged, rootlike leaves.
   a.   *Azolla*
   b.   *Salvinia*
   c.   *Marsilea*
   d.   *Botrychium*
   e.   *Ophioglossum*

True-False Questions

1. **Evolution of Vascular Plants; p. 425; easy; ans: T**
    The great height reached by some vascular plants was made possible by the
    evolution of the ability to synthesize lignin.

2. **Evolution of Vascular Plants; p. 425; moderate; ans: F**
    Unlike seed plants, most seedless vascular plants have gametophytes that are
    nutritionally dependent on the sporophyte.

3. **Organization of the Vascular Plant Body; p. 426; easy; ans: T**
    The shoot system of vascular plants consists of the stem and the leaves.

4. **Organization of the Vascular Plant Body; p. 427; easy; ans: F**
    The vascular cambium produces primary xylem and primary phloem.

5. **Organization of the Vascular Plant Body; p. 427; easy; ans: T**
    As a result of secondary growth, the periderm replaces the epidermis as the
    dermal tissue system.

6. **Organization of the Vascular Plant Body; p. 428; moderate; ans: F**
    Tracheids most likely evolved from vessel elements.

7. **Organization of the Vascular Plant Body; p. 429; moderate; ans: T**
    The eustele, found in most seed plants, evolved from a protostele.

8. **Reproductive Systems; pp. 430–431; moderate; ans: F**
    In homosporous plants, gametophytes develop inside the spore wall.

9. **The Phyla of Seedless Vascular Plants; p. 431; moderate; ans: T**
    Three of the seven phyla of seedless vascular plants—*Rhyniophyta*,
    Zosterophyllophyta, and *Trimerophytophyta*—had become extinct by the end of
    the Devonian period.

10. **Phylum *Rhyniophyta*; p. 434; moderate; ans: F**
    Aglaophyton major, formerly known as *Rhynia major*, was reclassified as a
    protracheophyte based on its lack of hydroids, which are characteristic of
    vascular plants.

11. **Phylum *Zosterophyllophyta*; p. 435; moderate; ans: T**
    The zosterophylls were most likely the ancestors of the lycophytes.

12. **Phylum *Lycophyta*; p. 435; moderate; ans: F**
    In *Lycopodium*, the sporangia are borne on microsporophylls and megasporophylls.

13. **Phylum *Lycophyta*; pp. 442–443; moderate; ans: T**
    Each leaf of *Isoetes* is a potential sporophyll, capable of bearing either megasporangia or microsporangia.

14. **Phylum *Trimerophytophyta*; p. 443; moderate; ans: T**
    The trimerophytes probably evolved from the rhyniophytes and gave rise to the ferns.

15. **Phylum *Psilotophyta*; p. 443; moderate; ans: T**
    The whisk ferns, genus *Psilotum*, are the only vascular plants that lack roots and leaves.

16. **Phylum *Sphenophyta*; p. 445; easy; ans: F**
    In *Equisetum*, strobili are clustered into sporangiophores.

17. **Phylum *Pterophyta*; p. 454; easy; ans: F**
    Most living ferns are heterosporous.

18. **Phylum *Pterophyta*; p. 455; moderate; ans: F**
    Ophioglossum is a leptosporangiate fern.

19. **Phylum *Pterophyta*; p. 455; moderate; ans: T**
    A frond is a structure of the sporophyte phase of *Filicales*.

20. **Phylum *Pterophyta*; pp. 455, 458; moderate; ans: F**
    Circinate vernation is a type of branching pattern in a megaphyll.

21. **Phylum *Pterophyta*; p. 462; moderate; ans: T**
    Azolla is a leptosporangiate fern.

Essay Questions

1. **Evolution of Vascular Plants; p. 425; moderate**
   Summarize the key events in the evolution of vascular plants from a *Coleochaete*-like organism.

2. **Organization of the Vascular Plant Body; pp. 426–427; moderate**
   Explain the differences between primary and secondary growth.

3. **Organization of the Vascular Plant Body; pp. 429–430; easy**
   Describe the evolution of megaphylls.

4. **Reproductive Systems; pp. 430–431; moderate**
   Explain how bisexual species of homosporous ferns can be "functionally unisexual."

5. **Reproductive Systems; pp. 430–431; moderate**
    Discuss the main differences between homospory and heterospory.

6. **The Phyla of Seedless Vascular Plants; pp. 431–435, 443–445; difficult**
    Discuss the evolution of primitive vascular plants by using representatives of
    the rhyniophytes, zosterophyllophytes, trimerophytes, and psilotophytes.

7. **Phylum *Rhyniophyta*; p. 434; easy**
    Does *Rhynia major* belong in the phylum *Rhyniophyta*? Why or why not?

8. **Phylum *Lycophyta*; pp. 435–443; moderate**
    Compare and contrast the three living families of lycophytes.

9. **Phylum *Sphenophyta*; pp. 445–449; moderate**
    Phylum *Sphenophyta* contains what is probably the oldest surviving plant
    genus. Name and describe this genus.

10. **Phylum *Pterophyta*; pp. 453–454; moderate**
    How is the development of a eusporangium different from that of a
    leptosporangium?

11. **Phylum *Pterophyta*; pp. 454–463; moderate**
    List the five principal orders of the phylum *Pterophyta* and describe the
    distinguishing features of each.

Chapter 20

Gymnosperms

Multiple-Choice Questions

1. **Evolution of the Seed; p. 468; easy; ans: a**
    A seed is composed of a(n) _____ and _____ .

    a. ovule; embryo
    b. ovule; megasporangium
    c. ovule; integuments
    d. megasporangium; embryo
    e. megasporangium; integuments

2. **Evolution of the Seed; p. 468; moderate; ans: d**
    In seed plants, the _____ is called the nucellus.

    a. young sporophyte
    b. megagametophyte
    c. megaspore
    d. megasporangium
    e. embryo

3. **Evolution of the Seed; p. 468; moderate; ans: b**
    Which of the following was NOT a step in the evolution of the ovule?

    a. retention of the megaspores in the megasporangium
    b. production of only four megaspore mother cells per megasporangium
    c. formation of a highly reduced endosporic megagametophyte
    d. production of only one functional megaspore per megasporangium
    e. development of the embryo within the megagametophyte

4. **Evolution of the Seed; p. 468; easy; ans: d**
    A micropyle is a(n):

    a. nutritive structure of the embryo.
    b. nutritive structure of the pollen grain.
    c. opening in the megaspore wall.
    d. opening in an integument.
    e. opening in the megasporangium wall.

5. **Evolution of the Seed; p. 468; moderate; ans: e**
   With the evolution of the ovule, the unit of dispersal shifted from the megaspore to the:

   a. sperm.
   b. egg.
   c. microspore.
   d. megaspore mother cell.
   e. seed.

6. **Evolution of the Seed; p. 468; moderate; ans: e**
   The fossil record indicates that the integument evolved through a gradual:

   a. thickening of the megasporangium wall.
   b. splitting of the megaspore wall.
   c. fusion of megaspore lobes.
   d. splitting of the ovule wall.
   e. fusion of integumentary lobes.

7. **Evolution of the Seed; p. 470; moderate; ans: c**
   Just before fertilization, a gymnosperm ovule contains:

   a. an embryo.
   b. antheridia.
   c. archegonia.
   d. megaspores.
   e. a seed coat.

8. **Evolution of the Seed; p. 470; easy; ans: c**
   Which of the following is NOT a gymnosperm phylum?

   a. *Ginkgophyta*
   b. *Cycadophyta*
   c. *Anthophyta*
   d. *Coniferophyta*
   e. *Gnetophyta*

9. **Evolution of the Seed; p. 470; moderate; ans: c**
   Seed plants evolved most directly from:

   a. lycophytes.
   b. ferns.
   c. progymnosperms.
   d. angiosperms.
   e. trimerophytes.

10. **Progymnosperms; p. 470; difficult; ans: c**
    The progymnosperms had characteristics intermediate between those of the _____ and those of the _____.

    a. Paleozoic ferns; gymnosperms
    b. Paleozoic ferns; angiosperms
    c. seedless vascular trimerophytes; seed plants
    d. seedless vascular trimerophytes; modern ferns
    e. seedless vascular trimerophytes; rhyniophytes

11. Progymnosperms; p. 471; moderate; ans: b
Which of the following is a genus of the phylum *Progymnospermophyta*?

a. *Elkinsia*
b. *Aneurophyton*
c. *Medullosa*
d. *Archaeosperma*
e. *Ginkgo*

12. Extinct Gymnosperms; p. 472; moderate; ans: c
Which extinct gymnosperms resembled modern cycads?

a. seed ferns
b. *Cordaitales*
c. *Bennettitales*
d. *Ophioglossales*
e. trimerophytes

13. Living Gymnosperms; p. 472; easy; ans: c
There are _____ phyla of gymnosperms with living representatives.

a. 2
b. 3
c. 4
d. 5
e. 6

14. Living Gymnosperms; pp. 472–473; moderate; ans: d
Which of the following statements about gymnosperms is FALSE?

a. Gymnosperm means "naked seed."
b. Ovules and seeds are exposed on the surface of sporophylls.
c. The female gametophyte produces several archegonia.
d. The male gametophyte produces several antheridia.
e. The male gametophyte is endosporic.

15. Living Gymnosperms; p. 472; moderate; ans: d
Polyembryony is common in gymnosperms because a(n):

a. seed produces several ovules.
b. ovule produces several megasporangia.
c. megasporangium produces several megaspores.
d. megagametophyte produces several archegonia.
e. archegonium contains several eggs.

16. Living Gymnosperms; p. 473; moderate; ans: a
_____ produce nonmotile sperm.

a. Conifers and gnetophytes
b. Conifers and cycads
c. Cycads and *Ginkgo*
d. Cycads and gnetophytes
e. *Ginkgo* and gnetophytes

17. Living Gymnosperms; p. 474; moderate; ans: c
In which of the following groups is the pollen tube haustorial?

a. conifers and gnetophytes
b. conifers and cycads
c. cycads and *Ginkgo*
d. cycads and gnetophytes
e. *Ginkgo* and gnetophytes

18. Living Gymnosperms; p. 474; moderate; ans: e
In which gymnosperms do multiflagellated sperm swim to an archegonium?

a. none
b. *Ginkgo* only
c. gnetophytes only
d. cycads only
e. *Ginkgo* and cycads

19. Phylum *Coniferophyta*; pp. 474, 475-476; easy; ans: b
Which of the following statements about conifers is FALSE?

a. They belong to the phylum *Coniferophyta*.
b. Their leaves have humidity-resistant features.
c. They are the most ecologically important gymnosperms.
d. The longest-lived tree is a conifer.
e. The tallest vascular plant is a conifer.

20. Phylum *Coniferophyta*; p. 474; moderate; ans: d
Which of the following statements about pine leaves is FALSE?

a. They have a thick cuticle.
b. They have sunken stomata.
c. They are spirally arranged and borne singly on the stem.
d. They are arranged in indeterminate bundles called fascicles.
e. Their vascular bundles are surrounded by transfusion tissue.

21. Phylum *Coniferophyta*; pp. 474–475; easy; ans: b
The likely function of the transfusion tissue of pine leaves is to:

a. protect leaf cells from physical damage.
b. conduct materials between the mesophyll and vascular bundles.
c. prevent water loss.
d. act as a barrier to the movement of toxic compounds.
e. give rise to indeterminate meristems.

22. Phylum *Coniferophyta*; p. 476; moderate; ans: b
In the stems of conifers, the vascular cambium produces _____ toward the outside and _____ toward the inside.

a. secondary xylem; secondary phloem
b. secondary phloem; secondary xylem
c. cortex; periderm
d. periderm; cortex
e. tracheids; sieve cells

23. Phylum *Coniferophyta*; p. 477; moderate; ans: c
In the pine life cycle, meiosis occurs in:

a. archegonia.
b. microspores.
c. microsporocytes.
d. pollen grains.
e. sporophylls.

24. Phylum *Coniferophyta*; p. 477; easy; ans: b
The daughter cells produced by a microsporocyte are:

a. microspore mother cells.
b. microspores.
c. pollen grains.
d. prothallial cells.
e. generative cells.

25. Phylum *Coniferophyta*; p. 477; moderate; ans: a
In pines, the immature male gametophyte consists of:

a. two prothallial cells, one generative cell, and one tube cell.
b. one prothallial cell, two generative cells, and one tube cell.
c. one prothallial cell, one generative cell, and two tube cells.
d. two prothallial cells, two generative cells, and one tube cell.
e. one prothallial cell, one generative cell, and one tube cell.

26. Phylum *Coniferophyta*; p. 480; moderate; ans: e
The seed-scale complex of an ovulate cone consists of:

a. a megasporophyll and two ovules.
b. a megasporophyll and one bract.
c. an ovuliferous scale and two ovules.
d. two ovules and two bracts.
e. an ovuliferous scale, two ovules, and one bract.

27. Phylum *Coniferophyta*; p. 481; easy; ans: c
In pines, a megaspore mother cell divides to give rise to four _____, of which
_____ disintegrate(s).

a. megaspores; one
b. megaspores; two
c. megaspores; three
d. megasporocytes; two
e. megasporocytes; three

28. Phylum *Coniferophyta*; p. 481; moderate; ans: d
In the pine life cycle, a pollen grain germinates shortly after it:

a. enters the pollen tube.
b. undergoes meiosis.
c. forms microspores.
d. comes in contact with the nucellus.
e. comes in contact with an archegonium.

29. **Phylum *Coniferophyta*; pp. 481–482; moderate; ans: b**
    In pines, the generative cell of the male gametophyte divides to produce:

    a.   two sperm cells.
    b.   one sterile cell and one spermatogenous cell.
    c.   one sterile cell and one sperm cell.
    d.   two spermatogenous cells.
    e.   two sterile cells.

30. **Phylum *Coniferophyta*; pp. 477, 481–482; difficult; ans: e**
    In pines, the mature male gametophyte consists of:

    a.   one tube cell and one generative cell.
    b.   one tube cell, one sterile cell, and one spermatogenous cell.
    c.   two prothallial cells, one tube cell, and one generative cell.
    d.   two prothallial cells, one tube cell, one sterile cell, and one
         spermatogenous cell.
    e.   two prothallial cells, one tube cell, one sterile cell, and two sperm.

31. **Phylum *Coniferophyta*; p. 482; difficult; ans: a**
    Which of the following statements about the embryogeny of pine is FALSE?

    a.   Three types of polyembryony occur in most species.
    b.   Four tiers of embryonic cells form at the lower end of the archegonium.
    c.   Suspensor cells force the embryos through the wall of the archegonium.
    d.   An embryo begins to develop from each of the four cells farthest from the
         micropyle.
    e.   The integument develops into a seed coat.

32. **Phylum *Coniferophyta*; p. 482; moderate; ans: b**
    In a conifer seed, the three different generations are represented by the:

    a.   integument, nucellus, and egg.
    b.   seed coat, embryo, and food supply.
    c.   seed coat, archegonium, and food supply.
    d.   integument, ovule, and embryo.
    e.   integument, megasporangium, and archegonium

33. **Phylum *Coniferophyta*; p. 482; moderate; ans: a**
    Which of the following statements about pine seeds is FALSE?

    a.   They are often shed from the cones during the first year following
         pollination.
    b.   They are often dispersed by the wind.
    c.   Some are dispersed only after the cones are scorched by fire.
    d.   Some are dispersed by birds.
    e.   Some are winged.

34. **Phylum *Coniferophyta*; p. 483; moderate; ans: c**
    In the _____, ovules are solitary and surrounded by an aril.

    a.   larches
    b.   hemlocks
    c.   yews
    d.   junipers
    e.   spruces

35. **Phylum *Coniferophyta*; p. 486; moderate; ans: d**
_____ is a conifer regarded as a "living fossil."

  a. *Tsuga*
  b. *Araucaria*
  c. *Sequoiadendron*
  d. *Metasequoia*
  e. *Sequoia*

36. **The Other Living Gymnosperm Phyla: *Cycadophyta*, *Ginkgophyta*, and *Gnetophyta*; p. 486; easy; ans: a**
Palmlike gymnosperms belong to the phylum:

  a. *Cycadophyta*.
  b. *Coniferophyta*.
  c. *Ginkgophyta*.
  d. *Progymnospermophyta*.
  e. *Gnetophyta*.

37. **The Other Living Gymnosperm Phyla: *Cycadophyta*, *Ginkgophyta*, and *Gnetophyta*; p. 486; moderate; ans: b**
Which of the following gymnosperms is a cycad native to the United States?

  a. *Gnetum*
  b. *Zamia*
  c. *Tsuga*
  d. *Taxus*
  e. *Metasequoia*

38. **The Other Living Gymnosperm Phyla: *Cycadophyta*, *Ginkgophyta*, and *Gnetophyta*; pp. 486–487; moderate; ans: c**
Which of the following statements about cycads is FALSE?

  a. They resemble palms.
  b. They exhibit secondary growth.
  c. They produce nonflagellated sperm.
  d. They are pollinated by insects.
  e. They harbor cyanobacteria.

39. **The Other Living Gymnosperm Phyla: *Cycadophyta*, *Ginkgophyta*, and *Gnetophyta*; p. 490; easy; ans: c**
Gymnosperms having fan-shaped deciduous leaves belong to the phylum:

  a. *Cycadophyta*.
  b. *Coniferophyta*.
  c. *Ginkgophyta*.
  d. *Progymnospermophyta*.
  e. *Gnetophyta*.

40. The Other Living Gymnosperm Phyla: *Cycadophyta, Ginkgophyta,* **and** *Gnetophyta*; **p. 492; easy; ans: d**
_____ is a gnetophyte characterized by two strap-shaped leaves growing from a massive woody disk.

a. *Rhopalotria*
b. *Zamia*
c. *Ephedra*
d. *Welwitschia*
e. *Gnetum*

41. The Other Living Gymnosperm Phyla: *Cycadophyta, Ginkgophyta,* **and** *Gnetophyta*; **p. 492; easy; ans: e**
The gymnosperms most closely resembling angiosperms belong to the phylum:

a. *Cycadophyta.*
b. *Coniferophyta.*
c. *Ginkgophyta.*
d. *Progymnospermophyta.*
e. *Gnetophyta.*

True-False Questions

1. Evolution of the Seed; p. 468; moderate; ans: T
In the evolution of the ovule, the apex of the megasporangium was modified to receive pollen grains.

2. Evolution of the Seed; p. 470; easy; ans: F
Gymnosperms, like the angiosperms, represent a single evolutionary line.

3. Progymnosperms; p. 470; moderate; ans: F
The progymnosperms were unique among the woody plants of the Devonian period in producing freely dispersed spores.

4. Extinct Gymnosperms; p. 472; easy; ans: T
The seed ferns, *Cordaitales*, and *Bennettitales* are all groups of extinct gymnosperms.

5. Living Gymnosperms; p. 474; moderate; ans: T
The pollen tube originally developed as a structure that permitted the male gametophyte to absorb nutrients during sperm formation.

6. Phylum *Coniferophyta*; **p. 475; moderate; ans: F**
The leaves of pines are usually replaced each year.

7. Phylum *Coniferophyta*; **p. 476; moderate; ans: T**
The xylem of conifers contains primarily tracheids, and the phloem has primarily sieve cells.

8. Phylum *Coniferophyta*; **pp. 476–477; easy; ans: F**
In *Pinus*, ovulate cones are usually borne on the lower branches of the tree.

9. **Phylum *Coniferophyta*; p. 477; moderate; ans: F**
   The microsporocytes of pines are haploid cells.

10. **Phylum *Coniferophyta*; p. 480; moderate; ans: T**
    In pine, the ovuliferous scales are modified determinate branch systems.

11. **Phylum *Coniferophyta*; pp. 481–482; moderate; ans: F**
    In pine, only one egg of all the archegonia in an ovule is fertilized.

12. **Phylum *Coniferophyta*; p. 482; easy; ans: T**
    In conifer seeds, the food supply is haploid.

13. **Phylum *Coniferophyta*; p. 483; moderate; ans: F**
    The tallest living plant is an individual of the species *Sequoiadendron giganteum*.

14. **The Other Living Gymnosperm Phyla: *Cycadophyta*, *Ginkgophyta*, and *Gnetophyta*; pp. 486–487, 490; moderate; ans: T**
    In both cycads and *Ginkgo*, male and female gametophytes are produced on separate plants.

15. **The Other Living Gymnosperm Phyla: *Cycadophyta*, *Ginkgophyta*, and *Gnetophyta*; p. 492; easy; ans: T**
    Members of the phylum *Gnetophyta* have strobili that resemble flower clusters.

16. **The Other Living Gymnosperm Phyla: *Cycadophyta*, *Ginkgophyta*, and *Gnetophyta*; p. 492; moderate; ans: T**
    Gnetum and *Welwitschia*, but not *Ephedra*, resemble angiosperms in lacking archegonia.

Essay Questions

1. **Introduction; p. 468; easy**
   In what way was the seed such an important adaptation in the evolution of plants?

2. **Evolution of the Seed; p. 468; moderate**
   List the key events that led to the evolution of the ovule.

3. **Progymnosperms; pp. 470–472; moderate**
   Discuss the relationship of the progymnosperms to the trimerophytes and seed plants. What evidence supports this relationship?

4. **Living Gymnosperms; pp. 472–474; moderate**
   Discuss the role and evolutionary development of the pollen tube.

5. **Phylum *Coniferophyta*; p. 474; moderate**
   Describe some of the drought-resistant features of pine leaves.

6. **Phylum *Coniferophyta*; pp. 476–482; difficult**
   Outline the main events of the pine life cycle, indicating the time frame for each.

7. **Phylum *Coniferophyta*; pp. 472, 481–482; moderate**
   What is polyembryony? Describe the types of polyembryony that occur in pines.

8. **Phylum *Coniferophyta*; p. 482; moderate**
   Describe the three different generations coexisting in a pine seed.

9. **The Other Living Gymnosperm Phyla: *Cycadophyta*, *Ginkgophyta*, and *Gnetophyta*; pp. 486–492; difficult**
   Compare and contrast the reproductive structures of cycads, *Ginkgo*, and gnetophytes.

10. **The Other Living Gymnosperm Phyla: *Cycadophyta*, *Ginkgophyta*, and *Gnetophyta*; p. 487; moderate**
    Describe the role of insects in the pollination of cycads.

11. **The Other Living Gymnosperm Phyla: *Cycadophyta*, *Ginkgophyta*, and *Gnetophyta*; p. 492; moderate**
    Describe the three living genera of gnetophytes. What features suggest that they are closely related to the angiosperms?

Chapter 21

Introduction to the Angiosperms

Multiple-Choice Questions

1. **Diversity in the Phylum *Anthophyta*; pp. 496, 498; moderate; ans: b**
    Which of the following statements about phylum *Anthophyta* is FALSE?

    a. It is the largest phylum of photosynthetic organisms.
    b. The two major classes are the monocots and dicots.
    c. Its members vary in size from 1 millimeter long to 100 meters tall.
    d. Some members are parasitic.
    e. Some members are saprophytic.

2. **The Flower; p. 499; moderate; ans: c**
    The stalk of an inflorescence is a:

    a. pedicel.
    b. receptacle.
    c. peduncle.
    d. carpel.
    e. perianth.

3. **The Flower; p. 499; moderate; ans: b**
    The receptacle of a flower is:

    a. a leaflike structure that contains the ovules.
    b. the part of the flower stalk to which flower parts are attached.
    c. the stalk of a flower or an inflorescence.
    d. the stalk of a flower in an inflorescence.
    e. the stalk that attaches an ovule to the ovary.

4. **The Flower; p. 501; easy; ans: a**
    The sterile parts of a flower are the:

    a. sepals and petals.
    b. sepals and stamens.
    c. petals and stamens.
    d. sepals and carpels.
    e. carpels and stamens.

5. **The Flower; p. 501; easy; ans: c**
   Collectively, the sepals form the:

   a.   corolla.
   b.   perianth.
   c.   calyx.
   d.   androecium.
   e.   gynoecium.

6. **The Flower; p. 501; moderate; ans: c**
   The perianth consists of all the _____ of a flower.

   a.   sepals
   b.   petals
   c.   sepals and petals
   d.   fertile parts
   e.   sepals, petals, stamens, and carpels

7. **The Flower; p. 501; moderate; ans: e**
   Which of the following statements about a stamen is FALSE?

   a.   It is a microsporophyll.
   b.   It usually consists of an anther and a filament.
   c.   It contains four microsporangia.
   d.   It contains pollen sacs.
   e.   It is part of the gynoecium.

8. **The Flower; p. 501; moderate; ans: a**
   The gynoecium consists of all the _____ of a flower.

   a.   carpels
   b.   ovules
   c.   ovaries
   d.   placentae
   e.   stigmas

9. **The Flower; p. 501; moderate; ans: c**
   The portion of a carpel that encloses the ovules is the:

   a.   style.
   b.   stigma.
   c.   ovary.
   d.   placenta.
   e.   funiculus.

10. **The Flower; p. 501; difficult; ans: e**
    The style connects the _____ to the _____.

    a.   anther; filament
    b.   calyx; corolla
    c.   androecium; gynoecium
    d.   ovules; ovary
    e.   ovary; stigma

11. The Flower; p. 501; moderate; ans: d
The part of the flower that receives the pollen is the:

a.   ovule.
b.   ovary.
c.   style.
d.   stigma.
e.   anther.

12. The Flower; p. 501; moderate; ans: a
A locule is a:

a.   chamber in an ovary.
b.   lobe of an anther.
c.   part of a stigma.
d.   whorl of sepals.
e.   cluster of microsporangia.

13. The Flower; p. 501; moderate; ans: b
In axile placentation, the ovules are borne:

a.   on the ovary wall.
b.   on a central column of tissue in a partitioned ovary.
c.   on a central column of tissue in an unpartitioned ovary.
d.   at the base of a unilocular ovary.
e.   at the top of a unilocular ovary.

14. The Flower; p. 501; easy; ans: e
A perfect flower contains:

a.   petals only.
b.   stamens only.
c.   carpels only.
d.   both petals and carpels.
e.   both carpels and stamens.

15. The Flower; p. 501; moderate; ans: d
A monoecious species has:

a.   flowers with all floral whorls.
b.   floral parts united with other members of the same whorl.
c.   floral parts united with members of other whorls.
d.   staminate and carpellate flowers on the same plant.
e.   staminate and carpellate flowers on different plants.

16. The Flower; pp. 501–502; moderate; ans: e
A flower that has only stamens and petals is:

a.   staminate and perfect.
b.   perfect and complete.
c.   perfect and incomplete.
d.   imperfect and complete.
e.   imperfect and incomplete.

17. **The Flower; p. 502; easy; ans: a**
    An example of connation is the union of stamens with:

    a. other stamens.
    b. petals.
    c. sepals.
    d. carpels.
    e. stigmas.

18. **The Flower; p. 502; moderate; ans: d**
    By definition, a flower with a polysepalous calyx has:

    a. no sepals.
    b. many sepals.
    c. sepals joined together.
    d. sepals not joined together.
    e. sepals joined with a flower part in a different whorl.

19. **The Flower; p. 502; moderate; ans: d**
    In flowers with sepals, petals, and stamens attached below the ovary, the ovary
    is said to be:

    a. hypogynous.
    b. epigynous.
    c. perigynous.
    d. superior.
    e. inferior.

20. **The Flower; p. 502; easy; ans: b**
    All radially symmetrical flowers are:

    a. hypogynous.
    b. regular.
    c. irregular.
    d. superior.
    e. inferior.

21. **The Angiosperm Life Cycle; p. 503; moderate; ans: b**
    In angiosperms, the mature male gametophyte consists of _____ cells.

    a. 2
    b. 3
    c. 5
    d. 7
    e. 9

22. **The Angiosperm Life Cycle; p. 503; moderate; ans: d**
    In angiosperms, the mature female gametophyte consists of _____ cells.

    a. 2
    b. 3
    c. 5
    d. 7
    e. 9

23. The Angiosperm Life Cycle; p. 503; easy; ans: c
In angiosperms, the ovary develops into a(n):

   a.   ovule.
   b.   seed.
   c.   fruit.
   d.   carpel.
   e.   perianth.

24. The Angiosperm Life Cycle; p. 504; moderate; ans: a
The sporogenous cells of angiosperms develop directly into:

   a.   microsporocytes.
   b.   microspores.
   c.   pollen grains.
   d.   tube cells.
   e.   generative cells.

25. The Angiosperm Life Cycle; p. 504; moderate; ans: e
The innermost layer of the pollen sac wall is the:

   a.   sporogenous layer.
   b.   sporopollenin layer.
   c.   exine.
   d.   intine.
   e.   tapetum.

26. The Angiosperm Life Cycle; p. 504; moderate; ans: c
Microsporocytes divide by _____, forming _____.

   a.   mitosis; haploid microsporocytes
   b.   mitosis; diploid microsporocytes
   c.   meiosis; haploid microspores
   d.   mitosis; diploid microspores
   e.   meiosis; haploid megaspores

27. The Angiosperm Life Cycle; p. 504; moderate; ans: c
Sporopollenin is the primary constituent of the:

   a.   integument.
   b.   nucellar wall.
   c.   exine.
   d.   intine.
   e.   tapetum.

28. The Angiosperm Life Cycle; p. 504; easy; ans: d
When the microspore of an angiosperm divides, it gives rise directly to:

   a.   two generative cells.
   b.   two tube cells.
   c.   two sperm cells.
   d.   a generative cell and a tube cell.
   e.   a sperm cell, a generative cell, and a tube cell.

29. **The Angiosperm Life Cycle; pp. 504, 506; moderate; ans: e**
Which of the following statements about pollen grains is FALSE?

    a. They vary considerably in size and shape.
    b. They differ in the number, arrangement, and shape of their apertures.
    c. They provide insights into past climates.
    d. They provide a means of identifying flowering plants.
    e. They are poorly represented in the fossil record.

30. **The Angiosperm Life Cycle; p. 506; easy; ans: b**
The funiculus is the:

    a. ovule-bearing region of the ovary wall.
    b. stalk of the ovule.
    c. wall of the megasporangium.
    d. inner integument.
    e. interior of the nucellus.

31. **The Angiosperm Life Cycle; p. 506; difficult; ans: b**
In the process of megasporogenesis, the _____ divides_____.

    a. megasporocyte; mitotically
    b. megasporocyte; meiotically
    c. megaspores; meiotically
    d. nucellus; mitotically
    e nucellus; meiotically

32. **The Angiosperm Life Cycle; p. 506; moderate; ans: b**
In the most common pattern of embryo sac development in angiosperms, how
many nuclei end up at the chalazal end?

    a. 1
    b. 3
    c. 5
    d. 7
    e. 9

33. **The Angiosperm Life Cycle; p. 506; moderate; ans: e**
The _____ are found at the chalazal end of the embryo sac.

    a. polar nuclei
    b. synergids
    c. eggs
    d. ovules
    e. antipodals

34. **The Angiosperm Life Cycle; p. 506; moderate; ans: b**
In embryo sac development, the egg apparatus contains the egg cell and the:

    a. polar nuclei.
    b. synergids.
    c. chalaza.
    d. central cell.
    e. antipodals.

35. **The Angiosperm Life Cycle; p. 506; moderate; ans: d**
In the center of the embryo sac are the:

   a.   ovules.
   b.   synergids.
   c.   eggs.
   d.   polar nuclei.
   e.   antipodals.

36. **The Angiosperm Life Cycle; p. 506; easy; ans: a**
The mature female gametophyte is called a(n):

   a.   embryo sac.
   b.   nucellus.
   c.   megasporangium.
   d.   endosperm.
   e.   ovule.

37. **The Angiosperm Life Cycle; p. 508; moderate; ans: b**
In the process of pollination, the:

   a.   anther dehisces, shedding its contents.
   b.   pollen passes from the anther to a stigma.
   c.   pollen tube grows through the style.
   d.   pollen tube enters the embryo sac.
   e.   sperm fertilizes an egg.

38. **The Angiosperm Life Cycle; p. 508; moderate; ans: d**
The transmitting tissue of angiosperms:

   a.   lines the inner wall of the pollen tube.
   b.   is a glandular tissue on the surface of the stigma.
   c.   secretes a sugary solution.
   d.   connects the stigmatic tissue with the ovule.
   e.   is absent in species with solid styles.

39. **The Angiosperm Life Cycle; p. 509; moderate; ans: c**
Commonly, the pollen tube releases its sperm and tube nucleus into:

   a.   the egg.
   b.   the central cell.
   c.   a synergid.
   d.   an antipodal cell.
   e.   a polar nucleus.

40. **The Angiosperm Life Cycle; p. 509; moderate; ans: b**
In the process of double fertilization, one sperm fuses with the _____, and the other sperm fuses with the _____.

   a.   egg; synergids
   b.   egg; polar nuclei
   c.   egg; antipodals
   d.   synergid; polar nuclei
   e.   synergid; antipodals

41. The Angiosperm Life Cycle; p. 509; easy; ans: c
In most angiosperms, the primary endosperm nucleus is:

    a.   *n*.
    b.   2*n*.
    c.   3*n*.
    d.   4*n*.
    e.   5*n*.

42. The Angiosperm Life Cycle; p. 509; moderate; ans: e
The process of double fertilization occurs:

    a.   only in angiosperms.
    b.   in all angiosperms and gymnosperms.
    c.   only in angiosperms and *Ephedra*.
    d.   only in angiosperms and *Gnetum*.
    e.   only in angiosperms, *Ephedra*, and *Gnetum*.

43. The Angiosperm Life Cycle; p. 510; easy; ans: c
The function of the endosperm is to provide _____ for the embryo.

    a.   support
    b.   oxygen
    c.   food
    d.   water
    e.   protection

44. The Angiosperm Life Cycle; p. 510; moderate; ans: d
In some angiosperms, the food store of the seed is perisperm derived from the:

    a.   integument.
    b.   embryo.
    c.   endosperm.
    d.   nucellus.
    e.   funiculus.

45. The Angiosperm Life Cycle; p. 510; difficult; ans: e
Unlike gymnosperms, the stored food in angiosperms is:

    a.   provided by the male gametophyte.
    b.   provided by the female gametophyte.
    c.   sporophyte.
    d.   formed before fertilization.
    e.   formed after fertilization.

46. The Angiosperm Life Cycle; p. 510; easy; ans: b
As the seed develops, the ovary wall becomes the:

    a.   perisperm.
    b.   pericarp.
    c.   endosperm.
    d.   nucellus.
    e.   integument.

47. The Angiosperm Life Cycle; pp. 510–511; moderate; ans: e
Which of the following conditions does NOT promote outcrossing?

a. Staminate and carpellate flowers occur on separate plants.
b. Stamens mature before carpels.
c. Carpels mature before stamens.
d. Stamens and stigmas are physically separated on the same flower.
e. The plant is genetically self-compatible.

48. The Angiosperm Life Cycle; p. 510; moderate; ans: b
In angiosperms, dichogamy refers to the condition in which:

a. sperm is directed to the egg.
b. stamens and carpels reach maturity at different times.
c. seed formation is speeded up.
d. fruit formation is speeded up.
e. polyembryony is facilitated.

49. The Angiosperm Life Cycle; pp. 510–511; moderate; ans: c
Gametophytic self-incompatibility refers to the condition in which the:

a. embryo does not develop from the zygote.
b. seed fails to germinate.
c. pollen fails to germinate because of its own genotype.
d. pollen fails to germinate because of its parent's genotype.
e. egg is not fertilized even though sperm are present.

50. The Angiosperm Life Cycle; p. 514; moderate; ans: c
Which of the following statements about self-pollination is FALSE?

a. It occurs in many angiosperms.
b. It is the transfer of pollen from an anther to a stigma of the same plant.
c. Self-pollinating flowers are usually larger than cross-pollinating flowers.
d. In some species it occurs in the bud.
e. Self-pollinating plants are often abundant where flower-visiting animals are rare.

True-False Questions

1. Diversity in the Phylum *Anthophyta*; pp. 496, 498; easy; ans: T
Most trees of phylum *Anthophyta* are eudicots.

2. The Flower; p. 499; easy; ans: T
A carpel is a leaflike structure that contains ovules.

3. The Flower; p. 501; easy; ans: T
A locule is a chamber in an ovary.

4. The Flower; p. 501; moderate; ans: F
If a single ovule is borne on a central column in an unpartitioned ovary, the placentation is described as parietal.

5. The Flower; p. 502; easy; ans: T
An example of adnation is the union of petals with stamens.

6. **The Flower; p. 502; moderate; ans: F**
   In a synsepalous flower, the sepals are joined not to each other but to another flower part.

7. **The Flower; p. 502; easy; ans: F**
   In a flower with a superior ovary, the perianth and stamens are epigynous.

8. **The Flower; p. 502; easy; ans: T**
   Bilaterally symmetrical flowers are said to be irregular.

9. **The Angiosperm Life Cycle; p. 503; moderate; ans: F**
   Angiosperms have archegonia but lack antheridia.

10. **The Angiosperm Life Cycle; p. 504; moderate; ans: T**
    The function of the tapetum is to provide food for developing microspores.

11. **The Angiosperm Life Cycle; p. 506 easy; ans: F**
    Pollen grains and spores are morphologically indistinguishable.

12. **The Angiosperm Life Cycle; p. 506; moderate; ans: F**
    Embryo sac development in *Lilium* is unique among angiosperms in having a single four-nucleate stage.

13. **The Angiosperm Life Cycle; p. 509; easy; ans: T**
    Endosperm may be an evolutionary derivative of a second embryo.

14. **The Angiosperm Life Cycle; p. 510; moderate; ans: F**
    In angiosperms, embryogeny begins with a free-nuclear stage.

15. **The Angiosperm Life Cycle; p. 510; easy; ans: T**
    Exocarp, mesocarp, and endocarp are the layers of the ovary wall in a fruit.

16. **The Angiosperm Life Cycle; p. 510; easy; ans: F**
    Dichogamy refers to a genetic incompatibility between two species.

17. **The Angiosperm Life Cycle; p. 511; moderate; ans: T**
    In gametophytic self-incompatibility, the ability of a pollen grain to germinate on a stigma is determined by the pollen's genotype.

18. **The Angiosperm Life Cycle; p. 514; moderate; ans: T**
    Self-pollinating plants usually form genetically uniform populations.

Essay Questions

1. **Diversity in the Phylum *Anthophyta*; p. 496; easy**
   List some of the types of plants found in the phylum *Anthophyta*.

2. **The Flower; pp. 498–499; moderate**
   Define the terms "angiosperm" and "flower." What is the relationship between these terms?

3. **The Flower; pp. 499, 501; easy**
   What is a carpel? Name the three parts of the carpel, and give the function of each.

4. **The Flower; pp. 501–502; easy**
   Can an incomplete flower be a perfect flower? Why or why not?

5. **The Angiosperm Life Cycle; p. 504; moderate**
   Summarize the events of microsporogenesis and microgametogenesis in angiosperms.

6. **The Angiosperm Life Cycle; pp. 504, 506; difficult**
   Compare and contrast the processes of sporogenesis and gametogenesis in the androecium and gynoecium.

7. **The Angiosperm Life Cycle; pp. 504, 506; easy**
   How would you distinguish a microspore from a pollen grain?

8. **The Angiosperm Life Cycle; p. 506; moderate**
   Summarize the main events of megasporogenesis and megagametogenesis in angiosperms.

9. **The Angiosperm Life Cycle; p. 509; easy**
   Describe the two fertilization events of "double fertilization."

10. **The Angiosperm Life Cycle; p. 509; easy**
    What is triple fusion in angiosperms? Name the product formed in triple fusion. What is the function of this product?

11. **The Angiosperm Life Cycle; p. 510; moderate**
    Compare and contrast the origin and ploidy of the stored food in angiosperms and gymnosperms.

12. **The Angiosperm Life Cycle; pp. 510–511; moderate**
    Discuss the factors that promote outcrossing in angiosperms.

13. **The Angiosperm Life Cycle; pp. 510–511; moderate**
    List some of the ways in which plants may be genetically self-incompatible.

14. **The Angiosperm Life Cycle; p. 514; moderate**
    Under what conditions is self-pollination advantageous?

Chapter 22

Evolution of the Angiosperms

Multiple-Choice Questions

1. **Introduction; p. 518; moderate; ans: d**
    Angiosperms first appear in the fossil record in the _____ period.

    a. Devonian
    b. Carboniferous
    c. Triassic
    d. Cretaceous
    e. Tertiary

2. **Relationships of the Angiosperms; p. 518; moderate; ans: d**
    Recent evidence supports the hypothesis that the seed plants most closely related to angiosperms are the:

    a. *Bennettitales* and *Ginkgo*.
    b. gnetophytes and *Ginkgo*.
    c. lycophytes and gnetophytes.
    d. *Bennettitales* and gnetophytes.
    e. *Bennettitales* and lycophytes.

3. **Origin and Diversification of the Angiosperms; p. 519; moderate; ans: b**
    Which of the following is NOT a unique characteristic of the angiosperms?

    a. double fertilization producing endosperm
    b. stamens with four pairs of pollen sacs
    c. a three-nucleate microgametophyte
    d. an eight-nucleate megagametophyte
    e. phloem having sieve tubes and companion cells

4. **Origin and Diversification of the Angiosperms; p. 519; easy; ans: a**
    Triaperturate pollen is characteristic of:

    a. eudicots.
    b. dicots.
    c. monocots.
    d. magnoliids.
    e. gymnosperms.

5. **Origin and Diversification of the Angiosperms; p. 519; easy; ans: e**
   The magnoliids comprise approximately _____ percent of living angiosperms.

   a.   97
   b.   63
   c.   50
   d.   10
   e.   3

6. **Origin and Diversification of the Angiosperms; p. 519; easy; ans: a**
   The magnoliids are the ancestors of:

   a.   both monocots and eudicots.
   b.   the monocots only.
   c.   the eudicots only.
   d.   the lycophytes and gnetophytes.
   e.   the *Bennettitales* and gnetophytes.

7. **Origin and Diversification of the Angiosperms; p. 519; moderate; ans: b**
   The magnoliids have traditionally been regarded as _____ but are now
   classified as _____ .

   a.   eudicots; monocots
   b.   dicots; eudicots
   c.   monocots; dicots
   d.   lycophytes; angiosperms
   e.   gymnosperms; angiosperms

8. **Origin and Diversification of the Angiosperms; p. 519; moderate; ans: e**
   _____ have single-pored pollen.

   a.   Only eudicots
   b.   Only monocots
   c.   Only magnoliids
   d.   Both eudicots and monocots
   e.   Both magnoliids and monocots

9. **Origin and Diversification of the Angiosperms; p. 519; easy; ans: c**
   Which plants are characterized by oil cells containing ethereal oils?

   a.   eudicots only
   b.   monocots only
   c.   magnoliids only
   d.   eudicots and monocots
   e.   magnoliids and monocots

10. **Origin and Diversification of the Angiosperms; p. 521; moderate; ans: d**
    The living paleoherbs include the family:

    a.   *Magnoliaceae.*
    b.   *Lauraceae.*
    c.   *Calycanthaceae.*
    d.   *Aristolochiaceae.*
    e.   *Asteraceae.*

11. Origin and Diversification of the Angiosperms; p. 521; moderate; ans: b
The ancestors of the monocots were most likely:

a. woody magnoliids.
b. paleoherbs.
c. eudicots.
d. similar to the *Magnoliaceae*.
e. similar to the *Lauraceae*.

12. Origin and Diversification of the Angiosperms; p. 522; moderate; ans: e
The earliest eudicots most likely had:

a. very large flowers.
b. brightly colored flowers.
c. complex flowers.
d. flowers resembling those of *Magnolia*.
e. petals and sepals that were not clearly differentiated.

13. Origin and Diversification of the Angiosperms; p. 522; moderate; ans: b
Which of the following was NOT an adaptive trait of early angiosperms to cold
and drought?

a. tough leaves
b. increased leaf size
c. deciduous habit
d. resistant seed coats
e. vessel elements

14. Origin and Diversification of the Angiosperms; p. 522; moderate; ans: a
Which of the following was NOT a development important to the early success
of the angiosperms?

a. resistance to warm, moist climates
b. chemical defenses
c. sieve-tube elements
d. mechanisms of pollination
e. mechanisms of seed dispersal

15. Origin and Diversification of the Angiosperms; p. 523; difficult; ans: c
At the time angiosperms were becoming abundant in the fossil record, the
central regions of West Gondwanaland were:

a. experiencing greater extremes of temperature.
b. experiencing greater extremes of humidity.
c. becoming milder.
d. becoming more arid.
e. becoming more subhumid.

16. Evolution of the Flower; p. 524; moderate; ans: a
In most angiosperms, petals are probably evolutionary derivatives of:

a. stamens.
b. sepals.
c. carpels.
d. receptacles.
e. peduncles.

17. **Evolution of the Flower; p. 524; moderate; ans: a**
Petals and stamens are similar in that both are supplied by:

  a.   a single vascular strand.
  b.   two vascular strands.
  c.   three vascular strands.
  d.   four vascular strands.
  e.   the same number of vascular strands as the leaves.

18. **Evolution of the Flower; p. 525; moderate; ans: d**
Which of the following statements about nectaries is FALSE?

  a.   They are glands.
  b.   They secrete nectar.
  c.   They secrete a sugary fluid.
  d.   Most are modified stamens.
  e.   Most are involved in attracting pollinators.

19. **Evolution of the Flower; p. 525; moderate; ans: a**
In archaic angiosperms, the carpels:

  a.   are leaflike.
  b.   are fused.
  c.   are completely closed.
  d.   lack ovules.
  e.   are sharply differentiated into stigma, style, and ovary.

20. **Evolution of the Flower; p. 527; moderate; ans: c**
Which of the following is NOT an evolutionary trend among flowers?

  a.   from radial symmetry to bilateral symmetry
  b.   from four floral whorls to fewer whorls
  c.   from an inferior ovary to a superior ovary
  d.   from an indefinite number of parts to a definite number of parts
  e.   from an undifferentiated perianth to one differentiated into a distinct calyx
       and corolla

21. **Evolution of the Flower; p. 528; moderate; ans: d**
In the *Asteraceae*, the pappus is a series of modified:

  a.   carpels.
  b.   stamens.
  c.   ovules.
  d.   sepals.
  e.   petals.

22. **Evolution of the Flower; p. 528; easy; ans: b**
Ray flowers and disk flowers are characteristic of the:

  a.   *Magnoliaceae.*
  b.   *Asteraceae.*
  c.   *Piperaceae.*
  d.   *Orchidaceae.*
  e.   *Aristolochiaceae.*

23. **Evolution of the Flower; p. 528; easy; ans: d**
The largest angiosperm family is the:

   a. *Magnoliaceae.*
   b. *Asteraceae.*
   c. *Piperaceae.*
   d. *Orchidaceae.*
   e. *Aristolochiaceae.*

24. **Evolution of the Flower; p. 528; moderate; ans: d**
In the *Orchidaceae*, the floral column is composed of a _____ fused with part of the _____ .

   a. carpel; sepal
   b. carpel; petal
   c. stamen; sepal
   d. stamen; carpel
   e. stamen; petal

25. **Evolution of the Flower; pp. 528–529; difficult; ans: b**
The pollinium consists of the:

   a. anther and filament.
   b. contents of an anther.
   c. stamen fused with the stigma and style.
   d. two lateral petals and the cuplike lip
   e. meristems used to clone orchids.

26. **Evolution of the Flower; p. 530; moderate; ans: e**
Which of the following was NOT an evolutionary adaptation of angiosperms in response to insects?

   a. closed carpels
   b. inferior ovary
   c. edible flower parts
   d. floral nectaries
   e. unisexual flowers

27. **Evolution of the Flower; p. 532; easy; ans: a**
The most important flower-visiting animals in angiosperm evolution are:

   a. bees.
   b. birds.
   c. beetles.
   d. bats.
   e. moths and butterflies.

28. **Evolution of the Flower; pp. 532–535; difficult; ans: c**
Which of the following is NOT an example of coevolution of bees and flowers?

   a. Bees have body parts adapted to collect and carry nectar.
   b. Bees have body parts adapted to collect and carry pollen.
   c. Flowers pollinated by bees are usually red.
   d. Flowers pollinated by bees have distinctive markings.
   e. Flowers pollinated by bees have showy, brightly colored petals.

29. Evolution of the Flower; pp. 532–535; difficult; ans: d
A type of bee that confines its visits to one or a few plant species:

a.   has fewer morphologic adaptations than bees that visit many species.
b.   usually lacks specialized mouthparts.
c.   is not an efficient pollinator.
d.   exerts a powerful evolutionary force on the plants it visits.
e.   exerts less evolutionary force on plants than do bees that visit many
     species.

30. Evolution of the Flower; pp. 532–535; easy; ans: d
Flowers with distinctive markings and complex passageways or traps are most
likely pollinated by:

a.   bats.
b.   birds.
c.   moths.
d.   bees.
e.   beetles.

31. Evolution of the Flower; pp. 535–536; moderate; ans: c
Flowers with a long corolla tube are most likely pollinated by:

a.   bats.
b.   birds.
c.   moths.
d.   bees.
e.   beetles.

32. Evolution of the Flower; p. 537; moderate; ans: b
Most flowers pollinated by birds:

a.   are blue or purple.
b.   are red or yellow.
c.   are white.
d.   have a strong odor.
e.   have landing platforms.

33. Evolution of the Flower; p. 538; easy; ans: e
Flowers that are visited regularly by large animals with high rates of energy
expenditure:

a.   have long corolla tubes.
b.   have complex passageways and traps.
c.   are odorless.
d.   have a dull color.
e.   produce copious nectar.

34. Evolution of the Flower; p. 539; moderate; ans: d
Dull-colored flowers with strong odors and producing copious nectar are most
likely pollinated by:

a.   beetles.
b.   bees.
c.   butterflies.
d.   bats.
e.   birds.

35. **Evolution of the Flower; p. 539; moderate; ans: a**
Flowers pollinated by _____ are most likely to produce pollen with the greatest amount of protein.

   a.   bats
   b.   bees
   c.   butterflies
   d.   moths
   e.   beetles

36. **Evolution of the Flower; p. 541; easy; ans: d**
Filiform pollen is common in _____ angiosperms.

   a.   wind-pollinated
   b.   self-pollinating
   c.   all aquatic
   d.   submerged aquatic
   e.   bird-pollinated

37. **Evolution of the Flower; p. 542; easy; ans: d**
The red, orange, and yellow pigments of plastids are:

   a.   anthocyanins.
   b.   flavonols.
   c.   flavonoids.
   d.   carotenoids.
   e.   betacyanins.

38. **Evolution of the Flower; p. 542; easy; ans: a**
The red and blue pigments stored in vacuoles in flowers are:

   a.   anthocyanins.
   b.   flavonols.
   c.   betalains.
   d.   carotenoids.
   e.   betacyanins.

39. **Evolution of the Flower; p. 543; moderate; ans: b**
_____ are flavonoids that vary in color with the pH of the cell sap.

   a.   Betacyanins
   b.   Anthocyanins
   c.   Betalains
   d.   Carotenoids
   e.   Chlorophylls

40. **Evolution of the Flower; p. 543; moderate; ans: e**
Ultraviolet reflectivity in flowers is related to the presence of:

   a.   anthocyanins.
   b.   flavonols.
   c.   chlorophylls.
   d.   betacyanins.
   e.   carotenoids.

41. Evolution of the Flower; p. 543; moderate; ans: d
The red color of beets is due to the presence of:

    a.   anthocyanins.
    b.   flavonols.
    c.   chlorophylls.
    d.   betacyanins.
    e.   carotenoids.

42. Evolution of Fruits; p. 543; moderate; ans: b
By definition an accessory fruit develops from:

    a.   a single ovary only.
    b.   an ovary plus additional flower parts.
    c.   several carpels of one gynoecium.
    d.   the gynoecia of more than one flower.
    e.   a receptacle only.

43. Evolution of Fruits; p. 543; easy; ans: e
By definition, a parthenocarpic fruit lacks:

    a.   petals.
    b.   carpels.
    c.   receptacles.
    d.   seeds.
    e.   stamens.

44. Evolution of Fruits; p. 543; moderate; ans: c
By definition, an aggregate fruit develops from:

    a.   a single ovary only.
    b.   an ovary plus additional flower parts.
    c.   several carpels of one gynoecium.
    d.   the gynoecia of more than one flower.
    e.   a receptacle only.

45. Evolution of Fruits; p. 543; moderate; ans: a
A simple fruit in which the inner layer of the fruit wall is fleshy is a(n):

    a.   berry.
    b.   drupe.
    c.   pome.
    d.   follicle.
    e.   achene.

46. Evolution of Fruits; p. 543; easy; ans: e
Peaches, cherries, and olives are:

    a.   capsules.
    b.   legumes.
    c.   pomes.
    d.   berries.
    e.   drupes.

47. **Evolution of Fruits; p. 544; moderate; ans: c**
A simple fruit in which the fleshy portion is derived largely from the base of the perianth is a(n):

    a. berry.
    b. drupe.
    c. pome.
    d. follicle.
    e. achene.

48. **Evolution of Fruits; p. 544; easy; ans: c**
By definition, a dehiscent fruit:

    a. is fleshy.
    b. has a single carpel.
    c. breaks open at maturity.
    d. has an inferior ovary.
    e. has a stony endocarp.

49. **Evolution of Fruits; p. 544; moderate; ans: d**
The fruit characteristic of the mustard family is a(n):

    a. achene.
    b. follicle.
    c. legume.
    d. silique.
    e. capsule.

50. **Evolution of Fruits; p. 545; easy; ans: a**
A winged achene is called a:

    a. samara.
    b. follicle.
    c. legume.
    d. silique.
    e. capsule.

51. **Evolution of Fruits; p. 545; moderate; ans: e**
Which of the following is NOT an indehiscent fruit?

    a. caryopsis
    b. nut
    c. samara
    d. schizocarp
    e. capsule

52. **Evolution of Fruits; p. 545; moderate; ans: c**
The grains typical of the grass family are a type of fruit known as a:

    a. samara.
    b. follicle.
    c. caryopsis.
    d. silique.
    e. capsule.

53. **Evolution of Fruits; p. 546; moderate; ans: b**
Which of the following is NOT an adaptation specific to wind-borne fruits or seeds?

a.   plumelike pappus
b.   tissue with large air spaces
c.   woolly hairs
d.   wings
e.   dustlike consistency

54. **Evolution of Fruits; p. 546; moderate; ans: d**
In _____, the seeds are discharged upward and outward from the plant.

a.   tumbleweed
b.   willows and poplars
c.   *Linaria*
d.   *Impatiens*
e.   the *Asteraceae*

55. **Evolution of Fruits; p. 548; easy; ans: a**
The function of elaiosomes is to:

a.   provide food for ants.
b.   produce noxious chemicals that deter herbivores.
c.   facilitate dispersal by wind.
d.   facilitate dispersal by water.
e.   adhere to the fur or feathers of animals.

56. **Biochemical Coevolution; p. 549; moderate; ans: d**
Which of the following is NOT a group of secondary plant products?

a.   glycosides
b.   alkaloids
c.   flavonoids
d.   carotenoids
e.   essential oils

57. **Biochemical Coevolution; pp. 549–550; easy; ans: e**
What is the apparent function of most secondary plant products?

a.   attracting pollinators
b.   facilitating fruit or seed dispersal
c.   serving as plant hormones
d.   serving as sources of energy for the plant
e.   restricting the palatability of the plant to herbivores

58. **Biochemical Coevolution; pp. 549–550; difficult; ans: d**
A relationship between plants of the mustard family *Brassicaceae* and larvae of the butterfly family *Pierinae* has developed such that the larvae:

a.   die when they eat the plants.
b.   escape predation by mimicry.
c.   provide food for the ants that live on the plants.
d.   feed only on those plants.
e.   protect the plants from herbivores.

59. Biochemical Coevolution; p. 550; easy; ans: b
Wild yams produce diosgenin, which is:

   a.   the active ingredient in marijuana.
   b.   similar to a human oral contraceptive.
   c.   the active ingredient in opium.
   d.   a heart poison.
   e.   a proteinase inhibitor.

60. Biochemical Coevolution; p. 551; moderate; ans: e
Which of the following statements about the coevolution of angiosperms and insects is FALSE?

   a.   Angiosperms exhibit a wide variety of pollination mechanisms.
   b.   Angiosperms exhibit a wide variety of fruit-dispersal mechanisms.
   c.   Whole families of plants are associated with major groups of plant-eating insects.
   d.   Whole families of plants can be characterized biochemically.
   e.   Many of the possible coevolution variants have evolved only once within a particular plant family.

True-False Questions

1. Origin and Diversification of the Angiosperms; p. 519; easy; ans: T
Like gymnosperms, the earliest angiosperms had pollen grains with a single pore.

2. Origin and Diversification of the Angiosperms; pp. 519, 521; moderate; ans: F
Archaeanthus linnenbergeri is a fossil representative of the paleoherbs.

3. Origin and Diversification of the Angiosperms; p. 521; moderate; ans: F
The woody magnoliids generally have flowers with fused carpels.

4. Origin and Diversification of the Angiosperms; p. 522; moderate; ans: F
The ancestor of the eudicots most likely had a flower resembling that of *Magnolia*.

5. Origin and Diversification of the Angiosperms; p. 522; easy; ans: T
Flowering plants achieved dominance first in the Northern Hemisphere, then in the Southern Hemisphere.

6. Origin and Diversification of the Angiosperms; p. 523; moderate; ans: F
Gondwanaland had split apart by the time the first angiosperms evolved.

7. Evolution of the Flower; p. 524; moderate; ans: F
In the earliest angiosperms, the perianth was sharply divided into calyx and corolla.

8. Evolution of the Flower; p. 525; moderate; ans: T
The earliest flowers were probably bisexual.

9. **Evolution of the Flower; p. 528; easy; ans: T**
   The composite head of the *Asteraceae* has the appearance of a single large flower.

10. **Evolution of the Flower; p. 529; moderate; ans: T**
    A pollinium is a pollen-dispersal unit characteristic of the *Orchidaceae*.

11. **Evolution of the Flower; p. 531; easy; ans: T**
    The diversification of bees and butterflies in the early Tertiary period was directly related to the diversification of angiosperms.

12. **Evolution of the Flower; p. 535; easy; ans: F**
    Some species of orchids attract pollinators by the resemblance of their flower to a female beetle.

13. **Evolution of the Flower; p. 537; easy; ans: T**
    In North and South America, the chief pollinators among the birds are hummingbirds.

14. **Evolution of the Flower; p. 539; easy; ans: F**
    Wind-pollinated angiosperms are the most primitive members of the *Anthophyta*.

15. **Evolution of the Flower; p. 542; easy: ans: F**
    The most important pigments in floral coloration are the carotenoids.

16. **Evolution of Fruits; p. 543; moderate; ans: F**
    The principal evolutionary role of fruits is to protect the seed from damage.

17. **Evolution of Fruits; p. 546; easy; ans: T**
    Either seeds or fruits may have wings or plumes for wind dispersal.

18. **Biochemical Coevolution; p. 550; easy; ans: T**
    Many insects that eat noxious plants are brightly colored.

Essay Questions

1. **Origin and Diversification of the Angiosperms; pp. 519–522; difficult**
   Discuss the characteristic features of the magnoliids. How is this group related to the monocots and eudicots?

2. **Origin and Diversification of the Angiosperms; pp. 521–522; moderate**
   What were the most likely ancestors of (a) the monocots and (b) the eudicots?

3. **Origin and Diversification of the Angiosperms; p. 522; moderate**
   To what type of climate were the early angiosperms adapted? Describe some of these adaptations.

4. **Origin and Diversification of the Angiosperms; p. 523; difficult**
   What changes occurred in Gondwanaland at the time angiosperms were becoming abundant in the fossil record? What were the accompanying changes in climate that influenced angiosperm evolution?

5. **Evolution of the Flower; pp. 524–525; moderate**
Discuss the evolution of the perianth, stamens, and carpels.

6. **Evolution of the Flower; pp. 527–529; moderate**
Compare and contrast the *Asteraceae* and the *Orchidaceae*.

7. **Evolution of the Flower; pp. 529–530; easy**
In what ways have plants "transcended their rooted condition" and become mobile?

8. **Evolution of the Flower; Evolution of Fruits; Biochemical Coevolution; pp. 530–551; difficult**
Discuss the coevolution of angiosperms and animals in terms of pollination, fruit dispersal, and secondary plant products.

9. **Evolution of the Flower; pp. 530–537; difficult**
What is the relationship between the color and scent of a flower and the type of insect that pollinates that flower?

10. **Evolution of the Flower; p. 538; difficult**
What is the correlation between flower shape and the rate of energy consumption of the animal that pollinates the flower?

11. **Evolution of the Flower; pp. 539–540; moderate**
Discuss the evolution of wind-pollinated flowers. What characteristic features are shared by wind-pollinated flowers?

12. **Evolution of the Flower; pp. 542–543; moderate**
List the different types of flower pigments and the colors for which they are responsible. Where in the plant cell is each located?

13. **Evolution of Fruits; pp. 543, 546; easy**
From an evolutionary point of view, what is the function of fruits? Give examples to support your answer.

14. **Evolution of Fruits; p. 543; easy**
Distinguish between simple, aggregate, and multiple fruits. Give an example of each.

15. **Evolution of Fruits; pp. 543–545; moderate**
Distinguish between fleshy fruits, dry indehiscent fruits, and dry dehiscent fruits. Give an example of each.

16. **Evolution of Fruits; pp. 547–548; moderate**
Discuss some of the evolutionary adaptations of fruits and seeds to dispersal by animals.

17. **Evolution of Fruits; p. 548; easy**
What is the predominant color among ripe fruits? What hypothesis has been suggested to explain this observation?

18. **Biochemical Coevolution; p. 549; difficult**
Discuss the role of secondary plant products in the coevolution of plants and animals.

Chapter 23

Early Development of the Plant Body

Multiple-Choice Questions

1. **Introduction; p. 556; moderate; ans: b**
    A plant's body plan consists of a(n) _____ and a(n) _____ pattern.

    a. apical-radial; basal
    b. apical-basal; radial
    c. basal-radial; apical
    d. apical; basal
    e. radial; apical

2. **Formation of the Embryo; p. 556; moderate; ans: c**
    When the zygote first divides, the two daughter cells are the _____ cell and the _____ cell.

    a. micropylar; chalazal
    b. root; shoot
    c. apical; basal
    d. proembryo; suspensor
    e. axis; proembryo

3. **Formation of the Embryo; p. 556; moderate; ans: d**
    During angiosperm embryogenesis, the basal cell gives rise to the:

    a. proembryo.
    b. embryo proper.
    c. axis.
    d. suspensor.
    e. apical cell.

4. **Formation of the Embryo; p. 556; difficult; ans: b**
    Which of the following statements about the polarity of an embryo is FALSE?

    a. It fixes the structural axis of the body.
    b. It is established only after the zygote has divided.
    c. It is essential to the development of all higher organisms.
    d. It refers to the condition in which one end is different from the other end.
    e. It is a key component of biological pattern formation.

5. Formation of the Embryo; pp. 557, 559; difficult; ans: a
Which of following lists the correct developmental sequence in eudicots, where
I is the globular stage; II, the heart stage; III, the proembryo; IV, the torpedo
stage; and V, the zygote?

    a.   V, III, I, II, IV
    b.   I, V, IV, II, III
    c.   III, V, II, IV, I
    d.   V, I, III, II, IV
    e.   V, III, II, I, IV

6. Formation of the Embryo; p. 557; moderate; ans: d
The procambium is the precursor of the:

    a.   epidermis.
    b.   ground tissues.
    c.   protoderm.
    d.   xylem and phloem.
    e.   primary meristems.

7. Formation of the Embryo; p. 557; moderate; ans: c
The primary meristems are the:

    a.   procambium and epidermis.
    b.   protoderm, xylem, and phloem.
    c.   procambium, protoderm, and ground meristem.
    d.   ground tissues, vascular tissues, and epidermis.
    e.   proembryo and embryo proper.

8. Formation of the Embryo; p. 557; moderate; ans: c
In embryogenesis in monocots, globular embryos next become:

    a.   two-lobed.
    b.   curved.
    c.   cylindrical.
    d.   heart-shaped.
    e.   spherical.

9. Formation of the Embryo; pp. 557, 559; moderate; ans: d
The root and shoot apical meristems first become discernible during the
transition between the _____ and _____.

    a.   proembryo; globular stage
    b.   torpedo stage; proembryo
    c.   heart stage; globular stage
    d.   globular stage; torpedo stage
    e.   torpedo stage; embryo proper

10. Formation of the Embryo; p. 559; moderate; ans: e
Where does the shoot apical meristem arise in eudicot and magnoliid embryos?

    a.   at the tip of a sheathlike extension
    b.   at the tip of a cotyledon
    c.   on one side of the single cotyledon
    d.   on one side of the two cotyledons
    e.   between the two cotyledons

11. **Formation of the Embryo; pp. 559–560; moderate; ans: d**
    Normal development of the _____ prevents formation of extra embryos by the
    _____.

    a.   embryo proper; apical meristems
    b.   primary meristems; embryo proper
    c.   suspensor; apical meristems
    d.   embryo proper; suspensor
    e.   suspensor; embryo proper

12. **Formation of the Embryo; p. 560; difficult; ans: c**
    During embryo-genesis in *Arabidopsis* mutants with the *twn* mutation:

    a.   no viable embryos are formed.
    b.   no food reserves are produced.
    c.   the suspensor forms secondary embryos.
    d.   the suspensor dies.
    e.   the suspensor loses its polarity.

13. **Formation of the Embryo; p. 560; moderate; ans: a**
    In *Arabidopsis*, embryonic pattern formation is known to be regulated by at
    least _____ genes.

    a.   50
    b.   75
    c.   100
    d.   250
    e.   500

14. **The Mature Embryo and Seed; p. 562; easy; ans: b**
    The funiculus joins the _____ and the _____.

    a.   endosperm; embryo
    b.   ovule; ovary wall
    c.   cotyledons; hypocotyl
    d.   hypocotyl; radicle
    e.   coleorhiza; coleoptile

15. **The Mature Embryo and Seed; pp. 562–563; moderate; ans: c**
    Which of the following CANNOT be part of the plumule?

    a.   shoot apical meristem
    b.   young leaves
    c.   radicle
    d.   epicotyl
    e.   stemlike axis

16. **The Mature Embryo and Seed; p. 562; easy; ans: d**
    The embryonic root is called the:

    a.   coleorhiza.
    b.   hilum.
    c.   plumule.
    d.   radicle.
    e.   epicotyl.

17. **The Mature Embryo and Seed; p. 563; moderate; ans: a**
In eudicots in which most of the endosperm is absorbed by the embryo, the cotyledons:

    a.   are large and fleshy.
    b.   are thin and membranous.
    c.   develop into the scutellum.
    d.   are absent.
    e.   absorb stored food during resumption of embryonic growth.

18. **The Mature Embryo and Seed; p. 563; easy; ans: a**
The cotyledon of grasses is called a(n):

    a.   scutellum.
    b.   hypocotyl.
    c.   epicotyl.
    d.   coleorhiza.
    e.   hilum.

19. **The Mature Embryo and Seed; p. 563; moderate; ans: a**
In monocots, the cotyledon can have all of the following functions EXCEPT:

    a.   protection of the plumule.
    b.   photosynthesis.
    c.   absorption of nutrients.
    d.   storage of nutrients.
    e.   transport of nutrients.

20. **The Mature Embryo and Seed; p. 563; easy; ans: d**
In monocot embryos, the plumule is enclosed by the:

    a.   coleorhiza.
    b.   scutellum.
    c.   funiculus.
    d.   coleoptile.
    e.   epicotyl.

21. **The Mature Embryo and Seed; p. 563; moderate; ans: a**
Which of the following statements about the seed coat is FALSE?

    a.   It develops from the ovary.
    b.   It protects the embryo.
    c.   It may be papery or very hard.
    d.   It may be impermeable to water.
    e.   The micropyle may be visible on the seed coat.

22. **The Mature Embryo and Seed; p. 563; easy; ans: c**
The scar left on the seed coat after the seed has separated from its stalk is called the:

    a.   micropyle.
    b.   funiculus.
    c.   hilum.
    d.   suspensor.
    e.   integument.

23. **Requirements for Seed Germination; pp. 564–565; difficult; ans: d**
Which of the following events is NOT associated with seed germination?

    a.   imbibition
    b.   activation of existing enzymes
    c.   synthesis of new enzymes
    d.   synthesis of food reserves
    e.   initiation of cell division and cell enlargement

24. **Requirements for Seed Germination; p. 565; moderate; ans: b**
When the seed coat is ruptured during germination, the seed:

    a.   switches to anaerobic glucose breakdown.
    b.   switches to aerobic respiration.
    c.   first begins to use glucose as a fuel molecule.
    d.   no longer uses glucose as a fuel molecule.
    e.   is no longer able to use oxygen.

25. **Requirements for Seed Germination; p. 565; moderate; ans: a**
The process of after-ripening involves:

    a.   enzymatic modification of a dormant seed so that it will germinate.
    b.   biochemical conversion of a germinating seed to a dormant seed.
    c.   cessation of the flow of nutrients from the parent plant to the ovule.
    d.   stimulation of the primary meristems to develop.
    e.   desiccation and hardening of the seed coat.

26. **Requirements for Seed Germination; p. 565; easy; ans: c**
In temperate regions of the world, after-ripening is triggered by:

    a.   drought.
    b.   high humidity.
    c.   low temperature.
    d.   high temperature.
    e.   low oxygen levels.

27. **Requirements for Seed Germination; p. 565; moderate; ans: b**
What induces the seeds of manzanita and other plants of the California chaparral to germinate?

    a.   light
    b.   fire
    c.   rainfall
    d.   activity of digestive enzymes
    e.   low temperature

28. **From Embryo to Adult Plant; p. 566; easy; ans: d**
The _____ is usually the first structure to emerge from a germinating seed.

    a.   epicotyl
    b.   hypocotyl
    c.   cotyledon
    d.   radicle
    e.   coleoptile

29. **From Embryo to Adult Plant; p. 566; easy; ans: b**
    In monocots, the root system commonly develops from:

    a.   branch roots.
    b.   adventitious roots.
    c.   the primary root.
    d.   the secondary root.
    e.   the taproot.

30. **From Embryo to Adult Plant; p. 566; easy; ans: c**
    The term "adventitious" describes structures that:

    a.   develop before a seed has germinated.
    b.   develop after a seed has germinated.
    c.   develop in unusual locations.
    d.   are involved in after-ripening.
    e.   are involved in seed dormancy.

31. **From Embryo to Adult Plant; p. 567; easy; ans: c**
    In epigeous germination, which structure emerges above ground first?

    a.   cotyledon
    b.   epicotyl
    c.   hypocotyl
    d.   coleoptile
    e.   radicle

32. **From Embryo to Adult Plant; p. 567; moderate; ans: d**
    In the pea (*Pisum sativum*), the _____ forms the hook that pushes to the soil surface during seed germination.

    a.   coleoptile
    b.   cotyledon
    c.   hypocotyl
    d.   epicotyl
    e.   radicle

33. **From Embryo to Adult Plant; p. 567; difficult; ans: a**
    Which of the following statements about seed germination in onion (*Allium cepa*) is FALSE?

    a.   Germination is hypogeous.
    b.   The plumule emerges from the cotyledon.
    c.   The cotyledon becomes photosynthetic.
    d.   The cotyledon forms the hook.
    e.   The stored food is found in the endosperm.

34. **From Embryo to Adult Plant; p. 568; moderate; ans: d**
    In maize (*Zea mays*), the first structure to emerge from the seed during germination is the:

    a.   epicotyl.
    b.   hypocotyl.
    c.   radicle.
    d.   coleorhiza.
    e.   coleoptile.

True-False Questions

1. **Formation of the Embryo; p. 556; easy; ans: F**
   During embryogenesis in angiosperms, the apical cell gives rise to the suspensor.

2. **Formation of the Embryo; pp. 557, 559; difficult; ans: F**
   The root and shoot apical meristems are two of the primary meristems.

3. **Formation of the Embryo; p. 557; moderate; ans: T**
   In the heart stage of development, the lobes of the "heart" are the cotyledons.

4. **Formation of the Embryo; pp. 559–560; easy; ans: F**
   The only role of the suspensor in angiosperms is to push the embryo into nutritive tissues.

5. **Formation of the Embryo; p. 560; moderate; ans: T**
   Regulatory genes affect the apical-basal pattern of embryonic development in *Arabidopsis*.

6. **The Mature Embryo and Seed; p. 562; moderate; ans: F**
   The first bud of the embryonic shoot of angiosperms is called the epicotyl.

7. **The Mature Embryo and Seed; pp. 562–563; moderate; ans: T**
   Seeds with large cotyledons typically have little or no endosperm.

8. **The Mature Embryo and Seed; p. 563; easy; ans: T**
   The scutellum is the cotyledon of grass embryos.

9. **The Mature Embryo and Seed; p. 563; easy; ans: F**
   The coleoptile and coleorhiza are typically found in monocots, eudicots, and magnoliids.

10. **Requirements for Seed Germination; p. 564; moderate; ans: T**
    Before a seed can germinate, it must first imbibe water.

11. **Requirements for Seed Germination; p. 565; moderate; ans: T**
    Before the seed coat ruptures during germination, glucose breakdown may be entirely anaerobic.

12. **Requirements for Seed Germination; p. 565; difficult; ans: F**
    In temperate regions, after-ripening ensures that seeds will germinate in the fall rather than in the spring.

13. **From Embryo to Adult Plant; p. 567; easy; ans: T**
    In castor bean (*Ricinus communis*), the stored food is found in the endosperm.

14. **From Embryo to Adult Plant; p. 567; easy; ans: T**
    In hypogeous germination, the cotyledons remain in the soil.

15. **From Embryo to Adult Plant; p. 568; moderate; ans: F**
    The period of the life cycle in which a plant is must susceptible to damage is between seed formation and germination.

Essay Questions

1. **Formation of the Embryo; p. 556; moderate**
Discuss the concept of polarity as it pertains to angiosperm embryogenesis.

2. **Formation of the Embryo; p. 557; moderate**
Define the term "primary meristem." Name the three primary meristems, and name the tissue(s) into which each develops.

3. **Formation of the Embryo; pp. 557–559; moderate**
List the principal stages of embryonic development in angiosperms, and describe the appearance of the embryo at each stage.

4. **Formation of the Embryo; pp. 559–560; moderate**
Discuss the role(s) of the suspensor during embryogenesis in angiosperms.

5. **Formation of the Embryo; pp. 559–560; difficult**
In flowering plants, how does the development of the embryo affect the development of the suspensor? Give evidence to support your answer.

6. **Formation of the Embryo; p. 560; moderate**
Give three examples of types of genes that govern *Arabidopsis* embryogenesis.

7. **The Mature Embryo and Seed; p. 562; moderate**
Name three tissues that accumulate food reserves within seeds.

8. **The Mature Embryo and Seed; pp. 562–563; moderate**
When is the term "hypocotyl-root axis" used instead of "radicle"?

9. **The Mature Embryo and Seed; pp. 562–563; moderate**
Describe the various roles played by cotyledons in angiosperms.

10. **The Mature Embryo and Seed; p. 563; moderate**
What is the relationship between the funiculus, the hilum, and the seed coat?

11. **Requirements for Seed Germination; p. 565; moderate**
Discuss the oxygen, temperature, and water requirements for the germination of most seeds.

12. **Requirements for Seed Germination; p. 565; difficult**
Define seed dormancy. Are the packaged seeds that you buy at the store likely to be dormant? Why or why not?

13. **Requirements for Seed Germination; p. 565; moderate**
What are the two most common causes of seed dormancy? Discuss some of the ways in which seeds can overcome dormancy.

14. **Requirements for Seed Germination; p. 565; moderate**
In what ways do dormancy and after-ripening have survival value for a plant?

15. **From Embryo to Adult Plant; pp. 566–567; moderate**
Define the terms "epigeous" and "hypogeous," and give examples of each type of germination.

16. **From Embryo to Adult Plant; p. 567; difficult**
During seed germination, how does the process of digestion and subsequent transport of stored food differ in the castor bean (*Ricinus communis*) and the garden bean (*Phaseolus vulgaris*)?

17. **From Embryo to Adult Plant; pp. 567–568; moderate**
Both onion (*Allium cepa*) and maize (*Zea Mays*) are monocots, but they differ in certain features of germination. Outline these differences.

Chapter 24

Cells and Tissues of the Plant Body

Multiple-Choice Questions

1. **Cells and Tissues of the Plant Body; p. 571; moderate; ans: d**
    Which of the following statements about the shoot and root apical meristems is FALSE?

    a. They are perpetually young tissues or cells.
    b. They are established during embryogenesis.
    c. It is through their activity that most plant development occurs.
    d. They lose the potential to divide soon after embryogenesis is complete.
    e. They generate cells that give rise to roots, stems, leaves, and flowers.

2. **Apical Meristems and Their Derivatives; p. 571; moderate; ans: c**
    Which of the following statements about primary growth is FALSE?

    a. It results in extension of the plant body.
    b. It involves the formation of primary tissues.
    c. It results in the thickening of the stem and root.
    d. It gives rise to the primary plant body.
    e   It results from activity of the root and shoot apical meristems.

3. **Apical Meristems and Their Derivatives; p. 571; easy; ans: c**
    Growth in plants is the counterpart of _____ in animals.

    a. reproduction
    b. differentiation
    c. motility
    d. irritability
    e. development

4. **Growth, Morphogenesis, and Differentiation; p. 571; moderate; ans: e**
    Most of the growth of a plant body is the result of:

    a. morphogenesis.
    b. embryogenesis.
    c. differentiation.
    d. cell division.
    e. cell enlargement.

5. **Growth, Morphogenesis, and Differentiation; p. 572; moderate; ans: b**
Morphogenesis refers to:

   a.   an irreversible increase in size.
   b.   the acquisition of a particular shape.
   c.   the sum of all the events that lead to formation of an organism's body.
   d.   the process by which cells become different from one another.
   e.   the fate of a plant cell.

6. **Growth, Morphogenesis, and Differentiation; p. 572; moderate; ans: b**
The process by which cells derived from meristematic cells become specialized is called:

   a.   derivation.
   b.   differentiation.
   c.   morphogenesis.
   d.   development.
   e.   interaction.

7. **Internal Organization of the Plant Body; p. 572; easy; ans: a**
The three tissue systems of vascular plants are:

   a.   the dermal, vascular, and ground tissue systems.
   b.   protoderm, procambium, and ground meristem.
   c.   parenchyma, collenchyma, and sclerenchyma.
   d.   epidermis, periderm, and protoderm.
   e.   xylem, phloem, and epidermis.

8. **Internal Organization of the Plant Body; p. 572; moderate; ans: d**
From which primary meristem does sclerenchyma develop?

   a.   parenchyma
   b.   collenchyma
   c.   procambium
   d.   ground meristem
   e.   protoderm

9. **Internal Organization of the Plant Body; p. 572; easy; ans: e**
The most common type of ground tissue is:

   a.   epidermis.
   b.   xylem.
   c.   sclerenchyma.
   d.   collenchyma.
   e.   parenchyma.

10. **Internal Organization of the Plant Body; p. 573; moderate; ans: b**
In a eudicot stem, _____ is the ground tissue external to the system of vascular strands and _____ is the ground tissue internal to these strands.

   a.   pith; cortex
   b.   cortex; pith
   c.   xylem; phloem
   d.   xylem; cortex
   e.   pith; phloem

11. **Internal Organization of the Plant Body; p. 573; easy; ans: d**
    _____ is a simple tissue, and _____ is a complex tissue.

    a. Xylem; phloem
    b. Phloem; xylem
    c. Parenchyma; collenchyma
    d. Collenchyma; xylem
    e. Xylem; sclerenchyma

12. **Ground Tissues; p. 574; moderate; ans: e**
    Which of the following statements about parenchyma cells is FALSE?

    a. They can photosynthesize.
    b. They can initiate adventitious roots.
    c. They are involved in secretion.
    d. They are capable of cell division.
    e. They lack secondary walls.

13. **Ground Tissues; p. 574; moderate; ans: b**
    The role of transfer cells is to:

    a. transport water and minerals throughout the plant.
    b. facilitate the movement of solutes over short distances.
    c. move sugars through the phloem.
    d. transfer solutes from the cortex to the pith via rays.
    e. increase the rate of water movement through stomata.

14. **Ground Tissues; p. 574; moderate; ans: a**
    _____ tissue is composed of cells having unevenly thickened primary walls.

    a. Collenchyma
    b. Xylem
    c. Parenchyma
    d. Phloem
    e. Sclerenchyma

15. **Ground Tissues; p. 575; moderate; ans: c**
    Which of the following statements about sclerenchyma cells is FALSE?

    a. They are ground-tissue cells.
    b. They often lack protoplasts at maturity.
    c. They strengthen plant parts that are still elongating.
    d. They are either fibers or sclereids.
    e. They may develop in any part of the primary and secondary plant bodies.

16. **Ground Tissues; p. 575; moderate; ans: b**
    Sclereids differ from fibers in that sclereids:

    a. are dead.
    b. are shorter.
    c. are more uniform in shape.
    d. have thick, lignified walls.
    e. occur singly.

17. Vascular Tissues; p. 576; easy; ans: a
_____ are types of tracheary elements.

a. Vessel elements and tracheids
b. Tracheids and xylem parenchyma
c. Vessel elements and xylem parenchyma
d. Xylem fibers and tracheids
e. Xylem fibers and xylem parenchyma

18. Vascular Tissues; p. 576; moderate; ans: c
_____ have pits but not perforations.

a. Parenchyma cells
b. Sclerenchyma cells
c. Tracheids
d. Vessel elements
e. Sieve elements

19. Vascular Tissues; p. 576; difficult; ans: b
Perforation plates are characteristic of the _____ of _____.

a. tracheids; angiosperms
b. vessel elements; angiosperms
c. tracheids; gymnosperms
d. vessel elements; gymnosperms
e. tracheids; seedless vascular plants

20. Vascular Tissues; p. 577; moderate; ans: e
One role of pit membranes in tracheids is to:

a. facilitate movement of air bubbles.
b. facilitate water movement.
c. facilitate solute transport.
d. provide support.
e. trap air bubbles.

21. Vascular Tissues; pp. 577–578; difficult; ans: b
By definition, metaxylem is _____ protoxylem.

a. formed bcfore
b. formed after
c. formed at the same time as
d. larger than
e. smaller than

22. Vascular Tissues; p. 578; easy; ans: b
Apoptosis is:

a. the blockage of vessel elements by air bubbles.
b. programmed cell death.
c. the storage of toxic materials.
d. the stretching of the secondary cell wall.
e. a type of solute transport.

23. **Vascular Tissues; pp. 579–580, 581–582; moderate; ans: d**
    In vascular plants, food is conducted through:

    a. companion cells only.
    b. sieve cells only.
    c. sieve-tube elements only.
    d. sieve cells and sieve-tube elements only.
    e. companion cells, sieve cells, and sieve-tube elements.

24. **Vascular Tissues; pp. 579–581; moderate; ans: b**
    Which of the following statements about sieve cells is FALSE?

    a. They transport food.
    b. They are interconnected to form sieve tubes.
    c. They are found in gymnosperms but not angiosperms.
    d. They are living cells at maturity.
    e. They have sieve areas.

25. **Vascular Tissues; p. 580; easy; ans: d**
    Callose is a _____ deposited in the _____ of sieve elements.

    a. protein; pits
    b. protein; pores
    c. carbohydrate; pits
    d. carbohydrate; pores
    e. fat; perforation plates

26. **Vascular Tissues; pp. 581–582; moderate; ans: e**
    Which of the following statements about P-protein is FALSE?

    a. It is found in the protoplasts of sieve-tube elements of magnoliids,
        eudicots, and some monocots.
    b. In undisturbed cells, it lines the sieve-plate pores.
    c. It may serve to seal the sieve-plate pores when the cell is wounded.
    d. It originates in P-protein bodies.
    e. The "P" stands for protection.

27. **Vascular Tissues; p. 582; difficult; ans: c**
    The parenchyma cells that are developmentally related to the _____ of
    angiosperms are called _____.

    a. sieve cells; companion cells
    b. sieve cells; albuminous cells
    c. sieve-tube elements; companion cells
    d. sieve-tube elements; albuminous cells
    e. sieve tubes; P-protein bodies

28. **Vascular Tissues; p. 582; moderate; ans: e**
    Albuminous cells are thought to have the same function as:

    a. vessel elements.
    b. tracheids.
    c. sieve cells.
    d. sieve-tube elements.
    e. companion cells.

29. Dermal Tissues; p. 583; moderate; ans: c
Stomata are the _____ between _____.

a.   cells; subsidiary cells
b.   cells; guard cells
c.   pores; guard cells
d.   pores; subsidiary cells
e.   pores; trichomes

30. Dermal Tissues; pp. 583–584; difficult; ans: a
Which of the following is NOT a function of trichomes?

a.   providing structural support
b.   defending against insects
c.   secreting salts
d.   absorbing water and minerals from the soil
e.   reflecting solar radiation

31. Dermal Tissues; pp. 586–587; moderate; ans: e
The cork cambium produces _____ on its outer surface and _____ on its inner surface.

a.   phelloderm; phellogen
b.   phellogen; phelloderm
c.   phellogen; phellem
d.   phelloderm; cork
e.   cork; phelloderm

True-False Questions

1.   Apical Meristems and Their Derivatives; p. 571; difficult; ans: T
Primary meristems eventually cease to be meristematic and begin to differentiate.

2.   Apical Meristems and Their Derivatives; p. 571; easy; ans: F
Primary growth refers to growth of the embryo; secondary growth refers to growth of the seedling.

3.   Apical Meristems and Their Derivatives; p. 571; moderate; ans: T
The unlimited growth of apical meristems is called indeterminate growth.

4.   Growth, Morphogenesis, and Differentiation; p. 571; easy; ans: T
Cell division does not necessarily increase the size of an organism.

5.   Growth, Morphogenesis, and Differentiation; p. 572; moderate; ans: F
The fate of a plant cell is determined solely by the genes in its chromosomes.

6.   Internal Organization of the Plant Body; pp. 572–573; moderate; ans: T
In general, vascular tissues are embedded within ground tissue, and the dermal tissue forms a surrounding layer.

7.   Internal Organization of the Plant Body; p. 573; easy; ans: T
Simple tissues are composed of only one cell type whereas complex tissues are composed of two or more cell types.

8. **Ground Tissues; p. 574; easy; ans: F**
Transfer cells are collenchyma cells specialized for short-distance solute transport.

9. **Ground Tissues; p. 575; easy; ans: T**
Fibers and sclereids are types of sclerenchyma cells.

10. **Vascular Tissues; p. 576; moderate; ans: F**
In the primary plant body, xylem develops from the vascular cambium.

11. **Vascular Tissues; p. 576; easy; ans: F**
Tracheids are tracheary elements found in angiosperms but not gymnosperms.

12. **Vascular Tissues; p. 577; moderate; ans: T**
Obstruction of water flow by air bubbles is more likely to occur in vessel elements than in tracheids.

13. **Vascular Tissues; pp. 577–578; difficult; ans: T**
The secondary walls of protoxylem, but not metaxylem, may have annual or helical thickenings.

14. **Vascular Tissues; p. 580; moderate; ans: F**
Sieve-tube elements lack sieve plates.

15. **Vascular Tissues; pp. 580–581; moderate; ans: T**
Unlike wound callose, definitive callose is deposited at the sieve areas and sieve plates of senescing sieve elements.

16. **Vascular Tissues; p. 581; easy; ans: T**
Sieve elements must be living in order to transport food.

17. **Vascular Tissues; p. 582; moderate; ans: T**
A companion cell and its associated sieve-tube element are derived from the same mother cell.

18. **Vascular Tissues; pp. 582–583; moderate; ans: F**
Albuminous cells are thought to have the same function as transfer cells.

19. **Dermal Tissues; p. 584; easy; ans: T**
The "bloom" on the surface of some leaves and fruits is caused by epicuticular wax.

20. **Dermal Tissues; pp. 586–587; moderate; ans: F**
The phellem produces both cork and phelloderm.

Essay Questions

1. **Apical Meristems and Their Derivatives; p. 571; easy**
What cell types comprise the apical meristems of plants? Is cell division limited to only one cell type in the apical meristem? Explain.

2. **Apical Meristems and Their Derivatives; p. 571; moderate**
    What is the difference between apical meristems and primary meristems?
    Between primary meristems and primary growth?

3. **Apical Meristems and Their Derivatives; p. 571; moderate**
    Discuss the idea that growth in plants is the counterpart of motility in animals.

4. **Growth, Morphogenesis, and Differentiation; pp. 571–572; difficult**
    Explain the differences among growth, morphogenesis, and differentiation.
    How does the interaction among these processes result in development?

5. **Growth, Morphogenesis, and Differentiation; p. 572; moderate**
    Discuss the various ways in which the fate of a plant cell is determined.

6. **Internal Organization of the Plant Body; pp. 572–573; moderate**
    How does the distribution of the three tissue systems vary in different parts of
    the eudicot plant body?

7. **Internal Organization of the Plant Body; p. 573; moderate**
    What is the difference between a simple tissue and a complex tissue? Give
    examples of each.

8. **Ground Tissues; pp. 573–576; moderate**
    Compare and contrast the structures and functions of the various cell types that
    comprise the ground tissue.

9. **Ground Tissues; p. 574; moderate**
    Discuss the roles of transfer cells in plants.

10. **Vascular Tissues; pp. 576–579; moderate**
    Compare and contrast the two main types of cells of xylem tissue in
    angiosperms.

11. **Vascular Tissues; pp. 576–577; moderate**
    In what way are vessel elements more efficient conductors of water than
    tracheids? In what way are they less safe?

12. **Vascular Tissues; pp. 577–578; difficult**
    In tracheary elements, what causes annual and helical thickenings to form
    instead of pits?

13. **Vascular Tissues; pp. 579–580; moderate**
    In what ways is a sieve cell less specialized than a sieve-tube member?

14. **Vascular Tissues; pp. 579–583; difficult**
    Compare and contrast the four main types of cells of phloem tissue in
    angiosperms. How do they differ from the phloem cells of gymnosperms?

15. **Dermal Tissues; pp. 585–586; moderate**
    What are trichomes? List some of their functions, and discuss the control of
    trichome development as deduced from studies of *Arabidopsis*.

16. Dermal Tissues; pp. 586–587; moderate
List the tissues that make up the periderm, and give the function of each type.

Chapter 25

The Root: Structure and Development

Multiple-Choice Questions

1. **Root Systems; p. 590; moderate; ans: d**
    In gymnosperms, magnoliids, and eudicots, the primary root is called the:

    a. fibrous root.
    b. adventitious root.
    c. lateral root.
    d. taproot.
    e. branch root.

2. **Root Systems; p. 590; moderate; ans: b**
    Which of the following statements about monocot roots is FALSE?

    a. The root system is generally shallower than in eudicots.
    b. The main root system has one prominent root.
    c. The roots form a fibrous root system.
    d. The main root system develops from adventitious roots.
    e. The primary root is short-lived.

3. **Root Systems; p. 590; moderate; ans: a**
    Feeder roots usually:

    a. occur in the upper meter of soil.
    b. occur at the tips of taproots.
    c. do not take up minerals.
    d. do not take up water.
    e. are coarser rather than nonfeeder roots.

4. **Origin and Growth of Primary Tissues; p. 591; moderate; ans: c**
    Which of the following statements about the rootcap is FALSE?

    a. It is covered by mucigel.
    b. Its cells are sloughed as it grows through the soil.
    c. Its sloughed cells are not replaced.
    d. It helps the root penetrate the soil.
    e. It protects the root apical meristem.

5. **Origin and Growth of Primary Tissues; p. 591; moderate; ans: b**
    Mucigel is:

    a. produced by the root apical meristem.
    b. secreted by the outer rootcap cells.
    c. a highly hydrated protein.
    d. responsible for stimulating cell division.
    e. a gravity sensor.

6. **Origin and Growth of Primary Tissues; pp. 591–592; difficult; ans: d**
    In a "closed type" of root apical organization, the:

    a. apical meristem lacks derivatives.
    b. cortex and vascular cylinder each have their own initials, but the rootcap and epidermis do not.
    c. cortex and rootcap each have their own initials, but the vascular cylinder and epidermis do not.
    d. rootcap, vascular cylinder, and cortex each have their own initials.
    e. rootcap, vascular cylinder, and cortex have common initials.

7. **Origin and Growth of Primary Tissues; p. 594; difficult; ans: a**
    Which of the following statements about the quiescent center is FALSE?

    a. It is located a short distance behind the apical meristem.
    b. It may play an important role in the organization and development of the root.
    c. It is a region that was mitotically active early in root development.
    d. It is a relatively inactive region of the mature root apical meristem.
    e. Quiescent centers isolated from maize can form whole roots in culture without first forming callus.

8. **Origin and Growth of Primary Tissues; p. 595; moderate; ans: c**
    Root hairs are produced in the:

    a. region of cell division.
    b. region of elongation.
    c. region of maturation.
    d. quiescent center.
    e. area between the regions of cell division and elongation.

9. **Origin and Growth of Primary Tissues; p. 595; easy; ans: c**
    The sequence of regions in a growing root, beginning immediately behind the rootcap, is:

    a. elongation, maturation, cell division.
    b. cell division, maturation, elongation.
    c. cell division, elongation, maturation.
    d. elongation, cell division, maturation.
    e. maturation, elongation, cell division.

10. **Origin and Growth of Primary Tissues, p. 595; moderate; ans: c**
    The _____ is also called the root-hair zone.

    a. region of cell division
    b. region of elongation
    c. region of maturation
    d. rootcap
    e. root apical meristem

11. **Primary Structure; p. 596; moderate; ans: c**
    Root hairs:

    a. are found mainly in the region of elongation.
    b. are absent from feeder roots.
    c. are relatively short-lived.
    d. are specialized cells.
    e. decrease the absorptive surface of the root.

12. **Primary Structure; p. 597; moderate; ans: c**
    Which of the following statements about mucigel is FALSE?

    a. It influences the availability of ions to the root.
    b. It may provide short-term protection from desiccation.
    c. It acts as a glue to hold several different roots together.
    d. It provides an environment favorable to beneficial bacteria.
    e. It helps to bind soil to the root.

13. **Primary Structure; p. 597; difficult; ans: b**
    The rhizosphere is the layer of:

    a. nitrogen-fixing bacteria surrounding the root.
    b. soil bound to the root.
    c. mucigel bound to the root.
    d. root hairs and mucigel.
    e. sloughed-off rootcap cells.

14. **Primary Structure; p. 598; easy; ans: e**
    The _____ occupies the greatest area of the primary root.

    a. epidermis
    b. xylem
    c. phloem
    d. pericycle
    e. cortex

15. **Primary Structure; p. 598; moderate; ans: c**
    Substances moving through the root cortex:

    a. follow a symplastic pathway only.
    b. follow an apoplastic pathway only.
    c. follow both symplastic and apoplastic pathways.
    d. are unable to travel via cell walls.
    e. are unable to pass through plasmodesmata.

16. Primary Structure; p. 598; easy; ans: d
The innermost layer of the cortex is the:

    a.   pericycle.
    b.   xylem.
    c.   epidermis.
    d.   endodermis.
    e.   pith.

17. Primary Structure; p. 598; moderate; ans: b
Which of the following statements about Casparian strips is FALSE?

    a.   They are found in the anticlinal walls of endodermal cells.
    b.   They are permeable to water but not to ions.
    c.   They are bandlike portions of the cell wall and middle lamella.
    d.   They contain suberin.
    e.   The do not contain plasmodesmata.

18. Primary Structure; p. 598; easy; ans: a
The presence of Casparian strips forces substances entering and leaving the vascular cylinder to pass through the protoplasts of _____ cells.

    a.   endodermal
    b.   epidermal
    c.   pericycle
    d.   xylem
    e.   phloem

19. Primary Structure; p. 598; moderate; ans: c
Suberin lamellae, such as those found in some older roots, are alternating layers of:

    a.   lignin and cellulose.
    b.   cellulose and wax.
    c.   suberin and wax.
    d.   suberin and lignin.
    e.   suberin and cellulose.

20. Primary Structure; p. 600; moderate; ans: e
Endodermal cells that do not become suberized in older roots are called _____ cells.

    a.   epidermal
    b.   exodermal
    c.   Casparian
    d.   transfer
    e.   passage

21. Primary Structure; p. 600; difficult; ans: e
Which of the following statements about the exodermis is FALSE?

    a.   It apparently reduces water loss from the root.
    b.   It may contain cellulose.
    c.   It contains suberin lamellae.
    d.   It contains Casparian strips.
    e.   It is part of the epidermis.

22. **Primary Structure; p. 600; moderate; ans: a**
The outermost layer of a root's vascular cylinder is the:

   a.   pericycle.
   b.   protoxylem.
   c.   metaxylem.
   d.   phloem.
   e.   endodermis.

23. **Primary Structure; p. 600; difficult; ans: c**
At the center of most eudicot roots is:

   a.   pith.
   b.   protoxylem.
   c.   metaxylem.
   d.   pericycle.
   e.   air.

24. **Primary Structure; p. 600; 25-18; moderate; ans: b**
Unlike eudicot roots, the roots of some monocots:

   a.   have a triarch xylem.
   b.   have a pith.
   c.   lack an endodermis.
   d.   lack a pericycle.
   e.   lack Casparian strips.

25. **Effect of Secondary Growth on the Primary Body of the Root; p. 600; moderate; ans: b**
Secondary growth depends on the activity of _____ and _____.

   a.   xylem; phloem
   b.   vascular cambium; cork cambium
   c.   vascular tissue; periderm
   d.   apical meristems; primary meristems
   e.   cork; phelloderm

26. **Effect of Secondary Growth on the Primary Body of the Root; p. 600; difficult; ans: e**
The vascular cambium is initiated by cells:

   a.   of the pericycle.
   b.   of the endodermis.
   c.   in the center of the vascular cylinder.
   d.   opposite the protoxylem poles.
   e.   between the primary xylem and primary phloem.

27. **Effect of Secondary Growth on the Primary Body of the Root; p. 601; moderate; ans: d**
In roots, the vascular cambium arises from the:

   a.   procambium only.
   b.   endodermis only.
   c.   pericycle only.
   d.   procambium and pericycle only.
   e.   procambium, endodermis, and pericycle.

28. **Effect of Secondary Growth on the Primary Body of the Root; p. 602; moderate; ans: a**
As the vascular cambium continues to divide during secondary growth of the root, most of the primary phloem:

   a.   is crushed.
   b.   is pushed inward.
   c.   remains where it was produced.
   d.   differentiates into pericycle.
   e.   differentiates into xylem.

29. **Effect of Secondary Growth on the Primary Body of the Root; p. 602; moderate; ans: e**
In roots, cork cambium arises in the:

   a.   procambium.
   b.   phloem.
   c.   xylem.
   d.   endodermis.
   e.   pericycle.

30. **Effect of Secondary Growth on the Primary Body of the Root; p. 602; easy; ans: b**
The function of lenticels in the periderm of roots is to:

   a.   serve as a barrier to the passage of minerals.
   b.   allow for gas exchange.
   c.   stimulate the vascular cambium.
   d.   stimulate phelloderm formation.
   e.   produce cork.

31. **Effect of Secondary Growth on the Primary Body of the Root; p. 602; difficult; ans: d**
In a woody root one meter in diameter, which tissue would NOT be present?

   a.   primary xylem
   b.   secondary xylem
   c.   secondary phloem
   d.   epidermis
   e.   cork

32. **Origin of Lateral Roots; p. 603; moderate; ans: a**
In angiosperms, cells of both the _____ and the _____ contribute to lateral root formation.

   a.   endodermis; pericycle
   b.   epidermis; pericycle
   c.   epidermis; endodermis
   d.   pith; cortex
   e.   primary phloem; secondary phloem

33. Aerial Roots and Air Roots; p. 605; moderate; ans: b
A pneumatophore is:
a. produced by belowground structures.
b. negatively gravitropic.
c. produced by trees growing in dry habitats.
d. produced by trees growing in well-drained soil.
e. also called an aerial root.

34. Aerial Roots and Air Roots; pp. 605–606; easy; ans: b
Which of the following plants has a velamen?
a. maize
b. epiphytic orchid
c. black mangrove
d. "flower pot plant"
e. banyan tree

35. Aerial Roots and Air Roots; p. 605; easy; ans: a
Velamen is a(n):
a. multiple epidermis.
b. multiple cortex.
c. epiphyte.
d. pneumatophore.
e. enzyme produced by lateral roots.

36. Adaptations for Food Storage: Fleshy Roots; p. 606; moderate; ans: d
Which of the following statements about supernumerary cambia in sugarbeet roots is FALSE?
a. They are types of additional cambia.
b. They are arranged in concentric rings.
c. They produce storage parenchyma.
d. They produce phloem toward the inside and xylem toward the outside.
e. They are responsible for most of the increase in thickness.

True-False Questions

1. Introduction; p. 590; easy; ans: T
Some hormones and secondary metabolites are synthesized in roots.

2. Root Systems; p. 590; moderate; ans: F
The root systems of eudicots are well suited to preventing soil erosion.

3. Root Systems; p. 590; easy; ans: T
The lateral spread of tree roots usually exceeds the spread of the crown of the tree.

4. Root Systems; p. 590; moderate; ans: F
Once feeder roots are damaged or destroyed, they cannot be replaced.

5. Origin and Growth of Primary Tissues; p. 591; moderate; ans: T
Roots frequently grow through spaces left in the soil by earlier roots that have died and decayed.

6. **Origin and Growth of Primary Tissues; p. 591; moderate; ans: T**
    The rootcap controls the response of the root to gravity.

7. **Origin and Growth of Primary Tissues; p. 594; difficult; ans: F**
    The quiescent center contains cells that cease to divide when bordering
    meristematic cells are injured.

8. **Origin and Growth of Primary Tissues; p. 595; moderate; ans: F**
    There is a gradual transition between the region of cell division and the region
    of maturation in the root.

9. **Primary Structure; p. 596; easy; ans: T**
    The root epidermis allows free passage of water and minerals into the root.

10. **Primary Structure; p. 597; easy; ans: T**
    In mycorrhizae, the fungal hyphae may extend into the soil well beyond the
    plant roots.

11. **Primary Structure; p. 598; moderate; ans: T**
    All substances entering the vascular cylinder must pass through living cells of
    the endodermis.

12. **Primary Structure; p. 598; moderate; ans: F**
    The development of suberin lamellae in the endodermis prevents the movement
    of substances across this layer of cells.

13. **Primary Structure; p. 600; moderate; ans: F**
    Pericycle originates from ground meristem.

14. **Primary Structure; p. 600; easy; ans: T**
    The terms "diarch" and "triarch" refer to the number of xylem ridges in a
    root's vascular cylinder.

15. **Effect of Secondary Growth on the Primary Body of the Root; p. 601;
    moderate; ans: F**
    The vascular cambium of roots produces secondary xylem toward the outside
    and secondary phloem toward the inside.

16. **Origin of Lateral Roots; p. 603; easy; ans: T**
    Lateral roots are endogenous structures.

17. **Aerial Roots and Air Roots; p. 603; easy; ans: F**
    Prop roots are types of lateral roots common in tropical trees.

18. **Aerial Roots and Air Roots; p. 605; moderate; T**
    Velamen provides mechanical protection and reduces water loss in some
    epiphytes.

19. **Adaptations for Food Storage: Fleshy Roots; p. 606; moderate; ans: F**
    Sweet potato roots have a single vascular cambium that produces storage
    parenchyma.

Essay Questions

1. **Root Systems; p. 590; easy**
Distinguish between the root systems of monocots and those of magnoliids and eudicots.

2. **Root Systems; pp. 590–591, 596–597; moderate**
When you transplant a small tree, how is the balance between root and shoot disrupted? What other damage can occur? How could you increase the chances that the transplant will be successful?

3. **Origin and Growth of Primary Tissues; p. 591; moderate**
List the functions of the rootcap.

4. **Origin and Growth of Primary Tissues; pp. 591–594; moderate**
Compare and contrast open and closed types of root apical organization.

5. **Origin and Growth of Primary Tissues; p. 594; difficult**
What is a quiescent center? Explain how a quiescent center can be inactive under certain conditions but active under others.

6. **Origin and Growth of Primary Tissues; p. 595; moderate**
List the three growth regions of the root, and describe the events that are characteristic of each. Give examples to show that the regions are not sharply delimited.

7. **Primary Structure; pp. 595–560; moderate**
Trace the movement of an ion from the soil to the center of a root, listing all tissues through which the ion must pass.

8. **Primary Structure; p. 598; difficult**
Explain the role of the endodermis in regulating the movement of substances into and out of the root's vascular cylinder. How does the structure of the endodermis change over time?

9. **Primary Structure; p. 600; moderate**
What is the pericycle? What are some of its functions?

10. **Effect of Secondary Growth on the Primary Body of the Root; pp. 600–601; difficult**
Describe the process by which the activity of the vascular cambium results in an increase in root diameter.

11. **Effect of Secondary Growth on the Primary Body of the Root; p. 602; moderate**
What is the function of the periderm? What tissues comprise the periderm, and how are they formed?

12. **Origin of Lateral Roots; p. 603; difficult**
Describe the process by which lateral roots arise. What is the advantage for the plant of lateral roots arising endogenously rather than exogenously?

13. Aerial Roots and Air Roots; pp. 603, 605; moderate
What is the difference between an aerial root and an air root? Give some
examples of each.

14. Adaptations for Food Storage: Fleshy Roots; p. 606; moderate
Describe some of the adaptations for food storage in roots.

Chapter 26

The Shoot: Primary Structure and Development

Multiple-Choice Questions

1. **Introduction; p. 611; easy; ans: c**
   The principal functions of the stem are:

   a. support and photosynthesis.
   b. photosynthesis and conduction.
   c. conduction and support.
   d. photosynthesis and storage.
   e. storage and absorption.

2. **Origin and Growth of the Primary Tissues of the Stem; p. 611; moderate; ans: b**
   Which of the following is(are) NOT produced by the shoot apical meristem?

   a. leaf primordia
   b. a protective covering
   c. bud primordia
   d. phytomeres
   e. cells of the primary plant body

3. **Origin and Growth of the Primary Tissues of the Stem; Fig. 26-4; moderate; ans: e**
   Which of the following is NOT part of a phytomere?

   a. axillary bud
   b. internode
   c. node
   d. leaf
   e. apical meristem

4. **Origin and Growth of the Primary Tissues of the Stem; p. 612; moderate; ans: d**
   Anticlinal divisions are different from periclinal divisions in that anticlinal divisions:

   a. occur in relatively large cells.
   b. occur solely in the corpus of the apical meristem.
   c. increase the number of cell layers.
   d. are perpendicular to the surface.
   e. are parallel with the surface.

5. **Origin and Growth of the Primary Tissues of the Stem; p. 612; difficult; ans: c**
    Which of the following statements about cells of the tunica is FALSE?
    a. Most divide anticlinally.
    b. Some divide periclinally.
    c. They form four layers in most angiosperms.
    d. Daughter cells from one layer may be displaced into another layer.
    e. They constitute the outermost cells of the shoot apex.

6. **Origin and Growth of the Primary Tissues of the Stem; p. 612; moderate; ans: d**
    In the angiosperm shoot apex, the bulk of the corpus corresponds to the:
    a. pith meristem.
    b. peripheral meristem.
    c. intercalary meristem.
    d. central mother cell zone.
    e. meristematic cap.

7. **Origin and Growth of the Primary Tissues of the Stem; p. 613; moderate; ans: b**
    Like the quiescent center of the root apical meristem, the _____ of the shoot apical meristem is a region of infrequent cell division.
    a. peripheral meristem
    b. central mother cell zone
    c. peripheral zone
    d. pith meristem
    e. tunica

8. **Origin and Growth of the Primary Tissues of the Stem; p. 613; difficult; ans: b**
    In the shoot apex, the procambium originates from the:
    a. pith meristem.
    b. peripheral meristem.
    c. intercalary meristem.
    d. ground meristem.
    e. meristematic cap.

9. **Origin and Growth of the Primary Tissues of the Stem; p. 614; moderate; ans: d**
    In palms, increase in stem thickness is due to the activity of a meristem located:
    a. at the stem tip.
    b. at the base of each internode.
    c. at the top of each internode.
    d. below the leaf bases
    e. above the bud primordia.

10. Origin and Growth of the Primary Tissues of the Stem; p. 614; easy; ans: e

A meristematic cap is a zone of _____ formation during primary growth of the stem.

a. bud
b. leaf
c. protoderm
d. ground meristem
e. procambium

11. Primary Structure of the Stem; pp. 614, 618; moderate; ans: c

Which of the following statements about the *Tilia* stem is FALSE?

a. The pith contains parenchyma and mucilage ducts.
b. The cortex contains collenchyma and parenchyma.
c. The xylem is located outside the phloem.
d. The epidermis is a single layer of cells.
e. The vascular bundles are separated by interfascicular parenchyma.

12. Primary Structure of the Stem; p. 618; difficult; ans: b

In *Tilia* and in the majority of stems, primary xylem develops from the _____ of the _____.

a. inner cells; vascular cambium
b. inner cells; procambium
c. outer cells; procambium
d. outer cells; vascular cambium
e. parenchyma cells; interfascicular regions

13. Primary Structure of the Stem; p. 618; difficult; ans: c

In the *Sambucus* stem, phloem differentiates toward the _____ and xylem differentiates toward the _____.

a. center; outside
b. outside; outside
c. outside; center
d. center; center and outside
e. center and outside; center

14. Primary Structure of the Stem; p. 619; easy; ans: d

In contrast to *Tilia* and *Sambucus*, *Medicago* and *Ranunculus* are:

a. monocots.
b. eudicots.
c. magnoliids.
d. herbaceous.
e. woody.

15. Primary Structure of the Stem; pp. 619, 621; moderate; ans: d

Which of the following statements about the *Ranunculus* stem is FALSE?

a. Its vascular bundles resemble those of monocots.
b. The procambium is lost at maturity.
c. Its vascular bundles are completely surrounded by sclerenchyma cells.
d. It has a vascular cambium.
e. It has closed vascular bundles.

16. **Primary Structure of the Stem; p. 621; moderate; ans: e**
The vascular bundles of *Zea* most closely resemble those of:

    a.   *Medicago* only.
    b.   *Medicago* and *Tilia*.
    c.   *Tilia* and *Sambucus*.
    d.   *Medicago* and *Ranunculus*.
    e.   *Ranunculus* only.

17. **Primary Structure of the Stem; p. 621; easy; ans: c**
_____ has a stem with scattered vascular bundles.

    a.   *Ranunculus*
    b.   *Sambucus*
    c.   *Zea*
    d.   *Medicago*
    e.   *Tilia*

18. **Relation between the Vascular Tissues of the Stem and the Leaf; p. 622; moderate; ans: a**
Extensions of vascular tissues into the leaves from the stem are called:

    a.   leaf traces.
    b.   leaf trace gaps.
    c.   branch traces.
    d.   stem bundles.
    e.   sympodia.

19. **Relation between the Vascular Tissues of the Stem and the Leaf; p. 622; difficult; ans: d**
A leaf trace gap consists of _____ in the _____.

    a.   xylem; ground tissue
    b.   phloem; ground tissue
    c.   ground tissue; leaf primordia
    d.   ground tissue; vascular cylinder
    e.   ground tissue; leaf trace

20. **Relation between the Vascular Tissues of the Stem and the Leaf; p. 622; moderate; ans: b**
A leaf trace extends between the leaf and a:

    a.   sympodium.
    b.   stem bundle.
    c.   leaf trace gap.
    d.   stem bundle gap.
    e.   bud primordium.

21. **Relation between the Vascular Tissues of the Stem and the Leaf; p. 622; moderate; ans: c**
A sympodium is:

    a.   all the branch traces in a stem.
    b.   all the leaf trace gaps in a stem.
    c.   a stem bundle and its associated leaf traces.
    d.   a branch trace and its associated branch trace gaps.
    e.   all the nodes and internodes in a stem.

22. **Relation between the Vascular Tissues of the Stem and the Leaf; p. 622; moderate; ans: b**
The vascular tissues that connect the buds with the stem are:

    a. bud traces.
    b. branch traces.
    c. leaf traces.
    d. leaf trace gaps.
    e. stem bundles.

23. **Relation between the Vascular Tissues of the Stem and the Leaf; p. 624; moderate; ans: d**
A plant with helical phyllotaxy has:

    a. opposite leaves.
    b. a decussate leaf arrangement.
    c. one leaf per node, arranged in two opposite ranks.
    d. one leaf per node, arranged spirally.
    e. three or more leaves per node.

24. **Relation between the Vascular Tissues of the Stem and the Leaf; p. 624; moderate; ans: c**
The field hypothesis of leaf arrangement states that leaf primordia:

    a. preferentially develop in a whorled phyllotaxy.
    b. cannot develop until sufficient space is available.
    c. are surrounded by a physiological field that prevents initiation of new primordia.
    d. contain vascular tissues organized in fieldlike rows.
    e. produce electrical fields that attract one another.

25. **Morphology of the Leaf; p. 624; easy; ans: e**
In magnoliids and eudicots, leaves generally consist of a _____ and a _____.

    a. blade; sheath
    b. sheath; petiole
    c. petiole; stipule
    d. stipule; blade
    e. blade; petiole

26. **Morphology of the Leaf; p. 624; easy; ans: e**
A sessile leaf, by definition, lacks a _____.

    a. blade.
    b. stipule.
    c. sheath.
    d. lamina.
    e. petiole.

27. **Morphology of the Leaf; p. 626; easy; ans: c**
A leaf having a rachis is a _____ leaf.

    a. simple
    b. palmately compound
    c. pinnately compound
    d. whorled
    e. sessile

28. Structure of the Leaf; pp. 626–627; moderate; ans: e
Plants that are characterized as hydrophytes:

a. are adapted to areas that are neither too wet nor too dry.
b. are adapted to dry habitats.
c. usually have stomata sunken in depressions.
d. usually have more stomata than other types of plants.
e. may lack stomata in some leaves.

29. Structure of the Leaf; p. 627; moderate; ans: b
Which of the following statements about the leaf epidermis is FALSE?

a. It is covered with a cuticle.
b. Its cells are loosely arranged.
c. Stomata may be sunken in depressions
d. Stomata may occur on one or both leaf surfaces.
e. Epidermal hairs may occur on one or both leaf surfaces.

30. Structure of the Leaf; pp. 630–631; moderate; ans: b
Which of the following statements about palisade parenchyma is FALSE?

a. It is lacking in the leaves of maize and other grasses.
b. It is usually located on the lower side of the leaf.
c. It consists of columnar cells.
d. It is where most of the photosynthesis in the leaf occurs.
e. It is part of the mesophyll.

31. Structure of the Leaf; p. 631; moderate; ans: e
Unlike the leaves of eudicots and magnoliids, most monocot leaves:

a. undergo secondary growth.
b. have veins containing both xylem and phloem.
c. have veins arranged in a branching pattern.
d. have netted venation.
e. have parallel venation.

32. Structure of the Leaf; p. 631; easy; ans: c
Most photosynthates are collected from leaf mesophyll cells by:

a. midribs.
b. major veins.
c. minor veins.
d. bundle-sheath cells.
e. bundle-sheath extensions.

33. Structure of the Leaf; pp. 631–632; moderate; ans: b
Which of the following statements about bundle-sheath extensions is FALSE?

a. They provide mechanical support for the leaf.
b. They interconnect all the bundle sheaths in a leaf.
c. They may conduct water from the xylem to the epidermis.
d. They consist of mesophyll cells.
e. They form connections between bundle sheaths and the epidermis.

34. **Grass Leaves; p. 632; difficult; ans: a**
Unlike C_3 grasses, C_4 grasses have leaves:

    a. with mesophyll and bundle-sheath cells in two concentric layers.
    b. with an interveinal distance of more than four cells.
    c. without Kranz anatomy.
    d. with a mestome sheath.
    e. with small bundle-sheath cells containing small chloroplasts.

35. **Grass Leaves; p. 632; moderate; ans: b**
Which of the following statements about bulliform cells is FALSE?

    a. They are found in monocots.
    b. They are large mesophyll cells.
    c. They are arranged in longitudinal rows.
    d. They become flaccid during excessive water loss.
    e. They play a role in leaf folding and unfolding.

36. **Development of the Leaf; pp. 632, 634; moderate; ans: a**
Which of the following statements about clonal analysis is FALSE?

    a. All the cells in the apical meristem under analysis are genetically identical.
    b. The offspring of marked cells are traced into the plant body.
    c. Chimeral meristems are produced.
    d. Genes involved in chlorophyll synthesis are used as markers.
    e. Genes involved in anthocyanin synthesis are used as markers.

37. **Development of the Leaf; p. 634; easy; ans: e**
Which layer(s) of the peripheral zone of the shoot apical meristem give(s) rise to leaf primordia?

    a. L1 only
    b. L2 only
    c. L3 only
    d. L1 and L2 only
    e. L1, L2, and L3

38. **Development of the Leaf; p. 634; difficult; ans: c**
The earliest structural evidence of leaf initiation is the development of a(n):

    a. midrib.
    b. sheath.
    c. leaf buttress.
    d. leaf primordium.
    e. intercalary meristem.

39. **Development of the Leaf; p. 634; moderate; ans: d**
Unlike the midrib, the leaf blade develops from:

    a. procambium.
    b. the central region of the primordium.
    c. founder cells.
    d. marginal bands of cells.
    e. the L1 layer of the tunica-corpus.

40. Development of the Leaf; p. 634; moderate; ans: e
In leaves, intercalary growth occurs by:

   a.   cell division alone throughout the petiole.
   b.   both cell division and cell enlargement throughout the petiole.
   c.   cell enlargement alone throughout the blade.
   d.   cell division alone throughout the blade.
   e.   both cell division and cell enlargement throughout the blade.

41. Development of the Leaf; p. 634; moderate; ans: d
Unlike determinate growth, indeterminate growth is:

   a.   characteristic of mesophyll.
   b.   characteristic of floral apices.
   c.   characteristic of leaves.
   d.   unlimited.
   e.   of short duration.

42. Development of the Leaf; pp. 634–635; moderate; ans: b
In magnoliids and eudicots, the major veins develop toward the _____ and the minor veins develop toward the _____.

   a.   leaf base; leaf margins
   b.   leaf margins; leaf base
   c.   leaf base; leaf base
   d.   leaf margins; leaf margins and leaf base
   e.   leaf base and leaf margins; leaf base

43. Development of the Leaf; p. 635; difficult; ans: a
Which of the following statements about the development of monocot leaves is FALSE?

   a.   The boundary between blade and sheath becomes distinct before the ligule develops.
   b.   Development of the sheath lags behind that of the blade.
   c.   Higher portions of the blade are more advanced in development than lower portions.
   d.   A basal intercalary meristem is involved in blade formation.
   e.   Early in development, the primordium acquires a hoodlike shape.

44. Sun and Shade Leaves; p. 636; difficult; ans: b
Compared with sun leaves, shade leaves:

   a.   have thicker-walled epidermal cells.
   b.   have a lower ratio of mesophyll surface area to leaf blade surface area.
   c.   have more extensive vascular systems.
   d.   are thicker.
   e.   are smaller.

45. Leaf Abscission; p. 636; difficult; ans: e
Which of the following events is NOT associated with leaf abscission?

   a.   formation of tyloses
   b.   formation of a suberized protective layer
   c.   cell division
   d.   enzymatic breakdown of cell walls
   e.   return of reusable substances to the leaf

46. **Transition between Vascular Systems of the Root and the Shoot;**
    pp. 636–637; difficult; ans: d
    Which of the following statements about the transition region and its formation is FALSE?
    a.   The transition region occurs between the root and the shoot.
    b.   The transition region is the region in which the vascular system changes locations.
    c.   Vascular transition is completed with the differentiation of the procambium in the seedling.
    d.   Vascular transition is initiated soon after seed germination.
    e.   The structure of the transition region is often very complex.

47. **Development of the Flower; p. 639; moderate; ans: a**
    Which of the following events does NOT occur in a shoot apex during flower development?
    a.   a change from determinate to indeterminate growth
    b.   elongation of the internodes
    c.   early development of lateral buds below the apex
    d.   an increase in mitosis
    e.   a change in shape of the apex to broad and domelike

48. **Development of the Flower; p. 639; moderate; ans: e**
    The homeotic mutation that results in "double flowers" in some rose varieties causes _____ to be converted into _____.
    a.   sepals; carpels
    b.   petals; stamens
    c.   stamens; carpels
    d.   petals; sepals
    e.   stamens; petals

49. **Development of the Flower; p. 639; moderate; ans: b**
    Studies of *Arabidopsis* mutants have shown three types of _____ that affect the identity of floral organs.
    a.   meristematic activity
    b.   homeotic genes
    c.   hormones
    d.   tunica-corpus organization
    e.   environmental factors

50. **Stem and Leaf Modifications; p. 641; easy; ans: c**
    Most tendrils are modified:
    a.   roots.
    b.   stems.
    c.   leaves.
    d.   flowers.
    e.   buds.

51. Stem and Leaf Modifications; p. 642; moderate; ans: d
The leaflike branches of asparagus are examples of:

a.   tendrils.
b.   prickles.
c.   spears.
d.   cladophylls.
e.   scales.

52. Stem and Leaf Modifications; p. 642; easy; ans: a
A true thorn is a modified:

a.   branch.
b.   leaf.
c.   epidermal hair.
d.   stem.
e.   flower.

53. Stem and Leaf Modifications; p. 642; easy; ans: b
The edible portion of an Irish potato is a:

a.   rhizome.
b.   tuber.
c.   corm.
d.   modified root.
e.   cladophyll.

54. Stem and Leaf Modifications; pp. 642–643; easy; ans: d
A stolon is different from a rhizome in that a stolon:

a.   is a leaf.
b.   is a stem.
c.   is a root.
d.   grows aboveground.
e.   grows underground.

55. Stem and Leaf Modifications; p. 643; easy; ans: d
A bulb is different from a corm in that a bulb:

a.   has thinner leaves.
b.   has a larger stem.
c.   has a fleshy stem
d.   stores food in its leaves.
e.   stores food in its stem.

56. Stem and Leaf Modifications; p. 643; easy; ans: e
A bulb is a type of:

a.   stolon.
b.   tuber.
c.   rhizome.
d.   corm.
e.   bud.

57. Stem and Leaf Modifications; p. 643; easy; ans: c
When you eat celery and rhubarb you are eating:

   a.   stems.
   b.   blades.
   c.   petioles.
   d.   flowers.
   e.   buds.

58. Stem and Leaf Modifications; p. 644; easy; ans: d
A succulent leaf is one that is specialized for:

   a.   growth in aqueous habitats.
   b.   chemical defense.
   c.   protection against predators.
   d.   water storage.
   e.   food storage.

True-False Questions

1. Origin and Growth of the Primary Tissues of the Stem; pp. 611–612; easy; ans: T
The shoot apical meristem produces leaf primordia and bud primordia.

2. Origin and Growth of the Primary Tissues of the Stem; pp. 611–612; moderate; ans: F
Lateral shoots are produced by leaf primordia.

3. Origin and Growth of the Primary Tissues of the Stem; p. 612; moderate; ans: F
Cells in the outermost layer of the tunica-corpus usually divide periclinally.

4. Origin and Growth of the Primary Tissues of the Stem; p. 613; moderate; ans: T
Cells of the peripheral zone of the angiosperm shoot apex divide frequently.

5. Origin and Growth of the Primary Tissues of the Stem; p. 613; moderate; ans: T
The central mother cell zone of the shoot is analogous to the quiescent center of the root.

6. Origin and Growth of the Primary Tissues of the Stem; p. 614; moderate; ans: F
Increase in stem length occurs largely by cell division.

7. Origin and Growth of the Primary Tissues of the Stem; p. 614; easy; ans: T
Palms are an example of plants that undergo extensive primary growth through the activity of a meristematic cap.

8. Primary Structure of the Stem; p. 614; moderate; ans: F
In most monocot but not eudicot stems, it is possible to distinguish cortex from pith.

9. **Primary Structure of the Stem; p. 618; moderate; ans: T**
In the *Tilia* and *Sambucus* stems, primary phloem fibers develop after internodal elongation is complete.

10. **Primary Structure of the Stem; p. 619; difficult; ans: T**
In the *Sambucus* and *Medicago* stems, vascular cambium develops partly from cells in the pith rays.

11. **Primary Structure of the Stem; pp. 619, 621; moderate; ans: F**
Stems with closed vascular bundles are capable of secondary growth.

12. **Primary Structure of the Stem; p. 621; moderate; ans: T**
In the *Zea* stem, the protoxylem and protophloem are destroyed during internodal elongation, forming a protoxylem lacuna.

13. **Relation between the Vascular Tissues of the Stem and the Leaf; p. 622; difficult; ans: F**
Procambial strands differentiate downward from the leaf primordium into the stem.

14. **Relation between the Vascular Tissues of the Stem and the Leaf; p. 622; easy; ans: T**
A stem bundle is commonly associated with several leaf traces.

15. **Relation between the Vascular Tissues of the Stem and the Leaf; p. 624; difficult; ans: T**
Decussate leaves can also be described as having an opposite phyllotaxy.

16. **Morphology of the Leaf; p. 624; easy; ans: F**
The petioles of monocot leaves are called stipules.

17. **Morphology of the Leaf; p. 626; easy; ans: T**
In a compound leaf, the blade is divided into leaflets.

18. **Structure of the Leaf; pp, 626–627; easy; ans: F**
Xerophytes can have both floating and submerged leaves.

19. **Structure of the Leaf; p. 627; easy; ans: T**
The stomata of most monocots are arranged in parallel rows, from leaf tip to leaf base.

20. **Structure of the Leaf; pp. 630–631; moderate; ans: F**
Spongy parenchyma usually contains more chloroplasts than palisade parenchyma.

21. **Structure of the Leaf; p. 631; easy; ans: T**
Most magnoliids and eudicot leaves have netted venation.

22. **Structure of the Leaf; p. 632; easy; ans: T**
Bundle-sheath extensions connect the bundle sheaths with the epidermis.

23. **Grass Leaves; p. 632; easy; ans: F**
    In a leaf having Kranz anatomy, the mesophyll and bundle-sheath layers are not concentrically arranged.

24. **Development of the Leaf; p. 634; moderate; ans: T**
    The white areas of green-white chimeras of ivy and geranium are produced by mutant cells that lack chlorophyll.

25. **Development of the Leaf; p. 634; moderate; ans: F**
    Founder cells of the shoot apical meristem give rise to bud primordia but not leaf primordia.

26. **Development of the Leaf; p. 634; easy; ans: T**
    Leaves exhibit determinate, rather than indeterminate, growth.

27. **Sun and Shade Leaves; p. 636; moderate; ans: T**
    Sun and shade leaves can be found on the same plant.

28. **Leaf Abscission; p. 636; moderate; ans: F**
    The abscission zone consists of three layers: the separation layer, the protective layer, and the layer of cell division.

29. **Transition between Vascular Systems of the Root and the Shoot; p. 637; easy; ans: F**
    In the transition region, there is a small but significant interruption between the vascular tissues of the root and the shoot.

30. **Development of the Flower; p. 639; easy; ans: T**
    The early stages of development are similar in floral parts and in leaves.

31. **Development of the Flower; p. 639; easy; ans: T**
    Snapdragon (*Antirrhinum*) and *Arabidopsis* have the same basic mechanisms for controlling floral organ identity.

32. **Stem and Leaf Modifications; pp. 642–643; easy; ans: F**
    A rhizome is a modified, aboveground root.

33. **Stem and Leaf Modifications; p. 644; moderate; ans: F**
    Succulent plants normally grow in habitats with abundant water.

Essay Questions

1. **Origin and Growth of the Primary Tissues of the Stem; pp. 611–612; easy**
   List the structures produced by the shoot apical meristem. In what ways is the shoot apical meristem different from the root apical meristem?

2. **Origin and Growth of the Primary Tissues of the Stem; p. 612; moderate**
   Describe the tunica-corpus organization of the angiosperm shoot apex.

3. **Origin and Growth of the Primary Tissues of the Stem; pp. 612–613; difficult**
   Describe the structure and function of the central mother cell zone, the peripheral meristem and the pith meristem of the shoot apex. What is their relationship to the tunica and corpus?

4. **Origin and Growth of the Primary Tissues of the Stem; pp. 613–614; moderate**
   Why can't the stem, like the root, be divided into regions of cell division, elongation, and maturation?

5. **Primary Structure of the Stem; p. 614; easy**
   Describe the three basic patterns of primary stem structure in seed plants.

6. **Primary Structure of the Stem; pp. 614–621; moderate**
   Compare and contrast the stem structures of *Tilia*, *Sambucus*, *Medicago*, *Ranunculus*, and *Zea*.

7. **Relation between the Vascular Tissues of the Stem and the Leaf; p. 622; difficult**
   Discuss how the arrangement of vascular tissues in the angiosperm stem is related to the arrangement of vascular tissues in the leaves.

8. **Relation between the Vascular Tissues of the Stem and the Leaf; p. 624; moderate**
   Compare and contrast the field hypothesis and the first available space hypothesis of leaf arrangement. Are these hypotheses mutually exclusive? Why or why not?

9. **Morphology of the Leaf; p. 626; easy**
   What two criteria would you use to determine whether a structure is a leaf or a leaflet?

10. **Structure of the Leaf; pp. 626–627; easy**
    Discuss some of the features that distinguish mesophytes, xerophytes, and hydrophytes.

11. **Structure of the Leaf; pp. 627–632; moderate**
    Compare and contrast the structure of monocot and eudicot leaves.

12. **Grass Leaves; p. 632; moderate**
    Discuss the differences and similarities between leaves from C_3 and C_4 grasses.

13. **Development of the Leaf; pp. 632–635; difficult**
    Compare and contrast the development of monocot and eudicot leaves.

14. **Sun and Shade Leaves; p. 636; difficult**
    List some differences between sun leaves and shade leaves. In what ways are these differences adaptive?

15. **Leaf Abscission; p. 636; moderate**
    Describe the sequence of events that culminates in leaf abscission.

5. **The Vascular Cambium; p. 648; moderate; ans: b**
    Which of the following statements about vascular rays is FALSE?

    a. They are composed largely of parenchyma cells.
    b. They serve as barriers to the movement of food substances and water.
    c. They store starch, proteins, and lipids.
    d. They synthesize secondary products.
    e. They are variable in length.

6. **The Vascular Cambium; p. 648; easy; ans: c**
    Strictly speaking, the vascular cambium refers to the:

    a. cambial zone.
    b. single radial file of derivatives.
    c. single radial file of initials.
    d. initials plus their immediate derivatives.
    e. initials, their immediate derivatives, and the most recent xylem and
        phloem.

7. **The Vascular Cambium; p. 649; moderate; ans: b**
    As the vascular cambium continues to divide, the cambial cells:

    a. are displaced inward.
    b. are displaced outward.
    c. remain in their original location.
    d. cease to divide periclinally.
    e. cease to divide anticlinally.

8. **Effect of Secondary Growth on the Primary Body of the Stem;**
    p. 650; moderate; ans: e
    Which of the following statements about secondary growth is FALSE?

    a. The secondary vascular tissues form a cylindrical shape.
    b. Vascular rays extend radially through the secondary xylem and secondary
        phloem.
    c. Primary phloem fibers remain intact longer than other primary phloem
        cells.
    d. The primary phloem is pushed outward.
    e. Most plants produce more secondary phloem than secondary xylem.

9. **Effect of Secondary Growth on the Primary Body of the Stem;**
    pp. 651–652; moderate; ans: d
    Secondary growth in the *Tilia* stem is different from that in the *Sambucus* stem
    because in the *Tilia* stem:

    a. more secondary xylem than secondary phloem is formed.
    b. more secondary phloem than secondary xylem is formed.
    c. only a small amount of secondary tissue is produced.
    d. dilated phloem rays are formed.
    e. phloem fibers are formed.

10. **Effect of Secondary Growth on the Primary Body of the Stem;**
    p. 653; easy; ans: e
    The periderm consists of:

    a.   cork only.
    b.   cork cambium only.
    c.   phelloderm only.
    d.   cork and cork cambium only.
    e.   cork, cork cambium, and phelloderm.

11. **Effect of Secondary Growth on the Primary Body of the Stem;**
    p. 653; moderate; ans: a
    In most woody plants, the first periderm usually arises in the:

    a.   cortex.
    b.   epidermis.
    c.   primary phloem.
    d.   primary xylem.
    e.   pith.

12. **Effect of Secondary Growth on the Primary Body of the Stem;**
    p. 654; moderate; ans: c
    Cells of the phelloderm differ from cells of the cortex in that phelloderm cells:

    a.   are dead at maturity.
    b.   have suberin lamellae on their inner wall surfaces.
    c.   are the innermost cells in the radial rows of the periderm.
    d.   are impermeable to water and gases.
    e.   have lignified cell walls.

13. **Effect of Secondary Growth on the Primary Body of the Stem;**
    p. 654; easy; ans: b
    Lenticels function primarily in:

    a.   water transport.
    b.   gas exchange.
    c.   mineral uptake.
    d.   protection.
    e.   hormone production.

14. **Effect of Secondary Growth on the Primary Body of the Stem;**
    p. 654; moderate; ans: a
    Which of the following statements about lenticels is FALSE?

    a.   They are found only on roots and stems.
    b.   They are portions of the periderm.
    c.   They contain numerous intercellular spaces.
    d.   In stems they generally arise below stomata.
    e.   On the surface of stems they appear as raised areas.

15. **Effect of Secondary Growth on the Primary Body of the Stem;**
    p. 654; easy; ans: c
    At the end of the first year's growth, bark is composed of:

    a.   periderm.
    b.   cork.
    c.   all tissues outside the vascular cambium.
    d.   only secondary tissues.
    e.   only primary tissues.

16. **Effect of Secondary Growth on the Primary Body of the Stem;**
    p. 655; moderate; ans: d
    After the first periderm is formed, additional periderms originate from _____
    cells.

    a.   epidermal
    b.   pith
    c.   cortical
    d.   phloem parenchyma
    e.   xylem parenchyma

17. **Effect of Secondary Growth on the Primary Body of the Stem;**
    p. 658; moderate; ans: e
    Unlike ring bark, scale bark:

    a.   is found in honeysuckle (*Lonicera*).
    b.   is found in grape (*Vitis*).
    c.   contains cork.
    d.   contains concentric rings of periderms.
    e.   contains discontinuous layers of periderms.

18. **Effect of Secondary Growth on the Primary Body of the Stem;**
    p. 659; difficult; ans: b
    Functional phloem:

    a.   is composed of primary tissue.
    b.   is part of the inner bark.
    c.   functions primarily to store foods.
    d.   consists of living or dead sieve elements.
    e.   makes up the majority of secondary phloem in older plants.

19. **The Wood: Secondary Xylem; p. 659; moderate; ans: a**
    "Softwood" is the name given to wood:

    a.   found in conifers.
    b.   found in eudicots.
    c.   found in magnoliids.
    d.   composed of primary xylem.
    e.   having no commercial value.

20. **The Wood: Secondary Xylem; p. 659; moderate; ans: c**
Which of the following statements about resin ducts is FALSE?

   a.   They may result from trauma.
   b.   They secrete resin that may protect against fungi and beetles.
   c.   They are intercellular spaces lined with collenchyma cells.
   d.   Their formulation may be stimulated by wounding.
   e.   In *Pinus* they occur in both the axial system and the rays.

21. **The Wood: Secondary Xylem; pp. 659–660; moderate; ans: d**
_____ is thought to block the movement of water or gases through a pit-pair in conifer tracheids.

   a.   Resin
   b.   The pit membrane
   c.   The middle lamella
   d.   A torus
   e.   A resin duct

22. **The Wood: Secondary Xylem; p. 662; difficult; ans: b**
Angiosperm wood differs from conifer wood in that angiosperm wood has:

   a.   tracheids.
   b.   vessels.
   c.   orderly radial files of cells.
   d.   smaller rays.
   e.   fewer cell types in the axial system.

23. **The Wood: Secondary Xylem; p. 664; difficult; ans: e**
Which of the following statements about growth rings is FALSE?

   a.   They may be absent in trees growing in the tropics.
   b.   They are caused by variations in the activity of the vascular cambium in each growing season.
   c.   They may occur in secondary phloem as well as in secondary xylem.
   d.   An annual ring represents one season's growth.
   e.   A false annual ring represents growth in an unusually short growing season.

24. **The Wood: Secondary Xylem; p. 665; difficult; ans: e**
In angiosperms, early wood _____ than late wood.

   a.   has narrower cells
   b.   has thicker cell walls
   c.   is denser
   d.   is produced later in the growing season
   e.   may have much larger pores

25. **The Wood: Secondary Xylem; p. 665; moderate; ans: b**
Ring-porous woods are different from diffuse-porous woods because in ring-porous woods:

   a.   the pores of late wood are larger than those of early wood.
   b.   the pores of early wood are larger than those of late wood.
   c.   the pores are fairly uniform in size throughout the growth layer.
   d.   almost all the water is conducted in the innermost growth layer.
   e.   vessels are lacking.

26. The Wood: Secondary Xylem; p. 665; moderate; ans: d
Which of the following statements about tyloses is FALSE?

a.   They may inhibit the spread of pathogens through the xylem.
b.   Their formation may be induced by plant pathogens.
c.   They may completely block a vessel.
d.   They are balloonlike outgrowths from pit membranes.
e.   They are formed when vessels become nonfunctional.

27. The Wood: Secondary Xylem; p. 666; difficult; ans: b
Compression wood:

a.   develops on the upper side of a leaning stem.
b.   is produced by increased cambial activity on the lower side of a leaning stem.
c.   is the reaction wood found in angiosperms.
d.   is characterized by wider portions of growth rings on the upper side of the stem.
e.   has less lignin and more cellulose than normal wood.

28. The Wood: Secondary Xylem; p. 667; easy; ans: b
The terms "straight," "spiral," and "interlocked" refer to the _____ of a wood.

a.   texture
b.   grain
c.   color
d.   density
e.   figure

29. The Wood: Secondary Xylem; p. 667; difficult; ans: c
By definition, a piece of wood with a coarse texture has:

a.   axial components parallel to the long axis.
b.   a decorative pattern on the longitudinal surface.
c.   wide bands of large vessels and broad rays.
d.   no visible difference between early and late woods.
e.   small vessels and narrow rays.

30. The Wood: Secondary Xylem; p. 667; moderate; ans: e
Unlike quartersawed boards, plainsawed boards:

a.   are radially cut.
b.   do not warp.
c.   are preferable for most uses.
d.   have growth rings appearing as parallel lines.
e.   have growth rings appearing as wavy bands.

31. The Wood: Secondary Xylem; p. 670; moderate; ans: d
In wood having a high specific gravity, the fibers have _____ walls and _____ lumens.

a.   thin; variable
b.   thin; narrow
c.   thin; wide
d.   thick; narrow
e.   thick; wide

True-False Questions

1. **Annuals, Biennials, and Perennials; p. 648; easy; ans: T**
   An annual completes its entire life cycle within one growing season.

2. **The Vascular Cambium; p. 648; moderate; ans: F**
   The meristematic cells of the vascular cambium contain dense cytoplasm and large nuclei.

3. **The Vascular Cambium; p. 648; moderate; ans: T**
   In secondary tissues, the cells of the axial system are oriented vertically and those of the radial system are oriented horizontally.

4. **The Vascular Cambium; pp. 649–650; moderate; ans: T**
   Unlike plants in temperate regions, many tropical plants exhibit continuous, year-round cambial activity.

5. **Effect of Secondary Growth on the Primary Body of the Stem; p. 650; easy; ans: T**
   Interfascicular cambium arises in parenchyma cells between the vascular bundles.

6. **Effect of Secondary Growth on the Primary Body of the Stem; p. 652; moderate; ans: F**
   The secondary plant body has a transition region similar to that of the primary plant body.

7. **Effect of Secondary Growth on the Primary Body of the Stem; p. 653; moderate; ans: T**
   The order of tissues in the periderm, from the outside of the plant toward the center, is phellem, phellogen, and phelloderm.

8. **Effect of Secondary Growth on the Primary Body of the Stem; p. 654; moderate; ans: F**
   The cortex of the stem is usually sloughed during the first year of growth.

9. **Effect of Secondary Growth on the Primary Body of the Stem; p. 654; easy; ans: T**
   Lenticels make it possible for stems to exchange gases with the surrounding air.

10. **Effect of Secondary Growth on the Primary Body of the Stem; p. 654; moderate; ans: T**
    Bark may contain primary tissues as well as secondary tissues.

11. **Effect of Secondary Growth on the Primary Body of the Stem; p. 658; moderate; ans: T**
    Unlike the outer bark, the inner bark contains living cells.

12. **Effect of Secondary Growth on the Primary Body of the Stem; p. 658; moderate; ans: F**
    In the cork oak, the cork produced by the first cork cambium has the greatest commercial value.

13. **Effect of Secondary Growth on the Primary Body of the Stem;**
    p. 659; difficult; ans: F
    Nonfunctional phloem is dead phloem tissue found in the inner bark.

14. **The Wood: Secondary Xylem; p. 659; easy; ans: T**
    Hardwood is another name for angiosperm wood.

15. **The Wood: Secondary Xylem; p. 659; moderate; ans: F**
    Conifer wood typically contains more parenchyma than angiosperm wood.

16. **The Wood: Secondary Xylem; p. 660; moderate; ans: T**
    Both a radial section and a tangential section are longitudinal sections.

17. **The Wood: Secondary Xylem; p. 662; moderate; ans: F**
    The rays of angiosperm woods are usually smaller than those of conifer woods.

18. **The Wood: Secondary Xylem; p. 664; easy; ans: F**
    Growing seasons with favorable environmental conditions produce narrower
    growth rings than seasons with unfavorable conditions.

19. **The Wood: Secondary Xylem; p. 664; easy; ans: T**
    Scientists have chronicled a continuous series of growth rings dating back more
    than 8200 years.

20. **The Wood: Secondary Xylem; p. 665; difficult; ans: T**
    In diffuse-porous woods, the pores are fairly uniform in size and distribution
    throughout the growth layers.

21. **The Wood: Secondary Xylem; p. 665; easy; ans: T**
    The proportion of heartwood to sapwood varies among species.

22. **The Wood: Secondary Xylem; p. 666; moderate; ans: F**
    Tension wood is a type of reaction wood found in gymnosperms.

23. **The Wood: Secondary Xylem; p. 667; difficult; ans: T**
    In cross-grained wood, the axial components are not aligned with the
    longitudinal axis of the wood.

24. **The Wood: Secondary Xylem; p. 667; moderate; ans: F**
    Plainsawing is more time-consuming and often more wasteful than
    quartersawing.

25. **The Wood: Secondary Xylem; pp. 668, 670; moderate; ans: T**
    The differences in specific gravity of woods depends on the proportion of cell
    wall substance to cell lumen.

Essay Questions

1. **Annuals, Biennials, and Perennials; p. 648; easy**
   Explain the differences between annuals, biennials, and perennials.

2. **The Vascular Cambium; pp. 648–649; difficult**
What two types of cells make up the vascular cambium? Explain why these cells must divide both anticlinally and periclinally.

3. **The Vascular Cambium; pp. 648–649; moderate**
Why do some botanists prefer the term "cambial zone" to "vascular cambium"?

4. **Effect of Secondary Growth on the Primary Body of the Stem; pp. 650–654; moderate**
Which tissues are produced by the vascular cambium and which by the cork cambium? Describe how the functioning of the cork cambium results from the activity of the vascular cambium.

5. **Effect of Secondary Growth on the Primary Body of the Stem; pp. 654–655; moderate**
What is bark? Explain how the composition of bark changes over time.

6. **Effect of Secondary Growth on the Primary Body of the Stem; pp. 654–655; moderate**
Explain why less secondary phloem than secondary xylem accumulates in the stem.

7. **Effect of Secondary Growth on the Primary Body of the Stem; pp. 653, 654, 658; moderate**
Distinguish among periderm, cork, and bark. What is the difference between the inner bark and the outer bark?

8. **Effect of Secondary Growth on the Primary Body of the Stem; p. 655; difficult**
Why is it advantageous for a plant to produce more than one periderm?

9. **Effect of Secondary Growth on the Primary Body of the Stem; pp. 658–659; easy**
Describe the manner in which commercial cork is obtained.

10. **The Wood: Secondary Xylem; pp. 659–662; moderate**
Compare and contrast angiosperm and conifer woods.

11. **The Wood: Secondary Xylem; p. 660; difficult**
Describe the appearance of conifer wood in transverse, radial, and tangential sections.

12. **The Wood: Secondary Xylem; pp. 664–665; moderate**
What causes growth rings to form? How can growth rings be used to determine past climates?

13. **The Wood: Secondary Xylem; pp. 665–667; moderate**
What is reaction wood? What are the characteristics of the two types of reaction wood?

14. **The Wood: Secondary Xylem; pp. 667–670; difficult**
Discuss the four factors that influence the gross features of wood.

15. **The Wood: Secondary Xylem; pp. 668, 670; moderate**
    Distinguish between the specific gravity and the density of wood. What do
    these properties tell us about the strength of wood?

Chapter 28

Regulating Growth and Development: The Plant Hormones

Multiple-Choice Questions

1. **Introduction; pp. 674–675; easy; ans: e**
   All plant hormones:

   a. are inorganic substances.
   b. act in the tissues where they are produced.
   c. are stimulatory.
   d. are active in large quantities.
   e. communicate information.

2. **Introduction; pp. 674–675; difficult; ans: d**
   Which of the following statements about plant hormones is FALSE?

   a. The same hormone can elicit different responses in different tissues.
   b. The same hormone can elicit different responses at different times in the same tissue.
   c. A tissue's response to a hormone depends on the sensitivity of the tissue to that hormone.
   d. All plant hormones can be classified into five groups.
   e. Many details of the mechanism of hormone action are unknown.

3. **Regulating Growth and Development: The Plant Hormones;**
   p. 675; moderate; ans: c
   _____ are recently discovered chemical signals that are related to animal steroids.

   a. Systemins
   b. Auxins
   c. Brassinolides
   d. Salicylic acids
   e. Jasmonates

4. **Auxins; p. 675; easy; ans: b**
    If a coleoptile tip is covered with a blackened glass tube then illuminated from the side, the coleoptile will:

    a. die.
    b. not bend.
    c. bend toward the light.
    d. bend away from the light.
    e. bend at right angles to the light.

5. **Auxins; p. 676; easy; ans: c**
    _____ is very similar to the amino acid tryptophan.

    a. Gibberellin
    b. Cytokinin
    c. Indoleacetic acid
    d. Abscisic acid
    e. Ethylene

6. **Auxins; p. 676; difficult; ans: c**
    Auxin is transported toward the _____ in stems, toward the _____ in leaves, and toward the _____ in roots.

    a. base; base; base
    b. tip; tip; tip
    c. base; base; tip
    d. base; tip; base
    e. tip; tip; base

7. **Auxins; p. 676; moderate; ans: e**
    Basipetal transport:

    a. refers to polar transport in general.
    b. refers to a type of nondirectional transport.
    c. can refer to transport in the stem from the soil surface toward the tip.
    d. can refer to transport in the root from the soil surface toward the tip.
    e. can refer to transport in the root from the tip toward the soil surface.

8. **Auxins; p. 676; moderate; ans: c**
    Auxin is transported through:

    a. vessels only.
    b. sieve tubes only.
    c. parenchyma cells only.
    d. both vessels and sieve tubes.
    e. both vessels and parenchyma cells.

9. **Auxins; p. 676; difficult; ans: c**
   If wounding causes the destruction of vascular tissues in a herbaceous eudicot stem:

   a. replacement vascular tissues will not form under any conditions.
   b. replacement vascular tissues will form if IAA is added to the stem just below the wound.
   c. replacement vascular tissues will form if the leaves and buds above the wound are intact.
   d. water will be transported in pith cells instead of in xylem.
   e. food will be transported in pith cells instead of in phloem.

10. **Auxins; p. 676; difficult; ans: e**
    Which of the following statements about the differentiation of vessel elements in isolated mesophyll cells of *Zinnia elegans* is FALSE?

    a. It does not require cytokinesis.
    b. It does not require DNA synthesis.
    c. It requires cytokinin.
    d. It requires auxin.
    e. It requires cell division.

11. **Auxins; p. 678; easy; ans: a**
    In apical dominance, the apical bud:

    a. inhibits the growth of lateral buds.
    b. stimulates the growth of lateral buds.
    c. stimulates the upward growth of the stem.
    d. stimulates the growth of leaves.
    e. inhibits the downward growth of the root.

12. **Auxins; pp. 678–679; moderate; ans: a**
    Which of the following statements about the use of auxin or synthetic auxin is FALSE?

    a. It is used to kill grasses.
    b. It is used to kill broad-leaf weeds.
    c. It inhibits the growth of already-growing roots.
    d. It stimulates the initiation of adventitious roots in cuttings.
    e. It stimulates the formation of parthenocarpic fruits.

13. **Cytokinins; p. 680; easy; ans: d**
    Folke Skoog and his colleagues found that _____ stimulated cells of tobacco to divide in culture.

    a. a vitamin
    b. a sugar
    c. a salt
    d. coconut milk
    e. auxin

14. Cytokinins; p. 680; moderate; ans: d
Which of the following statements about cytokinins is FALSE?

   a.   They stimulate cell division.
   b.   They are present in bleeding sap.
   c.   They are found primarily in actively dividing tissues and in root tips.
   d.   Kinetin is a naturally occurring cytokinin.
   e.   The most active naturally occurring cytokinin is zeatin.

15. Cytokinins; p. 681; moderate; ans: a
When tobacco pith callus is treated with equal concentrations of auxin and kinetin, _____ is(are) formed.

   a.   more callus
   b.   roots
   c.   buds
   d.   leaves
   e.   vascular tissue

16. Cytokinins; p. 681; difficult; ans: d
In studies of leaf senescence, when an excised leaf containing radioactive amino acids is spotted with a kinetin-containing solution, the spot:

   a.   turns yellow and becomes nonradioactive.
   b.   turns yellow and becomes more radioactive.
   c.   remains green and becomes nonradioactive.
   d.   remains green and becomes more radioactive.
   e.   turns brown and becomes nonradioactive.

17. Cytokinins; p. 682; easy; ans: d
Cytokinins are synthesized in ____ and transported from there to all other parts of the plant.

   a.   stems
   b.   shoot apical meristems
   c.   leaves
   d.   roots
   e.   fruits

18. Ethylene; p. 682; moderate; ans: b
Which of the following statements about ethylene is FALSE?

   a.   The final step in its synthesis in plants is catalyzed by enzymes on the tonoplast.
   b.   Its synthesis in plants begins with the amino acid tryptophan.
   c.   It exerts an influence on many aspects of plant growth and development.
   d.   It is the active component of illuminating gas that affects plant growth.
   e.   It has the formula $H_2C\!=\!CH_2$.

19. **Ethylene; p. 682; difficult; ans: c**
   In etiolated pea seedlings, ethylene causes _____ longitudinal growth, _____
   radial expansion of epicotyls, and _____ growth of epicotyls.

   a. increased; increased; horizontal
   b. increased; decreased; horizontal
   c. decreased; increased; horizontal
   d. decreased; increased; angular
   e. increased; decreased; angular

20. **Ethylene; p. 683; moderate; ans: e**
   All of the following are associated with fruit ripening EXCEPT:

   a. an increase in cellular respiration.
   b. the digestion of pectin.
   c. the metabolism of organic acids to sugars.
   d. the degradation of chlorophyll.
   e. the metabolism of sugars to starches.

21. **Ethylene; p. 683; easy; ans: d**
   The climacteric is the phase in which fruits undergo a(n):

   a. decrease in photosynthesis.
   b. decrease in respiration.
   c. increase in photosynthesis.
   d. increase in respiration.
   e. increase in pectin synthesis.

22. **Ethylene; p. 683; moderate; ans: c**
   Which of the following statements about abscission is FALSE?

   a. Ethylene triggers cell wall dissolution in leaves.
   b. Ethylene promotes abscission.
   c. Auxin increases the sensitivity of abscission zone cells to ethylene.
   d. Auxin prevents abscission.
   e. High concentrations of applied auxin may promote abscission.

23. **Abscisic Acid; pp. 683–684; moderate; ans: d**
   Which of the following statements about abscisic acid is FALSE?

   a. It is a growth inhibitor.
   b. It induces the closing of stomata.
   c. It prevents premature seed germination.
   d. It inhibits the production of seed storage proteins.
   e. It stimulates the yellowing of leaves.

24. **Gibberellins; p. 684; easy; ans: a**
   The highest concentrations of gibberellin are found in:

   a. immature seeds.
   b. germinating seeds.
   c. stems.
   d. leaves.
   e. fruits.

25. Gibberellins; pp. 684–685; easy; ans: e
Dwarf mutant plants are short because they:

   a.   synthesize too much gibberellin.
   b.   synthesize too much abscisic acid.
   c.   cannot synthesize ethylene.
   d.   cannot synthesize auxin.
   e.   cannot synthesize gibberellin.

26. Gibberellins; p. 685; moderate; ans: b
In germinating barley seeds, the _____ releases gibberellins, which then
diffuse to the _____ where they stimulate the synthesis of hydrolytic enzymes.

   a.   endosperm; aleurone layer
   b.   embryo; aleurone layer
   c.   aleurone layer; embryo
   d.   aleurone layer; endosperm
   e.   embryo; seed coat

27. Gibberellins; p. 686; easy; ans: d
Bolting is the process of:

   a.   flowering in short-day plants.
   b.   fruit development in cabbages.
   c.   root elongation.
   d.   stem elongation.
   e.   leaf elongation.

28. Gibberellins; pp. 686–687; moderate; ans: c
Which of the following statements about gibberellins is FALSE?

   a.   They induce bolting in long-day plants.
   b.   They induce early flowering in biennials.
   c.   They cause rosette formation in cabbages and carrots.
   d.   They cause development of parthenocarpic apples and currants.
   e.   They are used in the commercial production of seedless grapes.

29. The Molecular Basis of Hormone Action; p. 687; moderate; ans: d
Which of the following statements about hormone action is FALSE?

   a.   Hormones influence the rate of cell division.
   b.   Hormones influence the rate of cell expansion.
   c.   Hormones influence the direction of cell expansion.
   d.   All hormones act by stimulating gene expression.
   e.   All hormones act as chemical messengers.

30. The Molecular Basis of Hormone Action; p. 688; moderate; ans: a
The regulatory sequences of a eukaryotic gene:

   a.   border the coding sequence and regulate transcription.
   b.   are in the middle of the coding sequence and interrupt transcription.
   c.   specify the amino acid sequence of the gene's protein product.
   d.   specify the amino acid sequences transcription factors.
   e.   bind directly to the coding sequence, thereby activating or repressing it.

31. **The Molecular Basis of Hormone Action; p. 688; moderate; ans: c**
   In isolated barley aleurone cells, treatment with GA causes the levels of α-amylase to _____ and treatment with ABA causes the levels to _____.
   a.   increase; remain unchanged
   b.   decrease; remain unchanged
   c.   increase; decrease
   d.   decrease; increase
   e.   increase; increase

32. **The Molecular Basis of Hormone Action; p. 689; moderate; ans: b**
   Cell wall extensibility, and thus cell expansion, is increased by _____ but decreased by _____.
   a.   auxin and ethylene; GA and ABA
   b.   auxin and GA; ABA and ethylene
   c.   auxin and ABA; GA and ethylene
   d.   ABA and ethylene; auxin and GA
   e.   ABA and GA; auxin and ethylene

33. **The Molecular Basis of Hormone Action; p. 689; difficult; ans: d**
   In the acid-growth hypothesis of cell wall extensibility, which of the following events occurs FIRST?:
   a.   Extensins disrupt hydrogen bonds.
   b.   Protons are pumped into the cell wall.
   c.   Specific genes are activated.
   d.   A proton-pumping enzyme is activated.
   e.   Cross-links between noncellulosic polysaccharides are broken.

34. **The Molecular Basis of Hormone Action; p. 690; moderate; ans: e**
   If cell wall microfibrils are oriented transversely, the cell normally expands:
   a.   in all directions.
   b.   at an oblique angle.
   c.   first laterally and then longitudinally.
   d.   laterally.
   e.   longitudinally.

35. **The Molecular Basis of Hormone Action; p. 690; difficult; ans: c**
   Gibberellins promote a _____ orientation of the microtubules underlying the plasma membrane, and ethylene promotes a _____ orientation.
   a.   random; longitudinal
   b.   transverse; random
   c.   transverse; longitudinal
   d.   longitudinal; transverse
   e.   longitudinal; random

36. **Plant Biotechnology; p. 690; moderate; ans: b**
   Which of the following is NOT a target cell mechanism?
   a.   measuring hormone levels
   b.   activating genes
   c.   identifying hormones
   d.   transferring information via biochemical pathways
   e.   converting information into developmental changes

37. **The Molecular Basis of Hormone Action; p. 691; moderate; ans: b**
Which of the following statements about protein kinases is FALSE?

    a.   They may be activated by calcium ions.
    b.   They elevate calcium ion levels in the cytoplasm.
    c.   They are enzymes.
    d.   They transfer phosphate groups to specific proteins.
    e.   They alter the activity of target proteins.

38. **The Molecular Basis of Hormone Action; p. 691; moderate; ans: d**
Which of the following statements about second messengers is FALSE?

    a.   They amplify the signal produced by a hormone.
    b.   They contribute to the diversity of responses to a hormone.
    c.   They transfer information from the hormone-receptor complex to the target protein.
    d.   An example is protein kinase.
    e.   An example is calcium ion.

39. **The Molecular Basis of Hormone Action; pp. 692–693; easy; ans: c**
Of the following events, which one is mediated by ABA and occurs most rapidly?

    a.   opening stomata
    b.   closing stomata
    c.   changing the osmotic potential of guard cells
    d.   changing the turgor pressure in guard cells
    e.   causing water to enter guard cells

40. **The Molecular Basis of Hormone Action; p. 693; difficult; ans: a**
In the ABA-induced closing of stomata, the following ion fluxes occur across guard cell plasma membranes: first calcium ions flow _____, then anions flow _____, and finally potassium ions flow _____.

    a.   in; out; out
    b.   in; in; in
    c.   out; in; out
    d.   out; out; in
    e.   out; out; out

41. **Plant Biotechnology; p. 693; easy; ans: b**
By definition, hydroponics is the technique by which plant biologists:

    a.   produce clones of plants.
    b.   grow plants without soil.
    c.   produce somatic hybrids.
    d.   fuse protoplasts.
    e.   produce pathogen-free plants.

42. **Plant Biotechnology; p. 693; easy; ans: b**
_____ is a collection of methods for growing cells in a sterile environment.

    a.   Protoplast fusion
    b.   Tissue culture
    c.   Clonal propagation
    d.   Micropropagation
    e.   Genetic engineering

43. Plant Biotechnology; p. 693; easy; ans: a
Which of the following statements about micropropagation is FALSE?

   a.   It involves the fusion of protoplasts.
   b.   It is a type of plant multiplication.
   c.   It is also called clonal propagation.
   d.   It produces genetically identical cells.
   e.   It is an aspect of plant tissue culture.

44. Plant Biotechnology; p. 693; easy; ans: d
Which of the following does NOT involve the use of tissue culture?

   a.   micropropagation
   b.   clonal propagation
   c.   protoplast fusion
   d.   hydroponics
   e.   meristem culture

45. Plant Biotechnology; p. 693; easy; ans: e
Which of the following is MOST DIRECTLY involved in producing interspecific somatic hybrids?

   a.   micropropagation
   b.   hydroponics
   c.   clonal propagation
   d.   meristem culture
   e.   protoplast fusion

46. Plant Biotechnology; p. 695; moderate; ans: c
Most fertile interspecific somatic hybrids are produced by fusing cells from:

   a.   the same individual.
   b.   members of the same species.
   c.   members of different species but the same genus.
   d.   members of different genera.
   e.   members of different families.

47. Plant Biotechnology; p. 695; easy; ans: b
Members of a clone of plants are:

   a.   similar but not identical.
   b.   genetically identical.
   c.   dissimilar.
   d.   members of different species.
   e.   members of different genera.

48. Plant Biotechnology; p. 695; moderate; ans: c
Meristem culture is an effective technique for producing pathogen-free plants because meristems:

   a.   grow faster in culture than other organs.
   b.   grow more slowly in culture than other organs.
   c.   lack vascular tissues.
   d.   contain leaf primordia.
   e.   develop embryos.

49. Plant Biotechnology; p. 696; easy; ans: e
Which of the following statements about genetic engineering is FALSE?

a. It is an application of recombinant DNA technology.
b. It allows the manipulation of genetic material for practical purposes.
c. It allows genes to be inserted into organisms in a precise way.
d. It has potential for improving crops in the future.
e. It is recombination only between species that can hybridize in nature.

50. Plant Biotechnology; p. 696; moderate; ans: c
The Ti plasmid is used in genetic engineering to:

a. protect *Agrobacterium tumefaciens* against infection.
b. protect the plant against antibiotics.
c. transfer foreign genes into a cell.
d. remove genes from a plant.
e. stimulate tumor formation in plants.

51. Plant Biotechnology; p. 696; moderate; ans: a
In *Agrobacterium tumefaciens*, opines are used:

a. as sources of nitrogen and carbon.
b. as hormones.
c. as an antibiotic.
d. to facilitate transfer of the plasmid into a host cell.
e. to stimulate cell division.

52. Plant Biotechnology; p. 697; moderate; ans: b
Plants having the *BT* gene from *Bacillus thuringiensis* are resistant to damage by:

a. viruses.
b. caterpillars.
c. *Agrobacterium tumefaciens*.
d. several herbicides.
e. *Salmonella*.

53. Plant Biotechnology; p. 698; easy; ans: e
What happened when a cytokinin gene was fused to a gene expressed only in senescent leaves, and the resulting chimeric gene was transferred into tobacco plants?

a. Ripening was stimulated.
b. Wilting was delayed.
c. A tumor-like growth was formed.
d. Leaf senescence was stimulated.
e. Leaf senescence was delayed.

54. Plant Biotechnology; p. 699; moderate; ans: c
One limitation of the *Agrobacterium* gene-transfer system is that it cannot:

a. transfer reporter genes.
b. transform eudicots.
c. transform monocots.
d. transfer genes from *Escherichia coli*.
e. transfer the luciferase gene.

55. **Plant Biotechnology; p. 699; easy; ans: a**
Electroporation can be used in both monocots and eudicots to:

    a.   transfer a gene into a host cell.
    b.   monitor the transfer of a gene into a host cell.
    c.   bind a gene to a microprojectile for delivery to a host cell.
    d.   fuse protoplasts.
    e.   culture meristems.

56. **Plant Biotechnology; p. 699; easy; ans: b**
Sustainable agriculture:

    a.   is a form of "invade and conquer" farming.
    b.   minimizes the use of chemicals.
    c.   maximizes the use of fertilizers.
    d.   minimizes the use of genetic engineering.
    e.   maximizes the use of herbicides.

True-False Questions

1. **Introduction; pp. 674–675; easy; ans: F**
The only way a plant can vary the intensity of a hormone signal is by varying the concentration of hormone that is produced.

2. **Auxins; p. 676; moderate; ans: F**
Plants can synthesize auxin only from tryptophan.

3. **Auxins; p. 676; moderate; ans: F**
In roots, a substance transported in a basipetal direction is transported toward the root tip.

4. **Auxins; p. 676; moderate; ans: T**
The transport of auxin is acropetal in roots and basipetal in stems.

5. **Auxins; p. 676; easy; ans: F**
Growth regulators usually operate alone rather than in concert.

6. **Auxins; pp. 678–679; easy; ans: T**
Auxin stimulates cell division in the vascular cambium in woody plants.

7. **Auxins; p. 679; easy; ans: T**
An example of a synthetic auxin used as a herbicide is 2,4-D.

8. **Cytokinins; p. 680; easy; ans: F**
Application of cytokinin inhibits the growth of lateral buds.

9. **Cytokinins; p. 681; moderate; ans: F**
In cultured tobacco stem tissues, the addition of kinetin without IAA stimulates cell division.

10. **Cytokinins; p. 681; moderate; ans: T**
An excised leaf floated in a kinetin solution will remain green longer than a leaf floated in water.

11. **Cytokinins; p. 682; moderate; ans: T**
Cytokinins are synthesized in roots and transported in the xylem to other parts of the plant.

12. **Ethylene; p. 682; difficult; ans: F**
SAM and ACC are inhibitors of the biosynthetic pathway from methionine to ethylene.

13. **Ethylene; p. 682; easy; ans: T**
Ethylene stimulates internodal elongation of rice plants that are submerged during monsoon flooding.

14. **Ethylene; p. 683; easy; ans: T**
Ethylene is used commercially to ripen fruits.

15. **Ethylene; p. 683; moderate; ans: F**
At high concentrations, applied auxin inhibits ethylene production.

16. **Abscisic Acid; p. 683; moderate; ans: F**
One of the principal roles of abscisic acid is in leaf, flower, and fruit abscission.

17. **Abscisic Acid; p. 684; easy; ans: T**
Mutant maize embryos that cannot make ABA germinate directly on the cob.

18. **Gibberellins; p. 684; easy; ans: T**
Rice plants grow rapidly and become spindly due to the presence of gibberellic acid produced by a fungus.

19. **Gibberellins; pp. 685–686; moderate; ans: F**
In barley seeds, gibberellins stimulate the synthesis of enzymes that catalyze the reactions associated with embryonic cell division.

20. **Gibberellins; p. 686; easy; ans: T**
Gibberellins promote fruit development in some species in which auxin does not.

21. **The Molecular Basis of Hormone Action; p. 688; easy; ans: F**
A cortical cell and a mesophyll cell from the same plant most likely have different genes in their nuclei.

22. **The Molecular Basis of Hormone Action; p. 688; moderate; ans: T**
A regulatory transcription factor binds directly to the regulatory sequence of a gene.

23. **The Molecular Basis of Hormone Action; p. 688; moderate; ans: F**
Most plant hormones affect the turgor pressure of a cell but have little direct effect on the extensibility of the cell wall.

24. **The Molecular Basis of Hormone Action; p. 689; moderate; ans: T**
Extensins are proteins that disrupt hydrogen bonds between cell wall polysaccharides.

25. **The Molecular Basis of Hormone Action; p. 690; Fig. 28-23; moderate; ans: F**
All receptor proteins in plant cells are located in the plasma membrane.

26. **The Molecular Basis of Hormone Action; p. 691; moderate; ans: F**
Second messengers do not cross the plasma membrane and enter the cytosol.

27. **The Molecular Basis of Hormone Action; pp. 692–693; moderate; ans: T**
In *Arabidopsis thaliana*, the receiver protein of the ethylene pathway is a negative modulator of the pathway.

28. **The Molecular Basis of Hormone Action; p. 693; moderate; ans: F**
When potassium ions move out of guard cells, water follows by osmosis and the stomatal pore opens.

29. **Plant Biotechnology; p. 693; easy; ans: T**
In plant tissue culture, large numbers of cells are grown in in a sterile and controlled environment.

30. **Plant Biotechnology; p. 695; moderate; ans: F**
Fertile, intergeneric hybrids have been obtained by protoplast fusion.

31. **Plant Biotechnology; p. 696; easy; ans: T**
Agrobacterium tumefaciens transfers the T-region of its Ti plasmid to the host's DNA.

32. **Plant Biotechnology; p. 697; easy; ans: T**
A transgenic plant contains foreign genes.

33. **Plant Biotechnology; p. 698; moderate; ans: T**
Nutritionally improved soybean, maize, and potato crops have been produced by genetic engineering.

34. **Plant Biotechnology; pp. 698–699; easy; ans: F**
The luciferase gene does not require activation by a promoter.

35. **Plant Biotechnology; p. 699; moderate; ans: F**
Micropropagation techniques work equally well for monocots and eudicots.

36. **Plant Biotechnology; p. 699; easy; ans: T**
Particle bombardment can be used to transform monocots.

Essay Questions

1. **Regulating Growth and Development: The Plant Hormones; pp. 674–675; moderate**
What are some basic characteristics of hormones? Name the five traditional groups of plant hormones, and give four examples of chemical regulators that do not fit into these groups.

2. **Auxins; p. 676; difficult**
   What is the relationship between IAA, tryptophan, and indole? Discuss the evidence that questions this relationship.

3. **Auxins; p. 676; moderate**
   What is meant by "polar transport"? Describe the mechanism of the polar transport of auxin in plants.

4. **Auxins; pp. 676–679; difficult**
   Describe the effects of applied auxin on (a) the differentiation of vascular tissues, (b) the growth of lateral buds, (c) the formation of adventitious roots, (d) the growth of fruits, and (e) the growth of weeds.

5. **Cytokinins; p. 681; difficult**
   Give examples to show how the cytokinin/auxin ratio regulates the differentiation of cells.

6. **Cytokinins; p. 681; moderate**
   Discuss the effects of kinetin on leaf senescence in excised leaves.

7. **Ethylene; p. 682; moderate**
   Explain how ethylene may stimulate cell expansion under some conditions but inhibit it under others.

8. **Ethylene; p. 683; moderate**
   Describe the effects of applied ethylene on (a) fruit ripening, (b) abscission, and (c) sex expression in cucurbits.

9. **Abscisic Acid; pp. 683–684; difficult**
   List some of the effects of ABA. Use these examples to show that ABA can be classed as an inhibitor in some respects and a promoter in others.

10. **Gibberellins; pp. 684–685; moderate**
    How many different gibberellins have been identified? Do all of these gibberellins cause their effects directly? Use evidence obtained from mutant maize plants to support your answer.

11. **Gibberellins; pp. 685–686; moderate**
    Describe the mechanism by which gibberellins stimulate the germination of barley seeds.

12. **The Molecular Basis of Hormone Action; p. 688; difficult**
    Using barley aleurone tissues as an example, explain how hormones are able to switch genes on and off.

13. **The Molecular Basis of Hormone Action; p. 689; moderate**
    What determines the rate at which a plant cell expands? Use the acid-growth hypothesis to explain the mechanism of expansion.

14. **The Molecular Basis of Hormone Action; p. 690; difficult**
    Discuss the ways in which plant hormones affect the shape of cells.

15. **The Molecular Basis of Hormone Action; pp. 690–691; moderate**
Explain how receptor proteins are involved in the mechanism of action of hormones.

16. **The Molecular Basis of Hormone Action; p. 691; difficult**
Discuss the functioning of second messengers. What do you think "first messengers" are?

17. **The Molecular Basis of Hormone Action; pp. 692–693; difficult**
Discuss the mechanism by which ABA regulates the opening of stomata.

18. **Plant Biotechnology; pp. 693, 695; moderate**
Discuss the various techniques of plant tissue culture. What are the advantages and disadvantages of each?

19. **Plant Biotechnology; p. 696; easy**
What advantages does genetic engineering have over natural and somatic hybridization?

20. **Plant Biotechnology; p. 696; moderate**
Describe the mechanism by which *Agrobacterium tumefaciens* causes crown-gall disease.

21. **Plant Biotechnology; p. 697; moderate**
Give some examples to show why *Agrobacterium tumefaciens* is a natural genetic engineer.

22. **Plant Biotechnology; pp. 698–699; moderate**
What is a reporter gene? Use an example to demonstrate why reporter genes are useful in genetic engineering.

23. **Plant Biotechnology; p. 699; moderate**
What are the principal limitations of the *Agrobacterium* gene-transfer system? Describe two gene-transfer methods that offer improvements over the *Agrobacterium* system.

24. **Plant Biotechnology; p. 699; moderate**
Discuss the risks and benefits of genetic engineering in plants.

Chapter 29

External Factors and Plant Growth

Multiple-Choice Questions

1. **The Tropisms; p. 703; moderate; ans: e**
    Which of the following statements about tropisms is FALSE?

    a. They are growth responses.
    b. They are responses to an external stimulus.
    c. They are directional responses.
    d. Responses involve bending or curving.
    e. Positive responses are away from the stimulus.

2. **The Tropisms; p. 703; difficult; ans: e**
    Which of the following is NOT a finding from experiments on the role of auxin in phototropism of coleoptile tips?

    a. Light does not affect the total amount of auxin.
    b. Light does not destroy auxin.
    c. Auxin migrates from the lighted side to the shaded side.
    d. More auxin can be isolated from the shaded side of an intact tip than from the lighted side.
    e. If the tip is split and the two halves separated by a barrier, more auxin can be isolated from the shaded side than from the lighted side.

3. **The Tropisms; p. 704; moderate; ans: d**
    _____ light is the most effective in producing a phototropic response.

    a. Green
    b. Yellow
    c. Orange
    d. Blue
    e. Red

4. **The Tropisms; p. 704; easy; ans: c**
    The pigments that absorb the light necessary for the phototropic response are most likely:

    a. proteins.
    b. chlorophylls.
    c. flavins.
    d. carotenoids.
    e. anthocyanins.

5. **The Tropisms; p. 704; moderate; ans: a**
    According to the original hypothesis on negative gravitropism of shoots, auxin
    is redistributed to the _____ side where it _____ cell expansion.

    a. lower; stimulates
    b. lower; inhibits
    c. upper; stimulates
    d. upper; inhibits
    e. upper; first inhibits, then stimulates

6. **The Tropisms; p. 704; difficult; ans: b**
    Which of the following statements about gravitropism is FALSE?

    a. Auxin-activated transcription of certain genes occurs only on the rapidly
        growing side of the stem.
    b. Experiments have shown that redistribution of auxin to the lower side of a
        root inhibits cell expansion.
    c. In many eudicots, bulk redistribution of auxin has not been observed.
    d. Some *Arabidopsis* mutants have roots that do not respond to gravity.
    e. Movement of auxin over just a few cell layers could dramatically affect
        root growth.

7. **The Tropisms; p. 704; moderate; ans: c**
    In plants oriented horizontally, calcium moves toward the _____ surface of
    shoots and toward the _____ surface of roots before bending occurs.

    a. upper; upper
    b. lower; lower
    c. upper; lower
    d. lower; upper
    e. cortical; epidermal

8. **The Tropisms; pp. 704–705; moderate; ans: e**
    In roots, statoliths are localized in the:

    a. epidermal cells.
    b. vascular tissues.
    c. cells surrounding the vascular tissues.
    d. epidermis of the rootcap.
    e. columella.

9. **The Tropisms; p. 705; moderate; ans: a**
    When a growing root is oriented horizontally, the amyloplasts in the central
    column of the rootcap:

    a. slide downward.
    b. float upward.
    c. attach to the nuclear envelope.
    d. attach to the mitochondria.
    e. do not move.

10. **The Tropisms; p. 705; difficult; ans: c**
In contrast to the starch-statolith and the plasmalemma central control hypotheses, the hydrostatic pressure hypothesis proposes that gravity is sensed by:

a. statoliths.
b. amyloplasts.
c. the pressure of the protoplast on the cell wall.
d. tension exerted on the plasma membrane by the cytoskeleton.
e. tension exerted on the plasma membrane by the protoplast.

11. **The Tropisms; p. 705; difficult; ans: d**
According to the plasmalemma central control hypothesis, after calcium channels open in gravity-stimulated cells, all of the following events occur EXCEPT:

a. activation of protein kinases.
b. increase in cytosolic calcium levels.
c. activation of calmodulin.
d. rapid efflux of calcium ions.
e. release of auxin.

12. **The Tropisms; p. 706; moderate; ans: b**
In hydrotropism in roots, the moisture gradient is sensed by cells in the:

a. cortex.
b. rootcap.
c. xylem.
d. phloem.
e. epidermis.

13. **Circadian Rhythms; p. 706; easy; ans: c**
Circadian rhythms have cycles of approximately:

a. 12 minutes.
b. 1 hour.
c. 24 hours.
d. 1 month.
e. 12 months

14. **Circadian Rhythms; pp. 706–709; difficult; ans: d**
Which of the following statements about circadian rhythms is FALSE?

a. They are endogenous.
b. They can be entrained by light-dark cycles.
c. They can be entrained by temperature cycles.
d. They speed up as the temperature rises.
e. They enable the plant to measure changing daylength.

15. Circadian Rhythms; p. 707; moderate; ans: d
What is responsible for keeping a plant's circadian rhythm in step with the natural day-night cycle?

a. an endogenous timing mechanism
b. auxin
c. phototropism
d. the environment
e. the plant's biological clock

16. Photoperiodism; p. 709; easy; ans: a
Photoperiodism is the biological response to changes in:

a. daylength.
b. intensity of light.
c. wavelength of light.
d. seasons.
e. temperature.

17. Photoperiodism; pp. 709–710; moderate; ans: b
Short-day plants flower:

a. in the summer.
b. in early spring or fall.
c. when exposed to 8 hours of daylight.
d. when the light period is longer than a critical length.
e. without respect to daylength.

18. Photoperiodism; p. 711; easy; ans: c
Experiments on cocklebur have shown that the _____ perceives the photoperiod.

a. apical meristem
b. columella
c. leaf blade
d. petiole
e. stem

19. Photoperiodism; p. 711; moderate; ans: e
If a short-day plant receives a one-minute exposure to light in the middle of the dark period rather than continuous darkness, it will:

a. produce more flowers.
b. produce smaller flowers.
c. produce larger flowers.
d. flower at a lower temperature.
e. not flower.

20. Photoperiodism; p. 711; moderate; ans: b
Commercial growers of chrysanthemums can delay flowering by a short _____ treatment _____.

a. light; at the beginning of the night.
b. light; in the middle of the night.
c. light; at the end of the night.
d. dark; in the middle of the day.
e. dark; at the end of the day.

21. **Photoperiodism; p. 711; difficult; ans: a**
Which type(s) of plants, when grown under noninductive conditions, can be induced to flower if the light period is interrupted by darkness?

  a. only long-day plants
  b. only short-day plants
  c. only day-neutral plants
  d. both long-day and short-day plants
  e. both short-day and day-neutral plants

22. **Chemical Basis of Photoperiodism; p. 711; moderate; ans: d**
Light with a wavelength of about _____ nanometers is the most effective for interrupting the dark period of both short-day and long-day plants.

  a. 220
  b. 340
  c. 440
  d. 660
  e. 730

23. **Chemical Basis of Photoperiodism; p. 712; moderate; ans: c**
Which of the following statements about P_r and P_{fr} is FALSE?

  a. They are photoreceptors.
  b. They participate in photoconversion reactions.
  c. P_r is the biologically active form.
  d. P_r absorbs 660-nanometer light.
  e. P_{fr} absorbs 730-nanometer light.

24. **Chemical Basis of Photoperiodism; p. 712; moderate; ans: b**
When lettuce seeds are exposed to red light:

  a. P_{fr} is converted to P_r.
  b. P_r is converted to P_{fr}.
  c. 660-nanometer light is converted to 730-nanometer light.
  d. 730-nanometer light is converted to 660-nanometer light.
  e. they will not germinate.

25. **Chemical Basis of Photoperiodism; p. 712; moderate; ans: e**
After a plant has been exposed to noontime sunlight for a few minutes:

  a. all the P_r has been converted to P_{fr}.
  b. all the P_{fr} has been converted to P_r.
  c. P_r is converted to P_{fr} faster than P_{fr} is converted to P_r.
  d. P_{fr} is converted to P_r faster than P_r is converted to P_{fr}.
  e. the proportion of P_{fr} is greater than that of P_r.

26. **Chemical Basis of Photoperiodism; p. 712; difficult; ans: e**
A high level of _____ in the middle of the dark period will inhibit flowering in _____ plants that otherwise would have flowered.

  a. P_r; long-day
  b. P_r; short-day
  c. P_{fr}; day-neutral
  d. P_{fr}; long-day
  e. P_{fr}; short-day

27. Chemical Basis of Photoperiodism; pp. 713–714; moderate; ans: b
An etiolated eudicot seedling:

   a.   has a short stem.
   b.   has small leaves.
   c.   is green.
   d.   lacks plastids.
   e.   cannot undergo further growth.

28. Chemical Basis of Photoperiodism; p. 714; moderate; ans: b
When the tip of a germinating seedling emerges from the soil into the light, the plant undergoes a set of changes in growth and development called:

   a.   phototropism.
   b.   photomorphogenic responses.
   c.   photoperiodic responses.
   d.   photoconversion reactions.
   e.   etiolation.

29. Chemical Basis of Photoperiodism; p. 714; moderate; ans: c
Which of the following plants would NOT be etiolated?

   a.   a bean seedling grown entirely in the dark
   b.   a dark-grown bean seedling that receives 5 minutes of far-red light each day
   c.   a dark-grown bean seedling that receives 5 minutes of far-red light followed by 5 minutes of red light each day
   d.   a grass seedling grown entirely in the dark
   e.   a grass seedling that receives 5 minutes of far-red light each day

30. Chemical Basis of Photoperiodism; p. 714; difficult; ans: d
Compared with similar plants growing in full sunlight, plants growing in the shade of other vegetation:

   a.   receive more red light and less far-red light.
   b.   are usually shorter.
   c.   receive more wavelengths below 700 nanometers than above.
   d.   have a higher equilibrium ratio of P_r to P_{fr}.
   e.   receive more reflected light and less transmitted light.

31. Hormonal Control of Flowering; p. 715; moderate; ans: e
Which of the following statements about the floral stimulus is FALSE?

   a.   It has been called florigen.
   b.   It is transmitted from leaf to bud.
   c.   It is transported in the phloem.
   d.   It can pass from one plant to another through a graft.
   e.   It can pass from one plant to another through agar.

32. **Hormonal Control of Flowering; p. 715; difficult; ans: c**
According to the gibberellin-anthesin hypothesis, _____ plants produce
_____ under noninducing conditions.

   a. long-day; an inhibitor
   b. long-day; gibberellin but not anthesin
   c. long-day; anthesin but not gibberellin
   d. short-day; anthesin but not gibberellin
   e. short-day; an inhibitor

33. **Hormonal Control of Flowering; p. 715; easy; ans: d**
The current evidence suggests that flowering is controlled by:

   a. auxin and gibberellin.
   b. a single flowering hormone.
   c. florigen and anthesin.
   d. inhibitors and promoters.
   e. P_r and P_{fr}.

34. **Hormonal Control of Flowering; p. 715; moderate; ans: c**
Flowering in a day-neutral plant is inhibited by a graft union with a _____
plant kept on _____.

   a. long-day; long days
   b. short-day; short days
   c. long-day; short days
   d. short-day; short days
   e. day-neutral; short days

35. **Genetic Control of Flowering; p. 716; moderate; ans: b**
Which of the following statements about flowering in *Arabidopsis thaliana* is
FALSE?

   a. Studies of mutants reveal genes involved in the transition from vegetative
       apex to reproductive apex.
   b. Studies of mutants reveal genes that promote flowering but not genes that
       inhibit flowering.
   c. In plants with mutations in the *LEAFY* and *APETALA* genes, flowers are
       replaced by leaflike shoots.
   d. Some mutants flower regardless of the photoperiod.
   e. Some mutants exhibit delayed flowering.

36. **Dormancy; p. 717; easy; ans: d**
A dormant seed will germinate only when:

   a. the temperature becomes milder.
   b. water becomes available.
   c. oxygen becomes available.
   d. precise environmental cues are received.
   e. water and oxygen become available at the same time.

37. Dormancy; p. 717; easy; ans: a
Stratification is the process used by horticulturalists in which seeds are:

a. moistened and exposed to low temperature.
b. moistened and exposed to high temperature.
c. mechanically abraded.
d. soaked in alcohol.
e. dried.

38. Dormancy; p. 718; moderate; ans: b
Rainfall will break seed dormancy in certain desert species by:

a. abrading the seed coat.
b. leaching away inhibitory chemicals.
c. dissolving waxy substances.
d. making the seed coat permeable to nutrients.
e. making the seed coat permeable to oxygen.

39. Dormancy; p. 718; easy; ans: c
Using radiocarbon dating, the oldest seed known to have germinated was found to be approximately _____ years old.

a. 13
b. 130
c. 1300
d. 13,000
e. 130,000

40. Dormancy; p. 718; easy; ans: b
The primary purpose of seed banks is to:

a. provide a storehouse for seeds to be used in times of famine.
b. preserve the genetic characteristics of wild varieties and early cultivars of crop plants.
c. make seeds available to farmers who cannot afford to purchase them.
d. provide seeds to be used solely for genetic engineering.
e. store dormant seeds until ways can be found to germinate them.

41. Dormancy; p. 718; easy; ans: d
In many trees, bud dormancy is initiated in:

a. winter.
b. early spring.
c. late spring.
d. midsummer.
e. autumn.

42. Dormancy; p. 718; moderate; ans: e
Which of the following statements about bud scales is FALSE?

a. They may contain growth inhibitors.
b. They protect the bud against desiccation.
c. They restrict the movement of oxygen into the bud.
d. They insulate the bud from heat loss.
e. They have leaf primordia in their axils.

43. **Dormancy; pp. 718–719; moderate; ans: a**
In temperate regions, the acclimation of buds is initiated primarily by decreasing _____ and it produces _____.

a.   daylength; cold hardiness
b.   temperature; cold hardiness
c.   moisture; moisture resistance
d.   levels of growth inhibitors; early flowering
e.   numbers of bud scales; early flowering

44. **Dormancy; p. 719; moderate; ans: d**
Deciduous fruit trees such as apple and peach cannot be cultivated in tropical areas because in these areas their:

a.   roots would drown.
b.   stems could not grow at the high temperatures.
c.   leaves would wither and drop off.
d.   buds could not grow without a cold period.
e.   seeds would germinate too rapidly.

45. **Dormancy; p. 719; easy; ans: c**
In peach buds, dormancy can be broken by a combination of low temperature and treatment with:

a.   cytokinin.
b.   ethylene.
c.   gibberellin.
d.   auxin.
e.   abscisic acid.

46. **Cold and the Flowering Response; p. 719; moderate; ans: e**
Winter rye is usually planted in the _____, and it flowers in the _____.

a.   early spring; summer
b.   early spring; autumn
c.   early spring; winter
d.   summer; following spring
e.   autumn; following summer

47. **Cold and the Flowering Response; p. 719; moderate; ans: b**
In the process of vernalization, early _____ is stimulated by exposing the germinating seeds to _____ temperatures.

a.   dormancy; low
b.   flowering; low
c.   germination; low
d.   dormancy; high
e.   germination; high

48. Cold and the Flowering Response; p. 719; difficult; ans: e
Which of the following statements about cold treatments and flowering is FALSE?

a.   In winter rye, after cold treatment, plants require long days in order to flower.
b.   In spinach, cold treatment shortens the daylength required for flowering.
c.   In clover, cold treatment completely removes the daylength requirement for flowering.
d.   In henbane, gibberellin treatment can substitute for cold treatment in hastening flowering.
e.   In nonrosette long-day plants, gibberellin treatment can substitute for cold treatment in hastening flowering.

49. Nastic Movements; p. 719; moderate; ans: e
The direction of a nastic movement is always _____ the direction of the stimulus.

a.   away from
b.   toward
c.   at right angles to
d.   at an oblique angle to
e.   independent of

50. Nastic Movements; pp. 719–720; moderate; ans: d
In nyctinastic movements, leaves:

a.   twine around a support such as a fence post.
b.   have their stomata open during the night and closed during the day.
c.   are rolled during the day and unrolled at night.
d.   are oriented vertically in darkness and horizontally in light.
e.   are oriented horizontally in darkness and vertically in light.

51. Nastic Movements; p. 720; moderate; ans: c
Which of the following statements about a pulvinus is FALSE?

a.   It is a jointlike thickening at the base of a leaf or a leaflet.
b.   It is responsible for most nyctinastic leaf movements.
c.   It lacks vascular tissue.
d.   It contains parenchyma cells.
e.   Some of its cells undergo changes in turgor.

52. Nastic Movements; pp. 720–721; difficult; ans: b
In *Mimosa pudica*, movement of leaflets involves all of the following changes in cells of the pulvinus EXCEPT:

a.   accumulation of sucrose unloaded from the phloem.
b.   an influx of potassium ions from the apoplast.
c.   a change in turgor pressure.
d.   an efflux of water.
e.   a decrease in water potential.

53. **Nastic Movements; p. 721; moderate; ans: a**
    The rapid closure of leaves of the Venus flytrap is now thought to be due to:

    a. changes in turgor pressure in the mesophyll cells underlying the epidermis.
    b. changes in turgor pressure in the upper epidermis.
    c. changes in turgor pressure in cells of the pulvinus.
    d. acid-induced wall loosening of motor cells.
    e. unloading of sucrose from the phloem.

54. **Generalized Effects of Mechanical Stimuli on Plant Growth and Development: Thigmomorphogenesis; p. 721; moderate; ans: e**
    In contrast to thigmonastic movements, thigmomorphogenesis:

    a. is a response to touch.
    b. is a response to a mechanical stimulus.
    c. is independent of the direction of the stimulus.
    d. involves cells of the pulvinus.
    e. involves an altered growth pattern.

55. **Generalized Effects of Mechanical Stimuli on Plant Growth and Development: Thigmomorphogenesis; pp. 721–722; moderate; ans: c**
    Plants grown in a natural environment are _____ than plants grown in a greenhouse.

    a. greener
    b. thinner
    c. shorter
    d. more spindly
    e. less stocky

56. **Generalized Effects of Mechanical Stimuli on Plant Growth and Development: Thigmomorphogenesis; pp. 721–722; easy; ans: d**
    In *Arabidopsis thaliana*, touch induces the expression of genes that encode proteins related to _____ , suggesting a role for _____ in thigmomorphogenesis.

    a. an endogenous inhibitor; the biological clock
    b. amylase; starch
    c. the cytoskeleton; microtubules
    d. calmodulin; calcium
    e. chlorophyll; magnesium

57. **Solar Tracking; pp. 722–723; moderate; ans: e**
    Which of the following statements about heliotropism is FALSE?

    a. It is a diurnal movement involving pulvini.
    b. In paraheliotropism, the leaf blades remain parallel to the sun's rays.
    c. In diaheliotropism, the leaf blades remain perpendicular to the sun's rays.
    d. It is also called solar tracking.
    e. It involves a mechanism similar to that of stem phototropism.

True-False Questions

1. **The Tropisms; p. 703; easy; ans: T**
   In tropisms, a response toward the stimulus is said to be positive.

2. **The Tropisms; p. 703; moderate; ans: F**
   Experiments using ^{14}C-labeled auxin indicate that light causes the destruction of auxin in coleoptiles.

3. **The Tropisms; p. 704; moderate; ans: T**
   In shoot gravitropism, auxin is distributed to the lower side where it stimulates cell expansion.

4. **The Tropisms; p. 704; easy; ans: T**
   The role of calcium in gravitropism is mediated by calmodulin.

5. **The Tropisms; p. 704; moderate; ans: F**
   In gravitropism in intact maize roots, the effect of calcium is most likely at the cortex.

6. **The Tropisms; p. 705; easy; ans: F**
   An example of hydrotropism is the closing of stomata in response to water stress.

7. **The Tropisms; p. 706; moderate; ans: T**
   The twining of a tendril around a fence post is an example of thigmotropism.

8. **Circadian Rhythms; p. 706; easy; ans: T**
   Circadian rhythms persist even when all environmental conditions are kept constant.

9. **Circadian Rhythms; pp. 707–708; moderate; ans: T**
   Circadian rhythms that are free-running under certain artificially maintained conditions can be entrained by other conditions.

10. **Circadian Rhythms; p. 708; moderate; ans: F**
    As the temperature rises, the biological clock speeds up.

11. **Circadian Rhythms; p. 708; moderate; ans: T**
    The biological clock of *Neurospora* involves a feedback loop in which the protein products of genes inhibit their own transcription.

12. **Photoperiodism; p. 709; easy; ans: T**
    Photoperiodism pertains to events that are influenced by the relative lengths of light and dark.

13. **Photoperiodism; p. 709; easy; ans: F**
    There are two main types of plants in terms of flowering response: long-day plants and short-day plants.

14. **Photoperiodism; p. 709; easy; ans: T**
In order to flower, short-day plants must have a light period shorter than a critical length, and long-day plants must have a light period longer than a critical length.

15. **Photoperiodism; p. 711; moderate; ans: F**
Temperature has no effect on the photoperiodic response of plants.

16. **Photoperiodism; p. 711; easy; ans: T**
In experiments on cocklebur, a completely defoliated plant could not be induced to flower by short-day exposure.

17. **Photoperiodism; p. 711; easy; ans: T**
With respect to effect on flowering, the middle of the dark period is the part most sensitive to interruption by a flash of light.

18. **Chemical Basis Photoperiodism; pp. 711–712; moderate; ans: F**
The wavelength of light most effective in causing germination of lettuce seeds is 730 nanometers, in the far-red region.

19. **Chemical Basis of Photoperiodism; p. 712; moderate; ans: T**
Treatment of lettuce seeds with red light can nullify the effects of a previous exposure to far-red light.

20. **Chemical Basis of Photoperiodism; p. 712; easy; ans: F**
When plants are transferred from light to darkness, the level of P_{fr} remains constant for many hours.

21. **Chemical Basis of Photoperiodism; pp. 712–713; moderate; ans: T**
The phytochrome molecule consists of a protein portion and a chromophore, and it exists in two interconvertible forms.

22. **Chemical Basis of Photoperiodism; p. 714; easy; ans: T**
An example of photomorphogenic responses is the changes that take place when an etiolated plant is exposed to light.

23. **Chemical Basis of Photoperiodism; p. 714; moderate; ans: F**
The interconversion of P_r and P_{fr} is now known to control time measurement in photoperiodism.

24. **Hormonal Control of Flowering; p. 715; easy; ans: F**
Florigen has been isolated and chemically identified.

25. **Hormonal Control of Flowering; p. 715; moderate; ans: T**
Gibberellin can stimulate flowering in long-day biennials such as celery and cabbage.

26. **Hormonal Control of Flowering; p. 715; moderate; ans: T**
Experiments on Biloxi soybean suggest that leaves of noninduced plants produce a flowering inhibitor.

27. **Genetic Control of Flowering; p. 716; easy; ans: T**
In *Arabidopsis*, some genes promote, and others inhibit, flowering.

28. **Dormancy; p. 717; easy; ans: F**
Seed dormancy can usually be overcome by placing the seeds in a warm, moist environment.

29. **Dormancy; p. 717; easy; ans: T**
Stratification is the process in which moist seeds are exposed to low temperatures in order to break dormancy.

30. **Dormancy; p. 718; easy; ans: F**
In the process of scarification, seeds are treated with solvents that dissolve the waxes in the seed coat.

31. **Dormancy; p. 718; easy; ans: T**
The roles of bud scales are similar to those of the seed coat.

32. **Dormancy; p. 719; moderate; ans: F**
The bulbs of tulips and hyacinths can be "forced" by bringing them indoors before the temperature drops.

33. **Cold and the Flowering Response; p. 719; easy; ans: F**
Vernalization is the exposure of flower buds to low temperatures to promote early flowering.

34. **Nastic Movements; pp. 719–720; easy; ans: T**
The sleep movements of plants are an example of nastic movements.

35. **Nastic Movements; pp. 721–722; moderate; ans: T**
The seismonastic movements of *Mimosa pudica* involve changes in the pulvini.

36. **Generalized Effects of Mechanical Stimuli on Plant Growth and Development: Thigmomorphogenesis; p. 721; easy; ans: F**
Regular rubbing or bending of a stem stimulates its elongation.

37. **Solar Tracking; p. 721; easy; ans: T**
The mechanism underlying heliotropism involves pulvini at the base of leaves and leaflets.

Essay Questions

1. **The Tropisms; pp. 703–704; moderate**
Describe the mechanism by which coleoptiles bend toward the light. Support your answer with experimental evidence.

2. **The Tropisms; p. 704; moderate**
In what ways are auxin and calcium involved in the gravitropism of stems and roots?

3. **The Tropisms; pp. 704–705; moderate**
Compare and contrast the starch-statolith, the hydrostatic pressure, and the plasmalemma central control hypotheses of gravity sensing.

4. **The Tropisms; p. 706; moderate**
   Why is it difficult to study hydrotropism? How has this difficulty been overcome?

5. **The Tropisms; p. 706; moderate**
   What is thigmotropism? In what way is "memory" involved in this response?

6. **Circadian Rhythms; pp. 706–707; moderate**
   What is the difference between a free-running and an entrained circadian rhythm? In nature, how is entrainment achieved?

7. **Circadian Rhythms; pp. 707–709; easy**
   In what ways is the biological clock useful to plants? How is the affect of temperature an important aspect of this usefulness?

8. **Photoperiodism; pp. 709–711; moderate**
   Define photoperiodism. Distinguish between short-day, long-day, and day-neutral plants.

9. **Photoperiodism; p. 711; moderate**
   Why could short-day plants be called "long-night" plants, and long-day plants be called "short-night" plants? How can commercial growers take advantage of this observation?

10. **Chemical Basis of Photoperiodism; pp. 711–712; difficult**
    Describe the photoconversion reactions of phytochrome. Use these reactions to explain why a lettuce seed exposed first to red light and then to far-red light will not germinate.

11. **Chemical Basis of Photoperiodism; p. 713; moderate**
    Describe the two parts of the phytochrome molecule. What is known about the mechanism of action of phytochrome?

12. **Chemical Basis of Photoperiodism; pp. 713–714; moderate**
    What are the main characteristics of an etiolated plant? How are these characteristics affected by red and far-red light?

13. **Chemical Basis of Photoperiodism; p. 714; difficult**
    Using data obtained from studies of the *long hypocotyl* mutants of *Arabidopsis*, discuss the roles of specific photoreceptor genes in morphogenesis.

14. **Hormonal Control of Flowering; p. 715; moderate**
    What is the evidence for the existence of florigen and the manner in which it is transported in plants?

15. **Hormonal Control of Flowering; p. 715; difficult**
    Discuss the evidence for the existence of a flowering inhibitor.

16. **Genetic Control of Flowering; p. 716; difficult**
    How has an analysis of mutants advanced our understanding of the flowering process?

17. **Dormancy; pp. 717–718; moderate**
How is dormancy adaptive in an evolutionary sense? Describe some mechanisms that have evolved to break seed dormancy.

18. **Dormancy; p. 718; easy**
What is a seed bank, and why are seed banks important?

19. **Dormancy; pp. 718–719; moderate**
Discuss bud dormancy and the different ways in which it can be overcome. What is the difference between acclimation and cold-hardiness?

20. **Cold and the Flowering Response; p. 719; moderate**
Describe the procedure of vernalization. In what ways is it economically important?

21. **Nastic Movements; pp. 719–720; moderate**
How is a nastic movement different from a tropism?

22. **Nastic Movements; pp. 720–721; moderate**
What is a pulvinus? How do pulvini function in nastic movements? Give some examples.

23. **Generalized Effects of Mechanical Stimuli on Plant Growth and Development: Thigmomorphogenesis; pp. 721–722; moderate**
What is thigmomorphogenesis, and in what ways is it adaptive?

24. **Solar Tracking; pp. 722–723; moderate**
Describe the two types of heliotropism and their usefulness to plants. What mechanism has been proposed to explain this phenomenon?

Chapter 30

Plant Nutrition and Soils

Multiple-Choice Questions

1. **Essential Elements; p. 727, and Table 30-1; easy; ans: d**
    Which of the following is NOT an essential element?

    a. potassium
    b. zinc
    c. boron
    d. lead
    e. chlorine

2. **Essential Elements; p. 727; moderate; ans: a**
    Which of the following is NOT a criterion for an element to be considered essential for plant growth?

    a. It is present in plant tissues.
    b. In its absence the plant shows deficiency symptoms.
    c. It is part of a molecule that is essential to the plant.
    d. It is needed by the plant to produce viable seed.
    e. It is needed by the plant to complete its life cycle.

3. **Essential Elements; p. 727; easy; ans: c**
    A micronutrient is a plant nutrient required in concentrations equal to or less than _____ mg per kg of dry matter.

    a. 1
    b. 10
    c. 100
    d. 1000
    e. 10,000

4. **Essential Elements; p. 727 and Table 30-1; easy; ans: b**
    Which of the following is NOT a micronutrient?

    a. manganese
    b. potassium
    c. zinc
    d. copper
    e. nickel

5. **Essential Elements; p. 727; easy; ans: d**
A macronutrient is a plant nutrient required in concentrations equal to or greater than _____ mg per kg of dry matter.

    a. 1
    b. 10
    c. 100
    d. 1000
    e. 10,000

6. **Essential Elements; p. 727; moderate; ans: b**
Eudicots generally require greater amounts of _____ than do monocots.

    a. magnesium and calcium
    b. calcium and boron
    c. boron and manganese
    d. manganese and zinc
    e. zinc and cobalt

7. **Essential Elements; p. 729; moderate; ans: e**
Legumes grown in culture benefit from the addition of cobalt because the cobalt:

    a. inhibits the growth of pathogenic bacteria.
    b. stimulates cell division of the apical meristem.
    c. stimulates the growth of root cells.
    d. is required by mycorrhizae.
    e. is required by symbiotic nitrogen-fixing bacteria.

8. **Functions of Essential Elements; p. 729; easy; ans: a**
In plants, necrosis is the:

    a. localized death of tissues.
    b. yellowing of leaves.
    c. loss of chlorophyll.
    d. development of tumors.
    e. healing of wounds.

9. **Functions of Essential Elements; p. 729; difficult; ans: c**
In magnesium-deficient plants, older leaves become more severely chlorotic than younger leaves because:

    a. magnesium is less phloem-mobile than chlorophyll.
    b. magnesium is more phloem-mobile than chlorophyll.
    c. younger leaves withdraw magnesium from older leaves.
    d. older leaves withdraw magnesium from younger leaves.
    e. magnesium is more mobile in younger leaves than in older leaves.

10. **Functions of Essential Elements; p. 729; easy; ans: d**
Which of the following is NOT a phloem-mobile element?

    a. nitrogen
    b. potassium
    c. magnesium
    d. calcium
    e. phosphorus

11. **Functions of Essential Elements; p. 729; difficult; ans: e**
A deficiency of _____ is MOST likely to produce symptoms in younger
leaves before older leaves.

    a. copper
    b. potassium
    c. manganese
    d. zinc
    e. iron

12. **The Soil; p. 731; easy; ans: b**
Which of the following statements about rocks is FALSE?

    a. They consist of several different minerals.
    b. They are divided into groups based on the amount of weathering.
    c. Igneous rocks are derived directly from molten material.
    d. Sedimentary rocks are formed from components deposited in water.
    e. Metamorphic rocks are formed from the transformation of igneous and
        sedimentary rocks.

13. **The Soil; p. 731; difficult; ans: e**
Which of the following processes is NOT involved in the weathering of rocks?

    a. dissolution by acids
    b. scouring by water and wind
    c. heating and cooling
    d. freezing and thawing
    e. melting under heat and pressure

14. **The Soil; p. 732; easy; ans: b**
Humus consists mostly of:

    a. living organic matter.
    b. dead organic matter.
    c. sand.
    d. silt.
    e. clay.

15. **The Soil; pp. 731–732; moderate; ans: e**
Which of the following statements about soil is FALSE?

    a. The A horizon has the greatest physical, chemical, and biological activity.
    b. The A horizon is the topsoil.
    c. The B horizon is a region of deposition.
    d. The B horizon is the subsoil.
    e. The C horizon is part of the true soil.

16. **The Soil; p. 732; easy; ans: c**
Which of the following types of soil particles has, by definition, the smallest
diameter?

    a. loam
    b. silt
    c. clay
    d. fine sand
    e. coarse sand

17. **The Soil; p. 734; moderate; ans: d**
    What fraction of the total soil volume is pore space?

    a.   one-tenth
    b.   one-quarter
    c.   one-third
    d.   one-half
    e.   two-thirds

18. **The Soil; p. 734; moderate; ans: e**
    Which of the following soils can hold the greatest amount of water against the action of gravity?

    a.   silt
    b.   loam
    c.   coarse sand
    d.   fine sand
    e.   clay

19. **The Soil; p. 734; moderate; ans: b**
    Field capacity is the:

    a.   total amount of water present in a one-hectare field after a soaking rain.
    b.   percentage of water that a soil can hold against the action of gravity.
    c.   percentage of water remaining in a soil when plants undergo irreversible wilting.
    d.   total amount of fertilizer required for maximal plant growth in a one-hectare field.
    e.   total amount of mineral nutrients present in a one-hectare field after a soaking rain.

20. **The Soil; p. 735; moderate; ans: d**
    The permanent wilting percentage is the percentage of:

    a.   colloidal particles remaining in the soil after a heavy rain.
    b.   colloidal particles that would cause irreversible wilting.
    c.   colloidal particles that would cause reversible wilting.
    d.   water remaining in a soil when irreversible wilting occurs.
    e.   water remaining in a soil when reversible wilting occurs.

21. **The Soil; p. 735; easy; ans: e**
    A soil with a water potential of _____ megapascals is considered to be at its permanent wilting percentage.

    a.   0
    b.   0.5
    c.   −0.5
    d.   1.5
    e.   −1.5

22. **The Soil; p. 735; moderate; ans: d**
    Cation exchange is important because:

    a. cations are more crucial to the plant than anions.
    b. cations can exchange with anions that are used by the plant.
    c. cations prevent anions from binding to colloidal particles.
    d. exchangeable ions are not lost with leaching water.
    e. potential harmful exchangeable ions are leached away from the roots.

23. **The Soil; p. 735; easy; ans: a**
    An example of cation exchange is _____ replacing _____ on a clay particle.

    a. H^+; K^+
    b. H^+; NO_3^-
    c. NO_3^-; SO_4^{2-}
    d. OH^-; SO_4^{2-}
    e. HCO_3^-; Mg^{2+}

24. **The Soil; p. 735; difficult; ans: d**
    Which of the following ions is MOST likely be precipitated in alkaline soils?

    a. hydroxide
    b. bicarbonate
    c. sulfate
    d. iron
    e. potassium

25. **Nutrient Cycles; pp. 735–736; moderate; ans: b**
    Which of the following statements about nutrient cycles is FALSE?

    a. They are also called biogeochemical cycles.
    b. All nutrients recycled to the soil are available for plant use.
    c. Some cycles involve the atmosphere.
    d. Each element has a different cycle.
    e. Both macronutrients and micronutrients are recycled.

26. **Nutrient Cycles; p. 736; moderate; ans: b**
    Which of the following processes does NOT contribute to the "leakiness" of nutrient cycles?

    a. leaching
    b. recycling
    c. burning
    d. harvesting
    e. erosion

27. **Nitrogen and the Nitrogen Cycle; p. 736; easy; ans: e**
    The chief reservoir of nitrogen is:

    a. the soil.
    b. the ocean.
    c. living organisms.
    d. dead organic material.
    e. the atmosphere.

28. Nitrogen and the Nitrogen Cycle; p. 737; moderate; ans: a
As dead organic materials are broken down by bacteria and fungi, the nitrogen not used by these organisms is released as _____ in a process called _____.

    a. NH_4^+; ammonification
    b. NO_3^-; nitrification
    c. NO_2^-; nitrification
    d. N_2; denitrification
    e. N_2O; denitrification

29. Nitrogen and the Nitrogen Cycle; p. 738; moderate; ans: c
_____ releases energy that is used by bacteria to reduce carbon dioxide.

    a.   Ammonification
    b.   Nitrogen fixation
    c.   Nitrification
    d.   Conversion of nitrite to ammonium
    e.   Nitrogen mineralization

30. Nitrogen and the Nitrogen Cycle; p. 738; moderate; ans: b
Nitrite is oxidized to nitrate by:

    a.   *Nitrosomonas*.
    b.   *Nitrobacter*.
    c.   ammonifying bacteria and fungi.
    d.   denitrifying bacteria.
    e.   plant roots.

31. Nitrogen and the Nitrogen Cycle; p. 738; difficult; ans: a
Which of the following processes is anaerobic?

    a.   denitrification
    b.   nitrification
    c.   conversion of ammonium to nitrite
    d.   conversion of nitrite to nitrate
    e.   oxidation of ammonium

32. Nitrogen and the Nitrogen Cycle; p. 738; difficult; ans: d
Which of the following is(are) a product of denitrification?

    a. NH_4^+
    b. NO_3^-
    c. NO_2^-
    d. N_2
    e.   amino acids

33. Nitrogen and the Nitrogen Cycle; p. 738; moderate; ans: e
Which of the following is NOT a way in which nitrogen is lost from an ecosystem?

    a.   harvesting of plants
    b.   soil erosion
    c.   burning of plants
    d.   leaching
    e.   nitrification

34. Nitrogen and the Nitrogen Cycle; p. 738; moderate; ans: e

In the process of nitrogen fixation, _____ is converted to _____.

   a. NH_4^+; NO_3^-
   b. NO_3^-; NO_2^-
   c. N_2; NO_2^-
   d. N_2; NO_3^-
   e. N_2; NH_4^+

35. Nitrogen and the Nitrogen Cycle; p. 738; easy; ans: c

In the symbiotic association between bacterium and legumes, the bacteria provides the plant with:

   a.   an energy source.
   b.   carbon-containing molecules.
   c.   nitrogen.
   d.   phosphate.
   e.   sulfate.

36. Nitrogen and the Nitrogen Cycle; p. 739; easy; ans: b

Rhizobia enter legumes by invading the:

   a.   seeds.
   b.   root hairs.
   c.   stems.
   d.   leaves.
   e.   flowers.

37. Nitrogen and the Nitrogen Cycle; p. 739; difficult; ans: a

During rhizobial infection of a legume, an infection thread is formed by inward growth of cell walls of the:

   a.   root hair.
   b.   bacteroid.
   c.   envelope.
   d.   cortex.
   e.   nodule.

38. Nitrogen and the Nitrogen Cycle; pp. 739–740; difficult; ans: e

The nitrogen-fixing nodules of legumes consist of:

   a.   bacteroids only.
   b.   bacteroids and root hairs only.
   c.   bacteroids and vascular bundles only.
   d.   bacteroids, cortical cells, and vascular bundles only.
   e.   bacteroids, cortical cells, vascular bundles, and root hairs.

39. Nitrogen and the Nitrogen Cycle; p. 740; moderate; ans: c

In soybeans, nitrogen is exported from the nodule to the plant in the form of:

   a.   ammonium.
   b.   nitrate.
   c.   ureides.
   d.   nitrite.
   e. N_2.

40. **Nitrogen and the Nitrogen Cycle; p. 740; difficult; ans: c**
    Which of the following statements about leghemoglobin is FALSE?

    a. It acts as an O_2 carrier.
    b. It buffers the O_2 concentration in nodules.
    c. It is found in the cytosol of the bacteroid.
    d. Its heme portion is produced by the bacteroid.
    e. Its globin portion is produced by the plant.

41. **Nitrogen and the Nitrogen Cycle; p. 740; moderate; ans: b**
    The *nif* genes of rhizobia are involved specifically in:

    a. nodule formation.
    b. nitrogen fixation.
    c. plasmid formation.
    d. root-hair curling.
    e. the formation of infection threads.

42. **Nitrogen and the Nitrogen Cycle; p. 740; difficult; ans: d**
    The direct role of flavonoids in nodule formation is to:

    a. stimulate formation of infection threads.
    b. stimulate cell wall degradation.
    c. stimulate root-hair curling.
    d. activate the bacterial *nodD* gene.
    e. activate the plant *Nod* genes.

43. **Nitrogen and the Nitrogen Cycle; p. 740; difficult; ans: a**
    The role of nodulins is to:

    a. stimulate cortical cell division.
    b. stimulate root-hair curling.
    c. stimulate the formation of infection threads.
    d. activate the plant *Nod* genes.
    e. activate the bacterial *nodD* gene.

44. **Nitrogen and the Nitrogen Cycle; p. 740; moderate; ans: e**
    Which of the following statements about lectins is FALSE?

    a. They bind sugars.
    b. They may facilitate binding of bacteria to root-hair cell walls.
    c. They may interact with rhizobia.
    d. They are secreted by legume roots.
    e. They are carbohydrates.

45. **Nitrogen and the Nitrogen Cycle; p. 741; moderate; ans: c**
    The _____ symbiosis is an example of a nitrogen-fixing symbiosis that
    continues throughout the life cycle of the host.

    a. legume-*Rhizobium*
    b. legume-*Bradyrhizobium*
    c. *Azolla-Anabaena*
    d. *Alnus*-actinomycete
    e. *Ceanothus*-actinomycete

46. Nitrogen and the Nitrogen Cycle; p. 741; difficult; ans: e
Which of the following groups contains species that are nonsymbiotic, photosynthetic, nitrogen-fixing bacteria?

a.   genus *Azotobacter*
b.   genus *Azotococcus*
c.   genus *Beijerinckia*
d.   genus *Clostridium*
e.   cyanobacteria

47. Nitrogen and the Nitrogen Cycle; pp. 741–742; moderate; ans: c
Which of the following statements about industrial nitrogen fixation is FALSE?

a.   It has a high energy cost.
b.   It involves the reaction of N_2 with H_2 under high temperature and pressure.
c.   Its use has decreased steadily since 1914.
d.   It is used mostly in the production of fertilizers.
e.   Its energy-expensive component is derived from fossil fuels.

48. Assimilation of Nitrogen; p. 742; easy; ans: b
_____ is the principal source of nitrogen available to crop plants.

a.   Organic nitrogen
b.   Nitrate
c.   Ammonium
d.   Glutamine
e.   Glutamate

49. Assimilation of Nitrogen; p. 742; moderate; ans: c
In most herbaceous plants grown in nitrate-rich soil, the glutamine synthetase–glutamate synthase pathway occurs in:

a.   leaf vacuoles.
b.   leaf mitochondria.
c.   leaf chloroplasts.
d.   root mitochondria.
e.   root plastids.

50. Assimilation of Nitrogen; p. 742; moderate; ans: c
In soils where nitrogen is limiting, _____ is a major and direct source of nitrogen.

a.   nitrate
b.   ammonia
c.   organic nitrogen
d.   nitrite
e. N_2

51. **The Phosphorus Cycle; p. 742; easy; ans: a**
_____ is one of the main ways in which phosphorus is lost from terrestrial ecosystems.

   a.   Leaching
   b.   Evaporation
   c.   Burning
   d.   Recycling
   e.   The action of microorganisms

52. **Human Impact on Nutrient Cycles and Effects of Pollution; p. 743; moderate; ans: b**
The destruction of marshes and wetlands most notably affects the _____ stage of the nitrogen cycle.

   a.   ammonification
   b.   denitrification
   c.   nitrogen fixation
   d.   nitrification
   e.   assimilation

53. **Human Impact of Nutrient Cycles and Effects of Pollution; p. 744; easy; ans: c**
Acid rain results primarily from interaction of the oxides of _____ with atmospheric moisture.

   a.   phosphorus and sulfur
   b.   phosphorus and nitrogen
   c.   sulfur and nitrogen
   d.   nitrogen and carbon
   e.   nitrogen and magnesium

54. **Soils and Agriculture; pp. 744–745; moderate; ans: c**
In agricultural soils in which nutrients are constantly removed by the harvesting of crops, the remaining cations are likely to:

   a.   evaporate.
   b.   be rapidly leached away.
   c.   be in a form that is not available to plants.
   d.   react with soil moisture to form acids.
   e.   react with other nutrients to form toxic compounds.

55. **Plant Nutrition Research; p. 745; easy; ans: c**
Lime is added to soils in order to:

   a.   provide nutrients.
   b.   raise the pH.
   c.   remove excess salts.
   d.   provide ions for cation exchange.
   e.   kill insect pests.

56. **Plant Nutrition Research; pp. 746–747; moderate; ans: a**
    Which of the following statements about hyperaccumulators is FALSE?

    a. They are plants that are harmed by high concentrations of potentially toxic
        materials.
    b. They may concentrate trace elements, radionucleides, and/or heavy metals.
    c. They are being used in phytoremediation.
    d. They may concentrate toxins to produce toxic foliage.
    e. They include the Indian mustard plant and alpine pennycress.

57. **Plant Nutrition Research; p. 747; moderate; ans: e**
    Recently, endophytic nitrogen-fixing species of *Acetobacter* and *Herbaspirillum*
    were found in:

    a. maize roots only.
    b. maize stems only.
    c. maize leaves only.
    d. sugarcane stems only.
    e. sugarcane stems and leaves.

True-False Questions

1. **Introduction; p. 727; easy; ans: T**
    Plants can synthesize all their required amino acids and vitamins.

2. **Essential Elements; p. 727; easy; ans: T**
    Nickel is an essential element for plant growth.

3. **Essential Elements; p. 727; moderate; ans: F**
    The concentration of a mineral element in a plant cannot be greater than its
    concentration in the surrounding soil.

4. **Essential Elements; p. 727; moderate; ans: T**
    Different species of plants growing in the same nutrient medium may differ
    greatly in nutrient content.

5. **Functions of Essential Elements; p. 729; easy; ans: T**
    Some essential elements are enzyme activators.

6. **Functions of Essential Elements; p. 729; moderate; ans: F**
    Deficiency symptoms of phloem-immobile elements such as potassium and
    nitrogen appear first in older leaves.

7. **The Soil; p. 731; easy; ans: T**
    Minerals are inorganic compounds consisting of two or more elements.

8. **The Soil; p. 731; moderate; ans: F**
    Within the Earth, metamorphic and sedimentary rocks can be changed by heat
    and pressure into igneous rocks.

9. **The Soil; p. 732; easy; ans: F**
    The C horizon of soil is also called the subsoil.

10. **The Soil; p. 732; moderate; ans: T**
    Loam soils, the best agricultural soils, contain sand, silt, and clay particles.

11. **The Soil; p. 735; moderate; ans: T**
    Wilted plants in a soil at its permanent wilting percentage will not recover even when placed in a humid chamber.

12. **The Soil; p. 735; moderate; ans: F**
    Clay particles have an excess of surface negative charges, and humus has an excess of positive charges.

13. **The Soil; p. 735; moderate; ans: T**
    Mycorrhizae are especially important in the absorption and transfer of phosphorus in plants.

14. **Nutrient Cycles; p. 736; easy; ans: F**
    In a nutrient cycle, all the nutrients returned to the soil are generally available for plant use.

15. **Nitrogen and the Nitrogen Cycle; p. 736; moderate; ans: T**
    The primary limiting factor for plant growth is usually the amount of nitrogen in the soil.

16. **Nitrogen and the Nitrogen Cycle; p. 738; moderate; ans: F**
    Most commercial nitrogen fertilizers contain either nitrate or nitrite.

17. **Nitrogen and the Nitrogen Cycle; p. 738; moderate; ans: F**
    Denitrification occurs only in waterlogged soils.

18. **Nitrogen and the Nitrogen Cycle; p. 738; easy; ans: T**
    The most common nitrogen-fixing bacteria are *Rhizobium* and *Bradyrhizobium*.

19. **Nitrogen and the Nitrogen Cycle; pp. 739–740; moderate; ans: F**
    All cells in a nitrogen-fixing root nodule are infected with rhizobia.

20. **Nitrogen and the Nitrogen Cycle; p. 740; easy; ans: T**
    Nitrogenase is inhibited by molecular oxygen.

21. **Nitrogen and the Nitrogen Cycle; p. 740; moderate; ans: F**
    Bacteria that induce nodule formation in clover will also induce nodule formation in soybeans.

22. **Nitrogen and the Nitrogen Cycle; p. 741; moderate; ans: T**
    Some nitrogen-fixing symbioses involve plants other than legumes.

23. **Assimilation of Nitrogen; p. 742; easy; ans: F**
    In nitrate-rich soils, nitrate reduction in herbaceous plants takes place in root plastids.

24. **The Phosphorus Cycle; p. 742; easy; ans: T**
    Of the elements for which the Earth's crust is the primary reservoir, the one most likely to limit plant growth is phosphorus.

25. **Human Impact on Nutrient Cycles and Effects of Pollution; p. 743; easy; ans: T**
    When phosphorus is added to aquatic ecosystems through drainage from fertilized fields, massive growths of algae and flowering plants often result.

26. **Plant Nutrition Research; p. 745; easy; ans: T**
    Some plants accumulate toxins to such levels that their foliage becomes toxic.

27. **Plant Nutrition Research; p. 747; easy; ans: T**
    As a consequence of having increased nitrogen fixation, a plant might exhibit decreased shoot activity.

Essay Questions

1. **Essential Elements; p. 727; moderate**
   What three criteria are used to determine whether an element is essential to a plant?

2. **Essential Elements; pp. 727, 729; moderate**
   What is the difference between a micronutrient and a macronutrient? Name the macronutrients and micronutrients required by plants.

3. **Functions of Essential Elements; p. 729; moderate**
   How are essential elements classified on the basis of their function? What is one problem with this type of classification?

4. **Functions of Essential Elements; p. 729; difficult**
   What is the relationship between the mobility of a nutrient in the phloem and the age of leaves in which deficiency symptoms first appear?

5. **The Soil; p. 731; moderate**
   Compare and contrast igneous, sedimentary, and metamorphic rocks. How do rocks give rise to soils?

6. **The Soil; pp. 731–732; moderate**
   Describe the differences between the A, B, and C horizons of soil.

7. **The Soil; p. 734; difficult**
   What is the relationship between soil water content and the potential at which water is held in sandy, loam, and clay soils?

8. **The Soil; p. 735; moderate**
   Explain the principle of cation exchange in soils. What are the consequences of this for plant growth?

9. **Nutrient Cycles; pp. 735–736; moderate**
   Why are nutrient cycles also called biogeochemical cycles? In what way are such cycles "leaky"?

10. Nitrogen and the Nitrogen Cycle; pp. 737–738; difficult
Trace the pathway by which nitrogen is released from dead organic matter and
converted to the form in which it is assimilated by plants. Name all the
organisms involved.

11. Nitrogen and the Nitrogen Cycle; pp. 738–739; easy
What is the biological basis for crop rotation?

12. Nitrogen and the Nitrogen Cycle; pp. 739–740; difficult
Describe the sequence of events in which a nitrogen-fixing root nodule is
formed.

13. Nitrogen and the Nitrogen Cycle; p. 740; moderate
Why is it important to regulate the O_2 concentration in a root nodule? How is
this regulation achieved?

14. Nitrogen and the Nitrogen Cycle; p. 740; difficult
Describe the process by which genes are activated sequentially to produce
nodule formation in legumes.

15. Nitrogen and the Nitrogen Cycle; pp. 741–742; moderate
How is nitrogen fixation accomplished industrially? What are some problems
associated with this process?

16. Assimilation of Nitrogen; p. 742; difficult
Describe the pathway by which the nitrogen in nitrate is assimilated into
organic compounds. What differences are observed in plants growing in
ecosystems where nitrogen is limiting?

17. The Phosphorus Cycle; p. 742; moderate
Describe the steps in the phosphorus cycle. What are the major differences
between the phosphorus cycle and the nitrogen cycle?

18. The Phosphorus Cycle; p. 742; moderate
Discuss the ways in which phosphorus is (a) added to and (b) lost from an
ecosystem.

**19. Human Impact on Nutrient Cycles and Effects of Pollution; pp. 743–744;
moderate**
In what ways have humans upset the recycling of both nitrogen and phosphorus
in ecosystems?

**20. Human Impact on Nutrient Cycles and Effects of Pollution; p. 744;
moderate**
What is acid rain? List some of the effects of acid rain in ecosystems.

21. Soils and Agriculture; pp. 744–745; moderate
Explain how the process of cation exchange can be disrupted in agricultural
soils.

22. Plant Nutrition Research; p. 747; moderate
Describe how genetic engineering has been used to reduce aluminum toxicity in
plants.

23. **Plant Nutrition Research; p. 747; moderate**
    Describe some of the ways in which scientists are manipulating biological
    nitrogen fixation to increase its efficiency.

Chapter 31

The Movement of Water and Solutes in Plants

Multiple-Choice Questions

1. **Introduction; p. 751; moderate; ans: e**
    Which of the following statements about xylem and phloem is FALSE?

    a. They are closely associated spatially.
    b. They are closely associated functionally.
    c. They form a continuous vascular system.
    d. They are involved in long-distance transport.
    e. Xylem transports only water and phloem transports only nutrients.

2. **Movement of Water and Inorganic Nutrients through the Plant Body; p. 751; easy; ans: b**
    Transpiration is the:

    a. flow of water through the xylem.
    b. loss of water vapor from plants.
    c. absorption of water by roots.
    d. loss of carbon dioxide from leaves.
    e. entry of carbon dioxide into leaves.

3. **Movement of Water and Inorganic Nutrients through the Plant Body; p. 752; easy; ans: d**
    A plant loses the greatest amount of water through the:

    a. leaf cuticle.
    b. stem cuticle.
    c. root cuticle.
    d. stomata.
    e. lenticels.

4. **Movement of Water and Inorganic Nutrients through the Plant Body; p. 752; moderate; ans: a**
    Stomatal transpiration involves two steps:

    a. evaporation and diffusion.
    b. evaporation and active transport.
    c. osmosis and evaporation.
    d. osmosis and active transport.
    e. photosynthesis and respiration.

5. **Movement of Water and Inorganic Nutrients through the Plant Body;
    p. 752; moderate; ans: e**
    Approximately _____ percent of the water transpired by a plant is lost
    through its stomata.

    a. 1
    b. 10
    c. 25
    d. 60
    e. 90

6.

    **Movement of Water and Inorganic Nutrients through the Plant Body;
    p. 752; difficult; ans: b**
    Which of the following events is MOST DIRECTLY responsible for the
    opening of stomata?

    a. Water leaves the guard cells.
    b. Water enters the guard cells.
    c. Solutes enter the guard cells.
    d. Solutes leave the guard cells.
    e. Water enters the stomata.

7. **Movement of Water and Inorganic Nutrients through the Plant Body;
    p. 753; moderate; ans: d**
    Radial micellation refers to the _____ of guard cells.

    a. shape
    b. opening and closing
    c. orientation of proteins in the plasma membrane
    d. orientation of cellulose microfibrils in the cell walls
    e. mode of attachment between the ends

8. **Movement of Water and Inorganic Nutrients through the Plant Body;
    p. 753; moderate; ans: a**
    _____ is the most important factor affecting stomatal movements.

    a. Water loss
    b. Temperature
    c. The intensity of sunlight
    d. The length of the day
    e. The length of the night

9. **Movement of Water and Inorganic Nutrients through the Plant Body;
    pp. 753–754; moderate; ans: e**
    Which of the following statements about stomatal movements is FALSE?

    a. An increase in CO_2 levels usually causes stomata to close.
    b. Stomata usually open in the light and close in the dark.
    c. Blue light stimulates stomatal opening.
    d. Red light stimulates stomatal opening.
    e. Temperatures above 30° to 35°C usually stimulate stomatal opening.

10. **Movement of Water and Inorganic Nutrients through the Plant Body; p. 754; difficult; ans: b**

    The response of stomata to temperature appears to be due to changes in intercellular _____ levels.

    a. potassium
    b. carbon dioxide
    c. flavin
    d. chlorophyll
    e. oxygen

11. **Movement of Water and Inorganic Nutrients through the Plant Body; p. 754; difficult; ans: d**

    Plants with crassulacean acid metabolism:

    a. open their stomata when conditions are favorable to transpiration.
    b. convert CO_2 to organic acids during the day.
    c. convert organic acids to CO_2 during the night.
    d. are exemplified by members of the stonecrop family.
    e. have a pathway for carbon flow much different from that of C_4 plants.

12. **Movement of Water and Inorganic Nutrients through the Plant Body; p. 754; easy; ans: e**

    The rate of transpiration is affected by stomatal movements and:

    a. air currents only.
    b. humidity only.
    c. temperature only.
    d. temperature and air currents only.
    e. temperature, humidity, and air currents.

13. **Movement of Water and Inorganic Nutrients through the Plant Body; p. 754; moderate; ans: a**

    Unlike grassland plants, plants growing in shady forests:

    a. have large leaf surfaces.
    b. have thick cuticles.
    c. have sunken stomata.
    d. transpire at a rate that is unaffected by air currents.
    e. transpire at a rate that is unaffected by humidity.

14. **Movement of Water and Inorganic Nutrients through the Plant Body; p. 754; difficult; ans: e**

    A dry breeze increases the rate of transpiration mainly because it:

    a. cools the leaf.
    b. warms the leaf.
    c. increases the CO_2 gradient across the leaf surface.
    d. decreases the CO_2 gradient across the leaf surface.
    e. increases the vapor pressure difference across the leaf surface.

15. **Movement of Water and Inorganic Nutrients through the Plant Body;**
    p. 754; easy; ans: b
    When a cut stem is placed in colored water, the dye can be seen to move:

    a. upward in the phloem.
    b. upward in the xylem.
    c. downward in the phloem.
    d. downward in the xylem.
    e. upward in both the xylem and phloem.

16. **Movement of Water and Inorganic Nutrients through the Plant Body;**
    p. 755; moderate; ans: e
    Which of the following statements is NOT consistent with the cohesion-tension theory?

    a. A gradient of water potential exists between the stem and the root.
    b. Transpiration brings about a lowered water potential in the leaves.
    c. Water in the xylem is under tension.
    d. A gradient of water potential provides the driving force for the movement of water from the soil through the plant to the atmosphere.
    e. Root pressure is essential to the movement of water from roots to leaves.

17. **Movement of Water and Inorganic Nutrients through the Plant Body;**
    p. 756; moderate; ans: a
    In vascular plants, cavitation is the:

    a. rupture of water columns.
    b. expulsion of air from water columns.
    c. formation of air bubbles due to particulate matter.
    d. reduction of surface tension at the meniscus spanning pores in the pit membrane.
    e. filling of vessels with water vapor.

18. **Movement of Water and Inorganic Nutrients through the Plant Body;**
    p. 756; difficult; ans: d
    An air bubble in a vessel can be prevented from spreading to an adjacent vessel by:

    a. air seeding.
    b. transport upward in the transpiration stream.
    c. movement of a torus away from an aperture of a bordered pit-pair.
    d. the surface tension of the meniscus spanning the pores of the bordered pit-pair membrane.
    e. the filtering effect of the roots.

19. **Movement of Water and Inorganic Nutrients through the Plant Body;**
    pp. 756–757; moderate; ans: a
    All of the following can induce embolisms in a vessel or tracheid EXCEPT:

    a. an air bubble coming in contact with a torus blocking an aperture of a bordered pit-pair.
    b. air being sucked in through a pit-membrane pore.
    c. an insect bite that damages xylem tissue.
    d. freezing of the xylem sap.
    e. drought.

20. **Movement of Water and Inorganic Nutrients through the Plant Body;**
    pp. 757–758; difficult; ans: e
    A pressure chamber can be used to test the cohesion-tension theory by:

    a. measuring the tensile strength of water.
    b. following the movement of colored water in the xylem.
    c. measuring the shrinkage of a trunk caused by negative xylem pressure.
    d. detecting the rate at which heated water moves through a plant.
    e. determining the magnitude of the tension in a twig.

21. **Movement of Water and Inorganic Nutrients through the Plant Body;**
    pp. 757–758; moderate; ans: e
    Which of the following does NOT provide evidence supporting the cohesion-
    tension theory?

    a. Water has sufficient tensile strength.
    b. The water potential in a stem is sufficient.
    c. In the morning, heated water begins to flow close to the leaves.
    d. In the morning, the diameter of a tree decreases first in the upper trunk.
    e. In the evening, the diameter of a tree increases first in the lower trunk.

22. **Movement of Water and Inorganic Nutrients through the Plant Body;**
    p. 758; easy; ans: c
    In the morning, the xylem sap of a tree begins to flow first in the:

    a. upper trunk.
    b. lower trunk.
    c. twigs.
    d. lateral roots.
    e. main root.

23. **Movement of Water and Inorganic Nutrients through the Plant Body;**
    p. 759; difficult; ans: d
    Which of the following indicates the correct sequence of tissues through which
    water moves from the soil into the root?

    a. root hairs, endodermis, exodermis, cortical cells, vascular cylinder
    b . root hairs, endodermis, cortical cells, vascular cylinder, exodermis
    c. exodermis, endodermis, epidermis, vascular cylinder, cortical cells
    d. epidermis, exodermis, cortical cells, endodermis, vascular cylinder
    e. epidermis, endodermis, exodermis, cortical cells, vascular cylinder

24. **Movement of Water and Inorganic Nutrients through the Plant Body;**
    p. 759; moderate; ans: b
    Water moving into a root from the cell wall of an epidermal cell to the cell
    wall of a neighboring cortical cell follows the:

    a. symplastic pathway only.
    b. apoplastic pathway only.
    c. transcellular pathway only.
    d. symplastic and apoplastic pathways.
    e. apoplastic and transcellular pathways.

25. **Movement of Water and Inorganic Nutrients through the Plant Body; p. 759; moderate; ans: e**
    The major difference between the symplastic and the transcellular pathways of the root is that the transcellular pathway involves:

    a. plasmodesmata.
    b. protoplasts.
    c. plasma membranes.
    d. cell walls.
    e. vacuoles.

26. **Movement of Water and Inorganic Nutrients through the Plant Body; pp. 759–760; difficult; ans: c**
    When water moving through the cortex via the apoplastic pathway reaches the endodermis:

    a. it continues through the apoplast.
    b. it enters the cell walls.
    c. apoplastic movement is stopped by the Casparian strips.
    d. the symplastic pathway is blocked.
    e. the transcellular pathway is blocked.

27. **Movement of Water and Inorganic Nutrients through the Plant Body; p. 761; moderate; ans: e**
    In the process of guttation, all of the following events occur EXCEPT:

    a. dewlike drops of water form at the tips of grass.
    b. dewlike drops of water form at the tips of leaves.
    c. water exudes through stomata that lack the ability to open and close.
    d. water exudes through openings in structures called hydathodes.
    e. water forms from the condensation of water from the air.

28. **Movement of Water and Inorganic Nutrients through the Plant Body; p. 761; easy; ans: e**
    In vascular plants, hydraulic lift is the:

    a. movement of water from roots to leaves.
    b. flow of water against a water potential gradient.
    c. movement of water across the endodermis.
    d. transfer of water across Casparian strips.
    e. transfer of water by roots from moist to dry regions of the soil.

29. **Movement of Water and Inorganic Nutrients through the Plant Body; p. 761; moderate; ans: d**
    When transpiration is proceeding at a high rate:

    a. stomata usually close.
    b. guttation is at its greatest.
    c. root pressure can push water to the top of a tall tree.
    d. water is pulled into the roots by bulk flow.
    e. hydraulic lift forces water into roots.

30. **Movement of Water and Inorganic Nutrients through the Plant Body;**
    p. 762; moderate; ans: c
    An ion moving across the root via the symplastic pathway would NOT pass
    across the:

    a.   plasma membrane of the epidermis.
    b.   epidermal cell protoplast.
    c.   cell walls of the exodermis.
    d.   plasmodesmata in the epidermal-cortical cell walls.
    e.   endodermis.

31. **Movement of Water and Inorganic Nutrients through the Plant Body;**
    p. 762; easy; ans: b
    Mycorrhizae are especially important in the plant's absorption of:

    a.   zinc.
    b.   phosphorus.
    c.   manganese.
    d.   copper.
    e.   iron.

32. **Movement of Water and Inorganic Nutrients through the Plant Body;**
    p. 763; moderate; ans: d
    In roots, active transport of minerals is required for:

    a.   uptake by epidermal cells only.
    b.   secretion into the vessels only.
    c.   transport through the apoplast and the symplast only.
    d.   uptake by epidermal cells and secretion into the vessels only.
    e.   uptake by epidermal cells, transport through the symplast, and secretion
         into the vessels.

33. **Movement of Water and Inorganic Nutrients through the Plant Body;**
    p. 764; moderate; ans: e
    Transport of inorganic ions within the leaf occurs via the:

    a.   apoplastic pathway only.
    b.   symplastic pathway only.
    c.   transpiration stream only.
    d.   apoplastic and symplastic pathways only.
    e.   transpiration stream and the apoplastic and symplastic pathways.

34. **Assimilate Transport: Movement of Substances through the Phloem;**
    pp. 764–765; difficult; ans: b
    Which of the following statements about sources and sinks in assimilate
    movement is FALSE?

    a.   A plant part unable to meet its nutritional needs functions as a sink.
    b.   In seedlings, the cotyledons commonly act as the major sinks.
    c.   In mature plants, the upper leaves commonly act as sources for the shoot
         apex.
    d.   In mature plants, the lower leaves commonly act as sources for the roots.
    e.   Developing fruits are highly competitive sinks.

35. **Assimilate Transport: Movement of Substances through the Phloem;**
    p. 765; easy; ans: c
    As a consequence of girdling a tree:

    a.   its leaves stop exporting solutes.
    b.   its leaves stop importing solutes.
    c.   the bark above the girdle swells.
    d.   the bark above the girdle dies.
    e.   the bark begins to act as a source.

36. **Assimilate Transport: Movement of Substances through the Phloem;**
    pp. 765–766; moderate; ans: a
    Which of the following is NOT evidence supporting the role of phloem in
    sugar transport?

    a.   Honeydew contains amino acids but not sucrose.
    b. ^{14}C-labeled sucrose is transported in sieve tubes.
    c.   The exudate from the stylets of feeding aphids is mainly sucrose.
    d.   Radioactive assimilates are transported in the phloem.
    e.   When a tree is girdled, the bark above the ring becomes swollen.

37. **Assimilate Transport: Movement of Substances through the Phloem;**
    p. 766; easy; ans: b
    At least 90 percent of the dry matter of sieve-tube sap is composed of:

    a.   protein.
    b.   sugar.
    c.   one or more mineral ions.
    d.   one or more nucleotides.
    e.   one or more amino acids.

38. **Assimilate Transport: Movement of Substances through the Phloem;**
    p. 767; moderate; ans: d
    The driving force for assimilate transport is:

    a.   diffusion.
    b.   active transport.
    c.   transpiration.
    d.   osmosis.
    e.   pressure-flow.

39. **Assimilate Transport: Movement of Substances through the Phloem;**
    p. 767; easy; ans: a
    Assimilates are transported from sources to sinks along a gradient of:

    a.   turgor pressure.
    b.   sucrose.
    c.   glucose.
    d.   protons.
    e.   ATP.

40. Assimilate Transport: Movement of Substances through the Phloem; pp. 767–768; difficult; ans: c

Which of the following statements about apoplastic phloem loading is FALSE?

a. It is driven by a proton gradient.
b. The mechanism is sucrose-proton cotransport.
c. The sieve tube provides much of the energy for transport.
d. In some cases, active transport occurs across the plasma membrane of the sieve tube.
e. In some cases, active transport occurs across the plasma membrane of the companion cell.

41. Assimilate Transport: Movement of Substances through the Phloem; pp. 767–768; difficult; ans: d

The loading of sugars from sieve tube-companion cell complexes is:

a. always apoplastic.
b. always symplastic.
c. primarily symplastic in families of flowering plants in the boreal zone.
d. primarily apoplastic in families of flowering plants in temperate regions.
e. primarily apoplastic in families of flowering plants in the tropics.

42. Assimilate Transport: Movement of Substances through the Phloem; pp. 768–769; difficult; ans: e

Which of the following statements about phloem unloading and transport is FALSE?

a. Unloading can be apoplastic.
b. Unloading can be symplastic.
c. In young leaves and roots, unloading is probably passive.
d. In storage organs, energy is needed to accumulate sugars in sink cells.
e. In young leaves and roots, post-sieve-tube transport is passive.

True-False Questions

1. Introduction; p. 751; moderate; ans: T

The developing fruits of many crop plants obtain water principally via the phloem rather than the xylem.

2. Movement of Water and Inorganic Nutrients through the Plant Body; p. 751; easy; ans: F

Gaseous carbon dioxide can readily enter plant cells.

3. Movement of Water and Inorganic Nutrients through the Plant Body; p. 752; moderate; ans: T

The cuticle of leaves is largely impervious to both water and carbon dioxide.

4. Movement of Water and Inorganic Nutrients through the Plant Body; p. 752; moderate; ans: F

When the stomata are closed, photosynthesis ceases.

5. Movement of Water and Inorganic Nutrients through the Plant Body; p. 752; moderate; ans: F

Stomata open when the turgor pressure of the guard cells decreases.

6. **Movement of Water and Inorganic Nutrients through the Plant Body;**
   p. 753; moderate; ans: T
   The stomata of many species open in the morning and close at night regardless
   of the amount of water available to the plant.

7. **Movement of Water and Inorganic Nutrients through the Plant Body;**
   p. 753; easy; ans: T
   Mutants that are unable to synthesize ABA are permanently wilted.

8. **Movement of Water and Inorganic Nutrients through the Plant Body;**
   p. 754; moderate; ans: T
   Guard cell chloroplasts are believed to be involved in stomatal adaptation to
   sun, shade, and temperature.

9. **Movement of Water and Inorganic Nutrients through the Plant Body;**
   p. 754; moderate; ans: F
   The temperature of the leaf surfaces increases as rapidly as the air temperature
   increases.

10. **Movement of Water and Inorganic Nutrients through the Plant Body;**
    p. 754; moderate; ans: T
    The greater the vapor pressure difference between the intercellular spaces and
    the leaf surface, the faster is the rate of transpiration.

11. **Movement of Water and Inorganic Nutrients through the Plant Body;**
    p. 754; easy; ans: T
    If a cut stem is placed in water that is colored with dye, the dye moves upward
    in the vessels and tracheids.

12. **Movement of Water and Inorganic Nutrients through the Plant Body;**
    pp. 754–755; easy; ans: F
    Root pressure is now known to be crucial in moving water to the top of a tall
    tree.

13. **Movement of Water and Inorganic Nutrients through the Plant Body;**
    p. 756; moderate; ans: T
    Adhesion is just as important as cohesion in the cohesion-tension mechanism of
    water movement in plants.

14. **Movement of Water and Inorganic Nutrients through the Plant Body;**
    p. 756; moderate; ans: F
    It is the smallest pores in the walls or pit membranes of vessels or tracheids
    that are the most susceptible to air seeding.

15. **Movement of Water and Inorganic Nutrients through the Plant Body;**
    p. 758; easy; ans: T
    The diameter of a tree trunk decreases following the onset of rapid
    transpiration.

16. **Movement of Water and Inorganic Nutrients through the Plant Body;**
    p. 759; easy; ans: F
    Most of the water absorbed by a plant enters through older parts of the root
    system.

17. **Movement of Water and Inorganic Nutrients through the Plant Body;**
    p. 759; easy; ans: T
    Root hairs provide a large surface area for the uptake of water and minerals from the soil.

18. **Movement of Water and Inorganic Nutrients through the Plant Body;**
    p. 760; moderate; ans: F
    Once it has crossed the exodermis, water can pass into the root's vascular cylinder by the apoplastic pathway.

19. **Movement of Water and Inorganic Nutrients through the Plant Body;**
    p. 761; easy; ans: T
    In the process of guttation, water is exuded from leaves through hydathodes.

20. **Movement of Water and Inorganic Nutrients through the Plant Body;**
    p. 761; easy; ans: T
    The upward flow of water through the xylem is called the transpiration stream.

21. **Movement of Water and Inorganic Nutrients through the Plant Body;**
    p. 762; easy; ans: F
    When the soil is dry, roots grow faster in order to reach water.

22. **Movement of Water and Inorganic Nutrients through the Plant Body;**
    p. 763; moderate; ans: T
    Roots growing under anaerobic conditions are not able to take up nutrients.

23. **Movement of Water and Inorganic Nutrients through the Plant Body;**
    p. 764; moderate; ans: F
    Attempts at supplementing plant nutrition by applying fertilizers directly to leaves have not been successful.

24. **Assimilate Transport: Movement of Substances through the Phloem;**
    p. 764; easy; ans: T
    The assimilate stream transports sugars in the phloem.

25. **Assimilate Transport: Movement of Substances through the Phloem;**
    pp. 765–766; easy; ans: F
    When a tree is girdled, the swelling of the bark below the girdled ring provides evidence that sugars are transported in the phloem.

26. **Assimilate Transport: Movement of Substances through the Phloem;**
    p. 767; moderate; ans: T
    In the pressure-flow mechanism, sugars are transported passively through sieve tubes.

27. **Assimilate Transport: Movement of Substances through the Phloem;**
    p. 767; easy; ans: T
    According to the pressure-flow mechanism, sugars are transported in the phloem by flowing down a gradient of turgor pressure.

28. **Assimilate Transport: Movement of Substances through the Phloem;**
    p. 768; moderate; ans: F
    Apoplastic loading of sugar into the phloem evolved before symplastic loading.

Essay Questions

1. **Movement of Water and Inorganic Nutrients through the Plant Body; p. 751; moderate**
   Explain how transpiration and photosynthesis are closely linked.

2. **Movement of Water and Inorganic Nutrients through the Plant Body; p. 751; moderate**
   In what way is transpiration an "unavoidable evil" for a plant?

3. **Movement of Water and Inorganic Nutrients through the Plant Body; pp. 752–753; moderate**
   Describe the role of the guard cells in the opening and closing of stomata.

4. **Movement of Water and Inorganic Nutrients through the Plant Body; p. 753; moderate**
   Discuss the effect of ABA on stomatal movements.

5. **Movement of Water and Inorganic Nutrients through the Plant Body; pp. 753–754; difficult**
   Discuss the effects of CO_2, light, humidity, and temperature on stomatal movements.

6. **Movement of Water and Inorganic Nutrients through the Plant Body; p. 755; moderate**
   Describe the mechanism by which water moves from the soil to the top of a tree.

7. **Movement of Water and Inorganic Nutrients through the Plant Body; p. 756; difficult**
   What causes water to be under tension in the xylem? What are the roles of cohesion and adhesion in the process?

8. **Movement of Water and Inorganic Nutrients through the Plant Body; pp. 756–757; moderate**
   Why are air bubbles so hazardous to the movement of water in the xylem? What mechanisms does the plant have for protection against this hazard?

9. **Movement of Water and Inorganic Nutrients through the Plant Body; pp. 757–758; moderate**
   Outline the experimental evidence that supports the cohesion-tension theory.

10. **Movement of Water and Inorganic Nutrients through the Plant Body; p. 758; moderate**
    Why is "transpirational-pull" an unfortunate name for the cohesion-tension mechanism?

11. **Movement of Water and Inorganic Nutrients through the Plant Body; p. 759; moderate**
    Explain the differences between the apoplastic, symplastic, and transcellular pathways for water movement across a root.

12. **Movement of Water and Inorganic Nutrients through the Plant Body; pp. 759–760; moderate**
   What role do Casparian strips play in the entry of water into the vascular cylinder of the root?

13. **Movement of Water and Inorganic Nutrients through the Plant Body; pp. 760–761; difficult**
   Explain the phenomenon of root pressure. What is the relationship of root pressure to guttation?

14. **Movement of Water and Inorganic Nutrients through the Plant Body; pp. 762–764; difficult**
   Describe the pathways by which inorganic ions are taken up by roots and distributed in the plant.

15. **Assimilate Transport: Movement of Substances through the Phloem; pp. 764–765; moderate**
   Explain the concept of source and sink in assimilate transport. Describe how a plant organ can be a source under certain conditions and a sink under others.

16. **Assimilate Transport: Movement of Substances through the Phloem; pp. 765–766; moderate**
   Give evidence to support the hypothesis that assimilate solutes are transported in the phloem.

17. **Assimilate Transport: Movement of Substances through the Phloem; p. 766; easy**
   Discuss the use of aphids in investigating assimilate transport.

18. **Assimilate Transport: Movement of Substances through the Phloem; pp. 766–767; moderate**
   Explain the pressure-flow hypothesis. Which parts of this mechanism are active and which passive?

19. **Assimilate Transport: Movement of Substances through the Phloem; pp. 767–769; difficult**
   Discuss the mechanisms of phloem loading and unloading. In what ways is phloem unloading both active and passive?

Chapter 32

The Dynamics of Communities and Ecosystems

Multiple-Choice Questions

1. **Introduction; p. 774; easy; ans: b**
   A(n) _____ includes, by definition, all the organisms in a particular place together with their environment.
   a. community
   b. ecosystem
   c. population
   d. species
   e. mutualistic interaction

2. **Introduction; p. 774; moderate; ans: d**
   A community consists of all the _____ in a particular area.
   a. individuals of a single species
   b. plants
   c. plants and animals
   d. organisms
   e. organisms and their environment

3. **Interactions between Organisms; p. 774; easy; ans: b**
   Mutualism is an interaction between two species in which:
   a. one species benefits and the other is harmed.
   b. both species benefit.
   c. both species are harmed.
   d. one species benefits, and the other is neither harmed nor helped.
   e. one species is harmed, and the other is neither harmed nor helped.

4.   Interactions between Organisms; pp. 774–775; difficult; ans: e
Which of the following statements about mycorrhizae is FALSE?

a.   In many vascular plants, nonmycorrhizal individuals are rarely encountered.
b.   In most plants, mycorrhizal fungi are zygomycetes.
c.   The majority of species of herbs and shrubs form endomycorrhizal associations.
d.   Basidiomycetes and ascomycetes form ectomycorrhizal associations with some conifers and eudicot trees.
e.   Endomycorrhizae are characteristic of trees growing at high latitudes or at high elevations.

5.   Interactions between Organisms; p. 775; moderate; ans: e
The ants of genus *Pseudomyrmex* that swarm over a bull's-horn acacia tree:

a.   harm the plant.
b.   provide the plant with sugar.
c.   provide the plant with protein.
d.   store the plant's seeds.
e.   protect the plant against predators.

6.   Interactions between Organisms; p. 775; moderate; ans: a
When ants of the genus *Pseudomyrmex* were removed from bull's-horn acacia trees, the trees:

a.   usually died.
b.   were unaffected.
c.   grew taller.
d.   lived longer.
e.   produced thorns.

7.   Interactions between Organisms; p. 776; easy; ans: c
Competition occurs, by definition, when two individuals living in the same area:

a.   belong to the same species.
b.   belong to the same population.
c.   require the same limiting resource.
d.   photosynthesize at the same rates.
e.   allocate energy in the same way.

8.   Interactions between Organisms; p. 776; moderate; ans: d
Which of the following statements about the role of growth rate in competition is FALSE?

a.   Plants with a high growth rate often have an advantage over plants with a low growth rate.
b.   Plants with a low growth rate may survive if they are able to photosynthesize at low light intensities.
c.   Leaf arrangement, crown shape, and patterns of allocation of energy all affect growth rate.
d.   A single combination of traits usually produces a competitor that is best in all environments.
e.   Within a single community, different species may coexist because of adaptations to different microenvironments.

9. **Interactions between Organisms; p. 776; moderate; ans: e**
According to the principle of competitive exclusion, two species with similar requirements:

   a.   generally grow to the same height.
   b.   generally photosynthesize at the same rate.
   c.   cannot take up water equally well.
   d.   cannot photosynthesize equally well.
   e.   cannot coexist indefinitely in the same habitat.

10. **Interactions between Organisms; pp. 776–777; moderate; ans: a**
When the two species are grown together, *Lemna polyrhiza* is replaced by *Lemna gibba* because *L. gibba* outcompetes *L. polyrhiza* for:

   a.   light.
   b.   space.
   c.   carbon dioxide.
   d.   oxygen.
   e.   minerals.

11. **Interactions between Organisms; p. 777; moderate; ans: d**
In the chalk grasslands of England, what happened to the plants when a viral disease reduced the rabbit population?

   a.   All the plants died.
   b.   The average height increased.
   c.   The average height decreased.
   d.   Diversity decreased.
   e.   Diversity increased.

12. **Interactions between Organisms; p. 777; easy; ans: c**
When dandelions produce seeds that contain embryos that are genetically identical to the parents, it is an example of:

   a.   phytoalexin production.
   b.   plant-herbivore interactions.
   c.   clonal reproduction.
   d.   allelopathy.
   e.   competitive exclusion.

13. **Interactions between Organisms; p. 779; easy; ans: c**
In nature, the penicillin produced by *Penicillium chrysogenum* growing on seeds:

   a.   inhibits seed germination.
   b.   inhibits the growth of the fungus.
   c.   inhibits the growth of bacteria.
   d.   stimulates seed germination.
   e.   stimulates the growth of bacteria.

14. **Interactions between Organisms; p. 779; 2-15; moderate; ans: e**
    In allelopathy:

    a.   the alleles of a bacterium or fungus produce pathogenic effects in a plant.
    b.   some alleles of a plant are pathogenic to that same plant.
    c.   chemicals produced by a bacterium inhibit growth of a plant.
    d.   chemicals produced by a fungus inhibit growth of a plant.
    e.   chemicals produced by one plant inhibit growth of another plant.

15. **Interactions between Organisms; p. 779; moderate; ans: c**
    After the cactus moth was introduced in Australia, what was the effect on the
    prickly-pear cactus?

    a.   The cacti were able to flower.
    b.   The cacti produced toxic chemicals.
    c.   The cactus population was vastly reduced.
    d.   The cactus population was completely wiped out.
    e.   The cacti were more widely available to grazing animals.

16. **Interactions between Organisms; pp. 779–780; moderate; ans: d**
    _____ produced by plants are the most important factors preventing attack by
    herbivorous insects.

    a.   Thorns
    b.   Spines
    c.   Waxy cuticles
    d.   Secondary metabolites
    e.   Leathery leaves

17. **Interactions between Organisms; p. 780; moderate; ans: c**
    Which of the following statements about defense mechanisms in seaweeds is
    FALSE?

    a.   Some are too tough for herbivores to consume.
    b.   Some produce chemicals that make them distasteful to herbivores.
    c.   Their array of defenses is considerably greater than that of flowering
         plants.
    d.   Palatable species may grow intermixed with unpalatable species.
    e.   Some escape herbivores by growing in cracks and holes.

18. **Interactions between Organisms; pp. 780, 782; difficult; ans: a**
    Which of the following statements about phytoalexins is FALSE?

    a.   They are produced in response to bacterial and fungal proteins.
    b.   They are lipidlike compounds.
    c.   Their synthesis is stimulated by leaf damage.
    d.   They are produced in response to elicitors.
    e.   They are natural antibiotics.

19. **Interactions between Organisms; p. 782; difficult; ans: b**
    What are tannins?

    a.   sex attractants in insects
    b.   compounds that make plant proteins indigestible to insects
    c.   compounds produced by a plant when it is attacked
    d.   fungal substances that detoxify pisatin
    e.   elicitors produced by bacteria and fungi

20. Nutrient Cycling; p. 782; moderate; ans: d
Which of the following statements about nutrient cycling is FALSE?

a.   An ecosystem has the property of regulating the cycling of nutrients.
b.   An ecosystem is self-sustaining in terms of its nutrient supply.
c.   Nutrients are continuously cycled between organisms and the environment.
d.   In an ideal nutrient cycle, small amounts of nutrients are lost.
e.   In a nutrient cycle, the rate of flow differs among nutrients and habitats.

21. Nutrient Cycling; pp. 782–783; difficult; ans: d
In the undisturbed Hubbard Brook forest, scientists detected:

a.   considerable loss of nutrients by leaching.
b.   considerable addition of nutrients from dissolving bedrock.
c.   considerable inefficiency in conserving minerals.
d.   a slight annual gain of nitrogen and potassium.
e.   a slight annual gain of calcium.

22. Nutrient Cycling; p. 783; difficult; ans: c
After all plants in a selected portion of the Hubbard Brook forest were killed, scientists detected a(n):

a.   decrease in net calcium loss.
b.   decrease in net potassium loss.
c.   increase in net nitrogen loss.
d.   increase in the amount of nitrate in the soil.
e.   increase in the amount of potassium in the soil.

23. Trophic Levels; p. 783; difficult; ans: e
Which of the following statements about trophic levels is FALSE?

a.   They are feeding levels.
b.   They are part of a food chain.
c.   They are part of a food web.
d.   They include consumers and decomposers.
e.   They include producers and consumers.

24. Trophic Levels; p. 783; difficult; ans: d
A fungus is both a(n) _____ and a _____.

a.   heterotroph; primary producer
b.   autotroph; primary consumer
c.   heterotroph; secondary consumer
d.   heterotroph; decomposer
e.   autotroph; decomposer

25. Trophic Levels; p. 784; easy; ans: c
In an ecosystem, energy flows from:

a.   consumers to autotrophs to decomposers.
b.   decomposers to consumers to autotrophs.
c.   autotrophs to consumers to decomposers.
d.   consumers to decomposers to autotrophs.
e.   autotrophs to decomposers to consumers.

26. **Trophic Levels; p. 784; moderate; ans: a**
    During the transfer of energy between trophic levels, most of the energy is:
    a. dissipated as heat.
    b. absorbed by the soil.
    c. used by herbivores.
    d. used by carnivores.
    e. used by decomposers.

27. **Trophic Levels; p. 784; moderate; ans: a**
    Plants use approximately _____ percent of the incident light for biosynthesis.
    a. 1 to 3
    b. 5 to 8
    c. 10 to 20
    d. 25 to 30
    e. 35 to 40

28. **Trophic Levels; p. 785; easy; ans: b**
    Food chains usually have _____ links.
    a. 1 or 2
    b. 3 or 4
    c. 5 or 6
    d. 7 or 8
    e. 9 or 10

29. **Trophic Levels; p. 785 and Fig. 32-8; difficult; ans: c**
    Which of the following statements about an upright pyramid of energy is FALSE?
    a. It represents the energy relationships among trophic levels.
    b. The total energy of producers is represented at the base of the pyramid.
    c. The total energy of consumers is greater than the total energy of producers.
    d. The total energy of primary carnivores is greater than the total energy of secondary carnivores.
    e. The total energy decreases at successively higher trophic levels.

30. **Trophic Levels; p. 786; moderate; ans: c**
    The existence of more grass plants than herbivores in a grassland ecosystem is illustrated by a(n):
    a. pyramid of energy.
    b. pyramid of biomass.
    c. pyramid of numbers.
    d. pyramid of body size.
    e. inverted pyramid of biomass.

31. Development of Communities and Ecosystems; pp. 786, 788; moderate; ans: a
Which of the following statements about succession in a plant community is FALSE?

a. In its later stages it is a unidirectional process.
b. It is a progressive series of changes over time.
c. It occurs at variable rates in disturbed areas.
d. It may occur after an earthquake.
e. It is easiest to observe in its early stages.

32. Development of Communities and Ecosystems; p. 788; moderate; ans: d
Pioneer tree species that colonize forest gaps are characterized by _____ than late successional trees.

a. fewer crown layers
b. longer life spans
c. lower growth rates in the sun
d. more diffuse crowns
e. heavier, more durable wood

33. Development of Communities and Ecosystems; p. 788; moderate; ans: d
The transition from sugar pine to white fir forests in California was caused mainly by:

a. volcanic eruptions.
b. frequent earthquakes.
c. logging.
d. reduced numbers of forest fires.
e. reduced numbers of settlers.

34. Development of Communities and Ecosystems; p. 792; moderate; ans: b
Nonprairie species are prevented by _____ from encroaching on prairie ecosystems.

a. wind
b. fire
c. rainfall
d. herbivores
e. low temperature

True-False Questions

1. Introduction; p. 774; moderate; ans: F
Ecology is defined as the study of the interactions among organisms.

2. Introduction; p. 774; difficult; ans: T
An example of a community is an oak tree and all the organisms that live in and on it.

3. Interactions between Organisms; p. 774; easy; ans: T
An example of mutualism is nitrogen-fixing bacteria growing in the root nodules of legumes.

4. **Interactions between Organisms; p. 774; easy; ans: F**
    Only a small percentage of vascular plants have mycorrhizal associations.

5. **Interactions between Organisms; pp. 774–775; easy; ans: T**
    Ectomycorrhizal associations can be highly specific.

6. **Interactions between Organisms; p. 775; moderate; ans: T**
    In the bull's-horn acacia, Beltian bodies provide food for ants of the genus
    Pseudomyrmex.

7. **Interactions between Organisms; p. 775; moderate; ans: F**
    Bull's-horn acacia trees inhabited by ants of the genus *Pseudomyrmex* grow
    more slowly than acacia trees without ants.

8. **Interactions between Organisms; p. 776; easy; ans: T**
    Competition results when individuals living in the same area require the same
    limiting resource.

9. **Interactions between Organisms; p. 776; moderate; ans: F**
    A species that is dominant in a community in one climatic area will also be
    dominant in different climatic areas.

10. **Interactions between Organisms; p. 776; moderate; ans: T**
    A species that thrives when grown alone may not survive when grown with a
    similar species.

11. **Interactions between Organisms; p. 777; moderate; ans: F**
    The different requirements of Engelmann spruce and subalpine fir make it
    impossible for them to coexist in a community indefinitely.

12. **Interactions between Organisms; p. 777; moderate; ans: T**
    The diversity of plants in the chalk grasslands of England was much greater
    when rabbits were abundant than when the rabbit population was much
    reduced.

13. **Interactions between Organisms; p. 777; easy; ans: F**
    The diversity of species in an environment that is continuously disturbed is
    smaller than that in a more stable environment.

14. **Interactions between Organisms; p. 777; moderate; ans: T**
    The asexual reproduction of plants may make it difficult to measure the extent
    of competition among members of the same species in a particular area.

15. **Interactions between Organisms; p. 779; easy; ans: T**
    An example of allelopathy is the inability of white pine to grow near black
    walnut trees.

16. **Interactions between Organisms; p. 779; easy; ans: F**
    The relationship between the cactus moth and the prickly-pear cactus in
    Australia is an example of a pollinator-plant interaction.

17. **Interactions between Organisms; p. 780; easy; ans: T**
Helanalin, pyrethrum, and chromenes are chemicals produced by plants to deter insect herbivores.

18. **Interactions between Organisms; pp. 780, 782; moderate; ans: F**
Phytoalexins are compounds in the walls of certain bacteria and fungi that stimulate the production of elicitors in plants.

19. **Interactions between Organisms; p. 782; moderate; ans: T**
In response to insect attack, some plants produce leaves with higher tannin levels than normal.

20. **Interactions between Organisms; p. 782; easy; ans: T**
The pool of nutrients in an ecosystem is continually renewed.

21. **Nutrient Cycling; p. 783; moderate; ans: F**
In the experiments at Hubbard Brook forest, the investigators showed that when the plants were killed, the ecosystem was able to recover when seedlings emerged in the following spring.

22. **Trophic Levels; p. 783; easy; ans: T**
An herbivore is a primary consumer; a carnivore is a secondary consumer.

23. **Trophic Levels; p. 784; moderate; ans: F**
Energy is recycled in an ecosystem.

24. **Trophic Levels; p. 785; moderate; ans: F**
Cole found that in Lake Cayuga, more of the original energy is available to us if we eat trout rather than the smelt that are eaten by the trout.

25. **Trophic Levels; p. 785; moderate; ans: T**
A pyramid of biomass is inverted only when the primary producers have very high rates of reproduction.

26. **Development of Communities and Ecosystems; p. 786; easy; ans: T**
Succession is the progression of changes within a community.

27. **Development of Communities and Ecosystems; p. 788; easy; ans: F**
In forest ecosystems, gaps provide opportunities for plant species with relatively low light requirements to flourish.

28. **Development of Communities and Ecosystems; p. 788; easy; ans: F**
Efforts to prevent forest fires in California resulted in enhanced growth of sugar pines.

29. **Development of Communities and Ecosystems; p. 792; easy; ans: F**
An example of a climax community is the first community that arises following a volcanic eruption.

30. **Development of Communities and Ecosystems; p. 792; easy; ans: T**
"Restoration" ecology refers to efforts by humans to reestablish natural communities.

Essay Questions

1. **Introduction; p. 774; moderate**
   Explain the differences between a population, a community, and an ecosystem.

2. **Interactions between Organisms; pp. 774–775; difficult**
   What are mycorrhizae? In what sense are mycorrhizae examples of mutualism?

3. **Interactions between Organisms; pp. 775–776; moderate**
   Describe the interactions between ants and acacia trees. What evidence supports the hypothesis that this is an example of mutualism?

4. **Interactions between Organisms; pp. 776–777; moderate**
   Use the experimental data from studies of *Lemna* species to explain the principle of competitive exclusion.

5. **Interactions between Organisms; pp. 776–777; difficult**
   Does the example of the coexistence of Engelmann spruce and subalpine fir violate the principle of competitive exclusion? Why or why not?

6. **Interactions between Organisms; p. 777; difficult**
   Discuss the relationship between the number of organisms in a particular area and the extent of species diversity.

7. **Interactions between Organisms; p. 779; moderate**
   How is allelopathy a type of interaction between organisms? Give examples to support your answer.

8. **Interactions between Organisms; pp. 780, 782; moderate**
   What are phytoalexins, and under what conditions are they produced? How might they be used commercially?

9. **Interactions between Organisms; pp. 780, 782; difficult**
   How is the role of tannins and other phenolics different from that of phytoalexins? How are the roles similar?

10. **Nutrient Cycling; pp. 782–783; moderate**
    Explain how the experiments at Hubbard Brook forest provided information about the retention of nutrients in ecosystems.

11. **Trophic Levels; p. 783; easy**
    Name the trophic levels present in most ecosystems, and describe the types of organisms present at each level. How do primary producers relate to these trophic levels?

12. **Trophic Levels; p. 783; easy**
    What is the difference between a food chain and a food web? Which more accurately describes a natural ecosystem?

13. **Trophic Levels; p. 785; moderate**
    Why is it more energy-efficient for humans to eat plants instead of animals?

14. **Trophic Levels; p. 785; moderate**
Why do food chains typically consist of only a few links? What is the relationship between body size and the links in food chains?

15. **Trophic Levels; pp. 785–786; difficult**
Explain what is meant by pyramids of energy, biomass, and numbers. Under what circumstances are these pyramids upright, and under what circumstances are they inverted?

16. **Trophic Levels; p. 786; moderate**
Discuss the efforts by humans to develop renewable sources of energy from plants.

17. **Development of Communities and Ecosystems; pp. 786, 788; moderate**
Describe the process of succession beginning with bare rock and ending with a forest.

18. **Development of Communities and Ecosystems; p. 788; moderate**
Discuss the roles of gaps in succession and the maintenance of species diversity.

19. **Development of Communities and Ecosystems; p. 788; moderate**
Are forest fires always detrimental to ecosystems? Give reasons to support your answer.

20. **Development of Communities and Ecosystems; pp. 789, 792; moderate**
Discuss the speed and stages of succession following a volcanic eruption, using Krakatau or Mount St. Helens as an example.

21. **Development of Communities and Ecosystems; p. 792; moderate**
In what sense is the term "climax community" a misnomer?

22. **Development of Communities and Ecosystems; pp. 792–793; moderate**
Describe the attempts to restore prairie communities by restoration ecologists.

Chapter 33

Global Ecology

Multiple-Choice Questions

1. **Introduction; p. 797; moderate; ans: e**
   Which of the following statements about a biome is FALSE?

   a. It has distinctive vegetation.
   b. It occupies large areas of land surface.
   c. It is a terrestrial set of ecosystems.
   d. It is controlled by climate.
   e. It is usually limited to a single continent.

2. **Life on the Land; p. 797; moderate; ans: c**
   The three factors that most affect the distribution of biomes are the patterns and/or distribution of:

   a. heat, minerals, and moisture.
   b. oxygen, minerals, and moisture.
   c. heat, air currents, and mountains.
   d. humidity, carbon dioxide, and oxygen.
   e. humidity, air currents, and mountains.

3. **Life on the Land; p. 801; moderate; ans: b**
   An increase in elevation of _____ and an increase in latitude of _____ are generally accompanied by the same decrease in mean atmospheric temperature.

   a. 100 meters; 0.5°
   b. 100 meters; 1.0°
   c. 100 meters; 10.0°
   d. 100 kilometers; 1.0°
   e. 100 kilometers; 10.0°

4. **Life on the Land; p. 801; moderate; ans: d**
   In the middle latitudes of the Northern Hemisphere, the _____ slopes of mountains are usually the warmest.

   a. northern and eastern
   b. northern and western
   c. southern and eastern
   d. southern and western
   e. southern and northern

5. Life on the Land; pp. 801–802; moderate; ans: a
Along the West Coast of the United States, the "rain shadow" is the _____
area on the _____ slope of the Sierra Nevada.

  a. dry; eastern
  b. dry; western
  c. moist; eastern
  d. moist; western
  e. moist; southern

6. Rainforests; p. 803; moderate; ans: d
Which of the following statements about tropical rainforests is FALSE?

  a. Broad-leaved evergreen trees dominate the vegetation.
  b. The forest floor accumulates little organic debris.
  c. The world's greatest diversity of species is found there.
  d. The world's greatest number of individuals per species is found there.
  e. They have no dry season.

7. Rainforests; p. 804; easy; ans: c
A liana is a(n):

  a. epiphyte.
  b. herbaceous plant.
  c. woody vine.
  d. tropical tree.
  e. orchid.

8. Rainforests; p. 804; difficult; ans: e
Most rainforest trees:

  a. have bark that is thick and rough.
  b. have large, toothed leaves.
  c. branch close to the ground.
  d. are smaller than the trees of temperate forests.
  e. are homogeneous in appearance.

9. Rainforests; p. 805; moderate; ans: a
Which of the following statements about tropical soils is FALSE?

  a. They have a thick topsoil.
  b. They are relatively infertile.
  c. Most are deficient in mineral nutrients.
  d. Most are acidic.
  e. Many contain toxic levels of aluminum.

10. Savannas and Deciduous Tropical Forests; p. 805; difficult; ans: d
Savannas occur in all of the following locations EXCEPT:

  a. between prairies and taiga.
  b. between prairies and temperate deciduous forests.
  c. between tropical rainforests and deserts.
  d. in gallery forests.
  e. in the thorn forest of Brazil.

11. Savannas and Deciduous Tropical Forests; p. 806; moderate; ans: c
In savannas, most trees:

    a.   have larger leaves than tropical rainforest trees.
    b.   are taller than tropical rainforest trees.
    c.   have thick bark.
    d.   have few branches.
    e.   flower when they are in full leaf.

12. Savannas and Deciduous Tropical Forests; p. 806; difficult; ans: e
Monsoon forests are similar to savannas EXCEPT that monsoon forests:

    a.   contain mainly small-leaved trees.
    b.   contain deciduous trees.
    c.   are seasonally dry.
    d.   occur only in Asia.
    e.   have a higher density of trees.

13. Savannas and Deciduous Tropical Forests; p. 806; moderate; ans: a
Most of Florida is covered by:

    a.   subtropical mixed forests.
    b.   tropical forests.
    c.   thorn forests.
    d.   savannas.
    e.   monsoon forests.

14. Deserts; p. 807; easy; ans: b
The largest desert in the world is in:

    a.   Australia.
    b.   Africa.
    c.   South America.
    d.   North America.
    e.   Europe.

15. Deserts; pp. 807, 809–810; moderate; ans: e
Plants of the desert biome:

    a.   carry out only C_3 photosynthesis.
    b.   usually are perennial rather than annual.
    c.   usually have leaves that lack chlorophyll.
    d.   have low maximum photosynthetic rates.
    e.   have wide-ranging roots unless growing in washes.

16. Deserts; pp. 807, 809; moderate; ans: d
The leaves of desert plants:

    a.   have thinner cuticles than those of plants in less arid biomes.
    b.   have more stomata than those of plants in less arid biomes.
    c.   are larger than those of plants in less arid biomes.
    d.   are oriented to minimize heat absorption.
    e.   are densely covered with trichomes.

17. **Grasslands; pp. 810–811; moderate; ans: a**
    Which of the following statements about grasslands is FALSE?

    a.   They are found between taiga and temperate deciduous forests.
    b.   They are characterized by cold winters.
    c.   When disturbed, they change to deserts or forests.
    d.   Fire is important in their development.
    e.   Trees are generally absent, except along streams.

18. **Grasslands; pp. 810–811; moderate; ans: b**
    Shortgrass and tallgrass prairies differ in that shortgrass prairies:

    a.   are unaffected by fire.
    b.   exist under near-drought conditions.
    c.   have traditionally been exploited for agriculture.
    d.   have the most productive soils.
    e.   occur only in the Corn Belt of North America.

19. **Temperate Deciduous Forests; p. 812; easy; ans: d**
    The temperate deciduous forest biome:

    a.   is widespread in both the Northern and Southern Hemispheres.
    b.   has water available to plants all year.
    c.   is characterized by trees that retain their leaves all year.
    d.   has precipitation evenly distributed throughout the year.
    e.   has cool summers and relatively warm winters.

20. **Temperate Deciduous Forests; p. 812; moderate; ans: c**
    The winters of temperate deciduous forests are analogous, with respect to
    plants' access to water, to the:

    a.   winters of tropical rainforests.
    b.   winters of the taiga.
    c.   dry seasons of savannas.
    d.   summers of grasslands.
    e.   winters of deserts.

21. **Temperate Deciduous Forests; p. 812; difficult; ans: c**
    Spring ephemerals differ from early summer species in that spring ephemerals:

    a.   have broader, thinner leaves.
    b.   have smaller storage organs.
    c.   complete their photosynthetic activity earlier.
    d.   are herbaceous.
    e.   are leafless in winter.

22. **Temperate Deciduous Forests; p. 813; moderate; ans: c**
    The soils of temperate deciduous forests:

    a.   are more fertile than prairie soils.
    b.   retain their nutrients even after the forest is cleared.
    c.   are ill-suited for long-term cultivation.
    d.   have abundant nutrients.
    e.   are alkaline.

23. **Temperate Mixed and Coniferous Forests; pp. 815–816; moderate; ans: b**
Temperate mixed and temperate deciduous forests differ in that temperate mixed forests:

   a. have more deciduous trees.
   b. have more conifers.
   c. have warmer winters.
   d. have soils richer in nutrients.
   e. are found farther south.

24. **Mediterranean Scrub; p. 816; moderate; ans: d**
The Mediterranean scrub biome is characterized by _____ summers and _____ winters.

   a. warm; warm
   b. cool; cool
   c. cool, moist; hot, dry
   d. hot, dry; cool, moist
   e. hot, moist; cool, dry

25. **Taiga; p. 818; moderate; ans: d**
In northeastern Siberia, the dominant conifers of the taiga are:

   a. firs.
   b. spruces.
   c. pines.
   d. larches.
   e. poplars.

26. **Taiga; p. 818; moderate; ans: b**
Which of the following statements about the taiga is FALSE?

   a. Lakes, bogs, and marshes are common.
   b. The number of plant and animal species is relatively large.
   c. A snow cover persists throughout the winter.
   d. It is absent from the Southern Hemisphere.
   e. It is also called boreal forest.

27. **Taiga; p. 819; moderate; ans: c**
Which of the following groups of organisms is NOT commonly found in the taiga?

   a. lichens
   b. mosses
   c. annuals
   d. willows
   e. dewberries

28. **Arctic Tundra; p. 819; easy; ans: d**
Approximately _____ of the Earth's land surface is Arctic tundra.

   a. one-fiftieth
   b. one-twentieth
   c. one-tenth
   d. one-fifth
   e. one-third

29. **Arctic Tundra; p. 820; moderate; ans: c**
    Which of the following statements about the Arctic tundra is FALSE?

    a.   In the extreme north it forms polar deserts.
    b.   Fixed nitrogen is in short supply.
    c.   The evaporation rate is high.
    d.   The growing season is less than two months.
    e.   The soils are acidic and low in nutrients.

True-False Questions

1. **Life on the Land; p. 797; easy; ans: T**
    Carbon dioxide is more readily available to photosynthetic organisms living on land than to those living in the sea.

2. **Life on the Land; pp. 800–801; moderate; ans: F**
    The Southern Hemisphere as a whole has a higher mean annual temperature than the Northern Hemisphere.

3. **Life on the Land; p. 801; moderate; ans: F**
    Temperature variation is much greater at high-altitude locations at the equator than at analogous Arctic and Antarctic locations.

4. **Life on the Land; p. 802; easy; ans: T**
    Given sufficient light and moisture, lichens growing on trees in the Northern Hemisphere are better developed on the moister, northeastern side of the trees.

5. **Rainforests; p. 803; easy; ans: T**
    In tropical rainforests, organic debris is quickly broken down by decomposers and absorbed by mycorrhizae.

6. **Rainforests; p. 804; moderate; ans: F**
    Tropical rainforests are confined to Central and South America.

7. **Rainforests; p. 805; easy: ans: T**
    After several years of cultivation, tropical soils become useless for agricultural purposes.

8. **Savannas and Deciduous Tropical Forests; pp. 805–806; moderate; ans: F**
    In savannas, epiphytes and annual herbs are abundant.

9. **Savannas and Deciduous Tropical Forests; p. 806; moderate; ans: F**
    Areas of monsoon forest have high precipitation throughout the year.

10. **Deserts; p. 807; easy; ans: T**
    The cold deserts have just a few weeks of high temperatures each year.

11. **Deserts; p. 807; easy; ans: T**
    More types of annual plants are found in deserts than in any other biome.

12. **Deserts; p. 810; moderate; ans: F**
The juniper savannas of the United States occur in areas of low-lying, hot desert.

13. **Grasslands; p. 810; moderate; ans: T**
The glades within forested areas of the central and eastern United States contain plants of the grasslands biome.

14. **Grasslands; pp. 811–812; moderate; ans: T**
Fire favors the growth of plants with basal meristems.

15. **Temperate Deciduous Forests; p. 812; easy; ans: F**
Spring ephemerals usually emerge and mature more slowly than do late summer species.

16. **Temperate Deciduous Forests; p. 813; moderate; ans: T**
In temperate deciduous forests, species that ripen seeds in the fall are dispersed by birds.

17. **Temperate Deciduous Forests; p. 813; easy; ans: F**
The temperate deciduous forest plants of east Asia are quite different from those of eastern North America.

18. **Temperate Mixed and Coniferous Forests; p. 815; moderate; ans: T**
Temperate mixed forest, with conifers and deciduous trees, is characteristic of the Great Lakes–Saint Lawrence River region.

19. **Mediterranean Scrub; pp. 816–817; easy; ans: F**
In regions of Mediterranean-type vegetation, most plant growth occurs in the cool, moist summers.

20. **Mediterranean Scrub; p. 817; easy; ans: T**
Mediterranean scrub communities in different areas of the world tend to resemble each other even though their plant species are unrelated.

21. **Taiga; p. 818; moderate; ans: T**
In North America, taiga extends farther north along the western side of the continent than along the eastern side.

22. **Taiga; p. 818; moderate; ans: F**
In the northernmost taiga, most of the precipitation falls in winter.

23. **Arctic Tundra; p. 820; easy; ans: T**
The Arctic tundra is entirely underlain by permafrost.

24. **Arctic Tundra; p. 820; easy; ans: T**
Some areas of tundra, with very low precipitation, are true polar deserts.

25. **Arctic Tundra; p. 821; moderate; ans: F**
Perennials of the Arctic tundra characteristically reproduce sexually rather than vegetatively.

Essay Questions

1. **Life on the Land; pp. 797, 800; moderate**
   Discuss the advantages and disadvantages to plants of life on land.

2. **Life on the Land; p. 801; moderate**
   Discuss the similarities and differences between high-altitude and high-latitude habitats.

3. **Life on the Land; pp. 801–802; moderate**
   Which slope is usually the driest in mountains of the Northern Hemisphere? Of the Southern Hemisphere? Explain your answers.

4. **Rainforests; pp. 803–804; moderate**
   Discuss the types of plants that grow in tropical rainforests. In what ways are they adapted to the tropical habitat?

5. **Rainforests; pp. 804–805; moderate**
   What characteristics of the rainforests make them particularly vulnerable to human pressures?

6. **Savannas and Deciduous Tropical Forests; pp. 805–807; difficult**
   Explain the similarities and differences among savanna, monsoon forest, and subtropical mixed forest.

7. **Deserts; pp. 807–808; moderate**
   What are the characteristic features of the desert biome? Discuss the adaptations exhibited by desert plants.

8. **Grasslands; p. 810; moderate**
   How do the rainfall and temperature of the grassland biome compare with those of the two neighboring biomes?

9. **Grasslands; pp. 810–812; moderate**
   What are the characteristics of the plants of the grassland biome? What role does fire play in grasslands?

10. **Temperate Deciduous Forests; pp. 812–813; difficult**
    What types of plants characterize the temperate deciduous forest biome? How do these plants differ from one season to the next?

11. **Temperate Mixed and Coniferous Forests; pp. 815–816; moderate**
    Discuss the similarities and differences between the temperate mixed forest and the temperate deciduous forest biomes.

12. **Mediterranean Scrub; pp. 816–818; difficult**
    What are the characteristics of plants in Mediterranean scrub communities? How have these communities changed as a result of human habitation?

13. **Taiga; p. 819; moderate**
    How are the coniferous forests of the Pacific Northwest adapted to the winter-wet/summer-dry climate of that region?

14. **Arctic Tundra; pp. 819–821; moderate**
    Describe the physical environment of the Arctic tundra. How are tundra plants
    adapted to this environment?

Chapter 34

The Human Prospect

Multiple-Choice Questions

1. **Introduction; p. 824; easy; ans: e**
   Members of the genus *Homo* evolved from _____ about 2 million years ago.
   a. Neanderthals
   b. chimpanzees
   c. gorillas
   d. tool-using human beings
   e. members of the genus *Australopithecus*

2. **The Agricultural Revolution; p. 824; moderate; ans: d**
   When the glaciers began to retreat about 18,000 years ago:
   a. humans began to domesticate animals.
   b. large grazing animals increased in number.
   c. grasslands became more extensive.
   d. the human population was about 5 million.
   e. forests migrated southward across North America.

3. **The Agricultural Revolution; p. 825; moderate; ans: b**
   In contrast to cultivated wheats, the wild wheats have:
   a. a tougher rachis.
   b. seeds that are scattered more readily.
   c. a rachis that holds the seeds until they are harvested.
   d. seeds that can be gathered more easily by humans.
   e. seeds that can be planted more easily.

4. **The Agricultural Revolution; p. 825; moderate; ans: b**
   The domestication of plants in the Fertile Crescent began about _____ years ago.
   a. 1100
   b. 11,000
   c. 22,000
   d. 55,000
   e. 110,000

5. **The Agricultural Revolution; p. 825; moderate; ans: d**
   The first plants cultivated in the Fertile Crescent were:

   a.   olives and dates.
   b.   peas and olives.
   c.   barley and peas.
   d.   wheat and barley.
   e.   wheat and lentils.

6. **The Agricultural Revolution; p. 825; moderate; ans: c**
   Legumes have long been important in the human diet because they are:

   a.   low in fats.
   b.   low in carbohydrates.
   c.   high in protein.
   d.   high in fats.
   e.   high in carbohydrates.

7. **The Agricultural Revolution; p. 826; easy; ans: d**
   Which of the following animals was NOT first domesticated in the Near East?

   a.   cattle
   b.   dogs
   c.   goats
   d.   horses
   e.   sheep

8. **The Agricultural Revolution; p. 827; difficult; ans: e**
   Poi is a starchy food made from plants of the genera:

   a.   *Mangifera* and *Citrus*.
   b.   *Citrus* and *Musa*.
   c.   *Musa* and *Sorghum*.
   d.   *Sorghum* and *Colocasia*.
   e.   *Colocasia* and *Xanthosoma*.

9. **The Agricultural Revolution; p. 827; easy; ans: c**
   Plantains are a _____ variety of _____ first cultivated in tropical Asia.

   a.   starchy; taro
   b.   sweet; taro
   c.   starchy; banana
   d.   sweet; banana
   e.   starchy; sorghum

10. **The Agricultural Revolution; p. 827; moderate; ans: d**
    Which of the following crops was NOT initially cultivated in Africa?

    a.   sorghum
    b.   okra
    c.   yams
    d.   rice
    e.   cotton

11. **The Agricultural Revolution; pp. 827, 828, 830–831; moderate; ans: b**
Which of the following crops was NOT first cultivated in the New World and later carried to the Old World?

    a.   potatoes
    b.   bananas
    c.   tobacco
    d.   maize
    e.   kidney beans

12. **The Agricultural Revolution; pp. 828–829; easy; ans: b**
Which of the following statements about cotton is FALSE?

    a.   It was domesticated independently in the New and in the Old World.
    b.   It was domesticated earlier in Mexico than in Peru.
    c.   New World cottons are polyploid.
    d.   Old World cottons are diploid.
    e.   New World cottons soon replaced Old World cottons in Eurasia.

13. **The Agricultural Revolution; p. 828; moderate; ans: c**
Which of the following crops was NOT initially cultivated in the New World?

    a.   maize
    b.   kidney beans
    c.   barley
    d.   lima beans
    e.   peanuts

14. **The Agricultural Revolution; p. 829; easy; ans: c**
Plants were domesticated in Mexico beginning about _____ years ago.

    a.   12,000
    b.   11,000
    c.   9000
    d.   8000
    e.   6000

15. **The Agricultural Revolution; p. 831; moderate; ans: a**
The _____ was first cultivated in Central and South America then carried to New Zealand, Hawaii, and other Pacific islands.

    a.   sweet potato
    b.   white potato
    c.   sunflower
    d.   yam
    e.   pineapple

16. **The Agricultural Revolution; pp. 831–832; moderate; ans: a**
In contrast to Europe and Asia, the New World lacked:

    a.   large herds of domestic animals.
    b.   turkeys.
    c.   Muscovy ducks.
    d.   guinea pigs.
    e.   llamas.

17. The Agricultural Revolution; p. 832; moderate; ans: d
Herbs differ from spices in that most herbs are:

a. buds.
b. seeds.
c. roots.
d. leaves.
e. flowers.

18. The Agricultural Revolution; p. 833; moderate; ans: e
Which of the following spices originated in the New World?

a. mace
b. cinnamon
c. ginger
d. nutmeg
e. vanilla

19. The Agricultural Revolution; p. 833; difficult; ans: c
_____ is an herb that is a member of the mint family.

a. Anise
b. Dill
c. Sage
d. Caraway
e. Fennel

20. The Agricultural Revolution; p. 833; difficult; ans: d
_____ is an herb that is a member of the parsley family.

a. Thyme
b. Oregano
c. Basil
d. Coriander
e. Mint

21. The Agricultural Revolution; p. 833; moderate; ans: a
Which herb consists of the dried stigmas of *Crocus sativus*?

a. saffron
b. thyme
c. mustard
d. parsley
e. anise

22. The Agricultural Revolution; p. 833; difficult; ans: b
Which of the following statements about coffee is FALSE?

a. It is now cultivated throughout the warm regions of the world.
b. It was first cultivated in the mountains of subtropical Asia.
c. Its seeds are used in preparing the beverage.
d. A third of the world's supply comes from Brazil.
e. It is a major source of income for many tropical countries.

23. **The Agricultural Revolution; p. 833; easy; ans: e**
More than half the world's sunflowers are produced in:

a. the United States.
b. Mexico.
c. Thailand.
d. West Africa.
e. Russia.

24. **The Agricultural Revolution; pp. 834–835; moderate; ans: d**
The great majority of the world's food supply is based on _____ kinds of crop plants.

a. 2
b. 6
c. 8
d. 14
e. 20

25. **The Agricultural Revolution; p. 834; moderate; ans: e**
Which of the following is NOT one of the six crop plants that provide more than 80 percent of the total calories consumed by humans?

a. rice
b. wheat
c. potatoes
d. manioc
e. soybeans

26. **The Growth of Human Populations; p. 836; moderate; ans: d**
Which of the following was NOT a consequence of the development of agriculture?

a. the development of cities
b. the diversification of lifestyles
c. an increase in the population
d. an increase in the birth rate
e. a decrease in the death rate

27. **The Growth of Human Populations; p. 836; easy; ans: d**
At present, there are approximately _____ humans on Earth.

a. 6 million
b. 60 million
c. 600 million
d. 6 billion
e. 60 billion

28. **The Growth of Human Populations; p. 836; moderate; ans: b**
The growth rate of the human population is about _____ percent per year.

a. 0.2
b. 1.5
c. 5
d. 10
e. 20

29. The Growth of Human Populations; p. 837; moderate; ans: d
It is estimated that in 1996, _____ percent of the world's people were living in absolute poverty.

a. 2
b. 5
c. 10
d. 25
e. 30

30. The Growth of Human Populations; p. 837; difficult; ans: c
Which of the following statements about world food production is FALSE?

a. Most of the land that can be cultivated using current techniques is already in cultivation.
b. In Africa south of the Sahara, per capita food production has been falling.
c. Humans presently consume, waste, or divert about 20 percent of the land's total net photosynthetic productivity.
d. Since 1950, 25 percent of the world's topsoil has been lost.
e. In developing countries, people may spend 80 to 90 percent of their income on food.

31. Agriculture in the Future; p. 838; easy; ans: e
One of the great problems facing agriculture worldwide is:

a. too little research.
b. overuse of irrigation.
c. underuse of fertilizers.
d. inefficient machinery.
e. the displacement of workers.

32. Agriculture in the Future; p. 838; moderate; ans: e
The most promising approach to solving the world's food problems is:

a. increasing the use of pesticides.
b. increasing the use of fertilizers.
c. increasing the availability of water.
d. cultivating more land.
e. improving the existing crops.

33. Agriculture in the Future; p. 838; easy; ans: c
_____ of the 20 amino acids required by human adults must be obtained from food.

a. Two
b. Five
c. Nine
d. Eleven
e. All

34. Agriculture in the Future; p. 838; easy; ans: c
Crops can be improved by increasing each of the following EXCEPT:

a.   the quantity of proteins
b.   the quality of proteins.
c.   nitrogen requirements.
d.   disease resistance.
e.   storage ability.

35. Agriculture in the Future; p. 839; easy; ans: c
Which of the following is NOT an advantage of hybrid maize varieties over nonhybrid varieties?

a.   They can be selected for cultivation in particular localities.
b.   They require more people to harvest and thus stimulate the economy.
c.   They require less pesticides to produce greater yields.
d.   They require less water to produce greater yields.
e.   They require less fertilizer to produce greater yields.

36. Agriculture in the Future; p. 839; easy; ans: a
The Green Revolution involves the use of all of the following EXCEPT:

a.   algae as food.
b.   improved strains of crop plants
c.   more effective herbicides.
d.   more effective insecticides.
e.   more effective agricultural equipment.

37. Agriculture in the Future; pp. 839–840; easy; ans: d
Triticale is a hybrid that combines the high yield and quality of _____ with the disease resistance and cold hardiness of _____.

a.   manioc; wheat
b.   maize; rice
c.   rye; maize
d.   wheat; rye
e.   wheat; rice

38. Agriculture in the Future; pp. 839–841; moderate; ans: d
The Irish potato famine of 1846–1847 and the U.S. epidemic of southern leaf blight of maize in 1970 both resulted from:

a.   increases in the mutation rate.
b.   mistakes in genetic engineering.
c.   the overuse of chemical sprays.
d.   widespread cultivation of genetically uniform varieties that were susceptible to disease.
e.   an unusual abundance of numerous types of plant pathogens.

39. Agriculture in the Future; p. 841; easy; ans: e
The Irish potato famine of 1846–1847 was caused by an organism of the genus:

a. *Fusarium.*
b. *Cochliobolus.*
c. *Verticillium.*
d. *Secale.*
e. *Phytophthora.*

40. Agriculture in the Future; p. 841; moderate; ans: d
It is advantageous to use plants rather than commercial laboratories to produce commodities such as oils, drugs, and perfumes because:

a. commercial laboratories use nonrenewable energy sources.
b. commercial laboratories are less expensive.
c. commercial laboratories are now archaic.
d. plants require no energy other than the sun.
e. plants produce these items more quickly.

41. Agriculture in the Future; p. 842; moderate; ans: a
_____ is an alternative source of natural rubber that can be cultivated in the desert.

a. Guayule
b. The Para rubber tree
c. Jojoba
d. Tagweed
e. *Solanum*

42. Agriculture in the Future; pp. 842–843; difficult; ans: c
Unlike cereals, the seeds of grain amaranths have a high content of:

a. wax.
b. rubber.
c. lysine.
d. protein.
e. fat.

43. Agriculture in the Future; p. 843; moderate; ans: e
_____ is a particularly salt-tolerant plant that may prove useful in producing salt-tolerant hybrids.

a. *Simmondsia chinensis*
b. *Parthenium argentatum*
c. *Amaranthus*
d. *Solanum lycopersicum*
e. *Solanum cheesmanii*

44. Agriculture in the Future; p. 844; moderate; ans: c
Until supplies of the plant were virtually exhausted, birth-control pills and cortisone were manufactured from substances produced by:

a. pigweed.
b. jojoba.
c. wild yams.
d. wild tomatoes.
e. guayule.

45. Agriculture in the Future; p. 845; moderate; ans: b
Solutions to the problem of world hunger must involve all of the following EXCEPT:

a. creating jobs.
b. increasing population growth.
c. raising living standards.
d. introducing new agricultural practices.
e. providing farmers access to new crops.

46. Agriculture in the Future; p. 845; easy; ans: e
The first step in achieving solutions to society's problems must be:

a. increasing the use of genetic engineering.
b. providing food to the people of all countries.
c. overcoming malnutrition.
d. the amelioration of poverty.
e. stabilizing the human population.

True-False Questions

1. Introduction; p. 824; moderate; ans: T
The Neanderthal people disappeared about 34,000 years ago.

2. The Agricultural Revolution; p. 824; moderate; ans: T
Human beings migrated eastward from Europe and western Asia, colonized Siberia, and eventually reached North America.

3. The Agricultural Revolution; p. 825; moderate; ans: F
In the wild species of wheat, the rachis holds the seeds tightly.

4. The Agricultural Revolution; p. 825; easy; ans: F
Plants were first domesticated in Africa.

5. The Agricultural Revolution; p. 826; easy; ans: T
Since grazing animals were first domesticated into large herds, they have caused widespread ecological damage.

6. The Agricultural Revolution; p. 826; moderate; ans: T
Agriculture reached Britain by about 4000 B.C.

7. The Agricultural Revolution; p. 827; easy; ans: T
Wild bananas have large, hard seeds.

8. The Agricultural Revolution; p. 827; moderate; ans: F
Coffee was first cultivated in Brazil.

9. The Agricultural Revolution; p. 828; moderate; ans: T
Cotton was domesticated independently in the New World and the Old World.

10. The Agricultural Revolution; p. 828; moderate; ans: F
New World cottons are diploid, and Old World cottons are polyploid.

11. **The Agricultural Revolution; p. 829; moderate; ans: T**
Plants were first domesticated in Mexico about 9000 years ago.

12. **The Agricultural Revolution; p. 831; moderate; ans: F**
Manioc was first domesticated in the tropics of Asia.

13. **The Agricultural Revolution; p. 832; easy; ans: T**
Spices are usually derived from parts of a plant other than the leaves.

14. **The Agricultural Revolution; p. 833; moderate; ans: T**
In contrast to spices, many herbs originated in Europe and the Mediterranean region.

15. **The Agricultural Revolution; p. 833; easy; ans: F**
Sunflowers were first cultivated in the area that is now Russia.

16. **The Agricultural Revolution; p. 834; moderate; ans: T**
Rubber (*Hevea*) grows best in areas where it is not native.

17. **The Agricultural Revolution; p. 834; easy; ans: F**
The coconut palm is widespread today because of dispersal by humans.

18. **The Growth of Human Populations; pp. 835–836; easy; ans: T**
For a variety of reasons, populations that are dependent on hunting tend to remain small.

19. **The Growth of Human Populations. p. 836; moderate; ans: F**
Crop-based economies require an average area of 5 square kilometers to provide sufficient food for one family.

20. **The Growth of Human Populations; p. 836; easy; ans: F**
Since the seventeenth century, the human birth rate has increased dramatically.

21. **The Growth of Human Populations; p. 836; easy; ans: T**
At the present time, the world's population is approximately 6 billion people.

22. **The Growth of Human Populations; p. 836; easy; ans: F**
The world's population is growing at about 10 percent per year.

23. **The Growth of Human Populations; p. 837; easy; ans: T**
In 1996 the World Bank estimated that one out of every four people were living in absolute poverty.

24. **The Growth of Human Populations; p. 837; easy; ans: F**
In Africa south of the Sahara desert, food production per capita has been increasing.

25. **Agriculture in the Future; p. 837; moderate; ans: F**
Irrigation was first practiced in Mexico.

26. **Agriculture in the Future; p. 838; easy; ans: T**
Crop plants selected for higher protein content have higher nitrogen requirements than their less modified ancestors.

27. **Agriculture in the Future; p. 839; moderate; ans: F**
Between the 1930s and 1980s, maize production in the United States increased eightfold as plant breeders made full use of the plant's genetic diversity.

28. **Agriculture in the Future; p. 839; easy; ans: T**
One effect of the Green Revolution has been to accelerate the consolidation of small farms into a few large holdings.

29. **Agriculture in the Future; pp. 839–840; moderate; ans: F**
Triticale arose following a spontaneous tripling of the chromosome number in a sterile hybrid of wheat and rye.

30. **Agriculture in the Future; pp. 840–841; moderate; ans: F**
In the course of the agricultural use of plants, most of their variability has been lost.

31. **Agriculture in the Future; p. 841; easy; ans: T**
Plant breeders' use of genetic material from wild tomatoes to improve cultivated tomatoes has been extremely successful.

32. **Agriculture in the Future; p. 841; moderate; ans: T**
About 25 percent of commonly available prescription drugs are derived from plants.

33. **Agriculture in the Future; pp. 841–842; easy; ans: F**
Jojoba is native to tropical rainforests.

34. **Agriculture in the Future; p. 842; easy; ans: T**
Very little natural rubber is now obtained from Fara rubber trees (*Hevea*).

35. **Agriculture in the Future; pp. 842–843; easy; ans: T**
The protein content of amaranth seeds is as high as that of the cereals.

36. **Agriculture in the Future; pp. 843–844; easy; ans: F**
Drugs can usually be produced more cheaply by synthesis in the laboratory than by derivation from plant products.

37. **Agriculture in the Future; p. 845; easy; ans: T**
The rapid loss of the world's plant species is due in part to human population growth and poverty.

38. **Agriculture in the Future; p. 845; moderate; ans: F**
Genetic engineering has now replaced hybridization as the primary method of crop improvement.

39. **Agriculture in the Future; p. 845; easy; ans: F**
Most plants that can be cultivated for human use are already under cultivation.

Essay Questions

1. **Introduction; p. 824; easy**
Summarize the evolutionary history of *Homo sapiens*.

2. **The Agricultural Revolution; pp. 824–825; moderate**
    Describe the hypothesized sequence of events that resulted in agriculture.

3. **The Agricultural Revolution; pp. 825–826; moderate**
    Where and when were plants first domesticated? How did the domestication of plants influence other aspects of human culture?

4. **The Agricultural Revolution; pp. 826–828; moderate**
    Discuss the domestication of plants in China, tropical Asia, and Africa.

5. **The Agricultural Revolution; pp. 828–831; difficult**
    How did the domestication of plants in the New World differ from that in the Old World? Name any crops that were independently domesticated in both areas.

6. **The Agricultural Revolution; pp. 831–832; moderate**
    What effects did the domestication of animals have on the domestication of plants?

7. **The Agricultural Revolution; pp. 832–833; moderate**
    What is the difference between a spice and an herb? Name some examples of each, and indicate from which part of the plant they are derived.

8. **The Agricultural Revolution; pp. 833–834; moderate**
    Use specific examples to illustrate how important crops have spread throughout the world.

9. **The Agricultural Revolution; pp. 834–835; moderate**
    Name the world's six major food crops. Discuss their nutritional value and limitations.

10. **The Growth of Human Populations; pp. 835–836; difficult**
    What mechanisms limit the number of individuals in hunter-gatherer societies?

11. **The Growth of Human Populations; p. 836; moderate**
    Discuss the consequences for the human population of the change from hunting and gathering to agriculture.

12. **The Growth of Human Populations; pp. 836–837; moderate**
    List some of the potential problems associated with overpopulation.

13. **Agriculture in the Future; p. 838; moderate**
    Discuss some of the problems facing agriculture today.

14. **Agriculture in the Future; pp. 838–839; moderate**
    What is the most promising approach to alleviating the world's food problem? Give some examples of how this might be achieved.

15. **Agriculture in the Future; p. 839; moderate**
    What is meant by the Green Revolution? What have been its advantages and disadvantages?

16. **Agriculture in the Future; pp. 839–840; moderate**
    What is triticale? Explain the advantages and disadvantages of cultivating
    triticale.

17. **Agriculture in the Future; pp. 840–841; difficult**
    Why is it important to maintain the genetic diversity of crop plants? Give
    examples to support your answer.

18. **Agriculture in the Future; pp. 841–843; difficult**
    Describe the agricultural potential of (a) jojoba, (b) guayule, (c) grain
    amaranths, and (d) the wild tomato *Solanum cheesmanii*.

19. **Agriculture in the Future; pp. 843–844; moderate**
    Explain why plants will continue to be important sources of medicinal drugs,
    including those drugs that can be synthesized in the laboratory.

20. **Agriculture in the Future; pp. 844–845; moderate**
    In what ways will solving the problem of world hunger require an integrated
    approach?